CERTAIN PHILOSOPHICAL QUESTIONS

A portrait of the young Newton while a student at
Trinity College, Cambridge

CERTAIN PHILOSOPHICAL QUESTIONS: NEWTON'S TRINITY NOTEBOOK

J. E. McGUIRE
PROFESSOR OF HISTORY AND
PHILOSOPHY OF SCIENCE
UNIVERSITY OF PITTSBURGH

MARTIN TAMNY
ASSOCIATE PROFESSOR OF
PHILOSOPHY
THE CITY COLLEGE OF NEW YORK

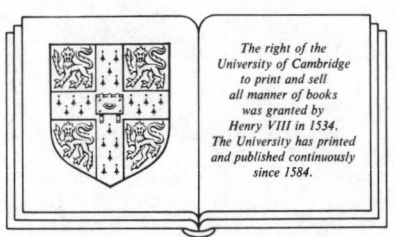

*The right of the
University of Cambridge
to print and sell
all manner of books
was granted by
Henry VIII in 1534.
The University has printed
and published continuously
since 1584.*

CAMBRIDGE UNIVERSITY PRESS
CAMBRIDGE
LONDON   NEW YORK   NEW ROCHELLE
MELBOURNE   SYDNEY

Published by the Press Syndicate of the University of Cambridge
The Pitt Building, Trumpington Street, Cambridge CB2 1RP
32 East 57th Street, New York, NY 10022, USA
10 Stamford Road, Oakleigh, Melbourne 3166, Australia

First published 1983
Reprinted 1985

Printed in Great Britain at the
University Press, Cambridge

Library of Congress Cataloging in Publication Data
McGuire, J. E.
Certain philosophical questions.
Bibliography: p.
Includes index.
1. Science–Philosophy.   2. Newton, Isaac, Sir,
1642–1727. Questiones quædam philosophicæ.   I. Tamny,
Martin.   II. Newton, Isaac, Sir, 1642–1727.
Questiones quædam philosophicæ. 1983.   III. Title:
IV. Title: Newton's Trinity notebook.
Q175.M417   1983   501   82-22200
ISBN 0 521 23164 7

Photographs (Plates 1 through 3) and text
from Add. 3996 and Add. 3975
appear by permission of the University Library, Cambridge.

For Jean, and in memory of my father
J.E.M.

For my parents
M.T.

CONTENTS

PART II. TRANSCRIPTION AND EXPANSION
OF *QUESTIONES QUÆDAM PHILOSOPHICÆ*

PREFACE

Some four years ago D. T. Whiteside brought the present writers together to begin what has been a stimulating collaboration. Prior to that occasion, each of us, in ignorance of one another's activities, had been engaged in preliminary research with a view to producing, at some future time, an edition of Newton's *Questiones quædam philosophicæ* (1664–5). This early manuscript has been much discussed by scholars, and its importance has often been stated, but the measure of its significance had never been established. In studying a thinker of Newton's complexity, the need for an understanding of his beginnings is clear. If he did undergo an *annus mirabilis* in 1666, the *Questiones* should surely enable us to assess its significance, because of its proximity in time. At the beginning of our collaboration, a working hypothesis restricted the significance of the *Questiones* to the period of 1666. As a result of our inquiries, a recognition has emerged of the extent to which many of the same problems and a similar philosophical outlook are present in Newton's later thought. The man grew, to be sure, but in large measure his progress can be gauged by the increase in the power of his thought within a metaphysical framework he adopted at the age of twenty.

When we began our Commentary, we envisaged a number of short essays on the major points of interest in the manuscript. As our work continued, the breadth and depth of the relationship between Newton's thought and that of his contemporaries and intellectual forebears became abundantly clear. At the same time, the significance of the *Questiones* for his later thought required us to make frequent reference to his later writings. In the end, the brief essays

xi

we had planned were expanded into the current Commentary, which far exceeds the length of the *Questiones* itself.

The following libraries and their staffs have aided our research considerably, and we wish to express our gratitude: University Library, Cambridge; the North Library of the British Library; the Bodelian Library; the Wertheim Study of the New York Public Library; the Morris Raphael Cohen Library of The City College of New York; the library of the Warburg Institute; the library of the Institute for Advanced Study, Princeton. We are specially indebted to the library of Trinity College, Cambridge, without whose resources this study could not have been completed. Special thanks are due to Trevor Kaye and the rest of the staff of the library for the unfailing courtesy and help that they extended to two visiting scholars.

We owe much to a number of individuals for the various ways in which they have helped our endeavor. We thank Richard Ziemacki, who followed us from Cambridge to New York, for his support in helping to bring our book to completion. We owe thanks to Miriam Drabkin for making available to us her paleographic skills. For checking the important Greek transcriptions, we are indebted to Jacob Stern, and thanks are due to Mary Louise Gill for listening to translations from Newton's Latin. Thanks are also due to Frank D. Grande for his generous help with some of the photographs we have used, and to Mary Hohman for the cheerful manner in which she produced early drafts of the Commentary. Also, we should like to thank Carolyn France for typing the final copy of the Commentary under less than ideal circumstances. Our wives have our love and gratitude.

Finally, we wish to acknowledge our indebtedness to all those students of Newton from whose careful scholarship we have benefited.

J. E. McGUIRE
M. TAMNY

PART I

COMMENTARY

INTRODUCTION

It is a happy circumstance when the private writings of a great man are available for posterity, especially when they provide a record of his early interests and activities. Fortunately, many of Isaac Newton's early papers still remain, thus enabling scholars to deepen our understanding of the development of his thought. Newton's earliest sustained observations on matters of science and philosophy are recorded in a notebook that he began during his first year at Cambridge. That small notebook is the subject of this volume. Needless to say, its value is immense. It allows us to understand the beginnings of an intellectual journey that culminated with the publication of the great *Principia Mathematica* in 1686 and the *Opticks* in 1704.

1. Description of the manuscript

In 1661, the eighteen-year-old Isaac Newton, having become a subsizer and then sizer of Trinity College, Cambridge, began to keep systematic notes pertaining to his studies. These he recorded in the notebook that has come down to us as Add. 3996 of the University Library, Cambridge. It is a leather-bound book whose covers are now worn, exposing the boards beneath at several points. The spine was replaced by John P. Gray & Son, Ltd., of Cambridge, in March of 1963. There have been no other repairs to the binding. The original boards are exposed on the inside, and on the verso of the front board is written "principū 9^{d} (?) est; quod operatur per principiu quo," as well as some proportions comparing the degrees of refraction of glass in what is certainly Newton's hand. The notebook was doubtless pur-

3

chased in its bound form by Newton when he reached Cambridge in June of 1661. The covers are 9.4 cm × 14.6 cm, and the bound pages are 9.4 cm × 14.3 cm.

On the front flyleaf we find, in Newton's hand, the inscription

Isaac Newton
Trin: Coll Cant
1661

The page bears several notes in pencil by the university staff, "Add. 3996," "(Repaired by Gray, Cambridge, March, 1963)," and "Isaac Newton." There also appears a signed comment by Thomas Pellet, written September 26, 1727: "Not fit to be printed." Pellet, as Newton's executor, had been empowered on Newton's death, March 20, 1727, to go through his manuscripts and decide what might be published posthumously from the great man's literary estate. As this volume attests, Mr. Pellet's judgment has finally been overturned after a period of more than 250 years.

Beginning from this flyleaf, there appear consecutive numbers written in pencil on the upper right-hand corner of the recto of each folio until we reach folio 30, at which point the notebook was turned upside down and begun from the other end, the new flyleaf being numbered 31 and subsequent pages numbered consecutively from that point. This numbering, almost certainly done by the University Library staff, attempts to follow the apparent order of Newton's entries, which begin with Greek notes from Aristotle's *Organon* on folios 3^r through 15^r, continue with Latin notes from Johannes Magirus's *Physiologiae peripateticae,* and end with astronomical notes in English on folios 26^v through 30^v. From this point, Newton's notes continue from the other end of the notebook, with additional Greek and Latin notes on folios 34^r through 81^v, followed by notes on Descartes in English on folios 83^r and 83^v. Finally, from folios 87^r to 135^r there are those English notes that Newton titled *'Questiones quædam Philosophicæ,"*[1] which form the major subject of this

---

[1] There are three acceptable Latin forms for the term "questions." These are "quaestiones," "quæstiones," and "questiones." Although the first two were the more common forms during the Medieval and Renaissance periods, the third was also deemed acceptable. We have therefore decided to follow Newton's usage and

book. The penciled folio numbers continue, however, for an additional 11 pages to folio 140^v, which faces folio 30^v, at which point the two sequences meet.

The only pages numbered by Newton himself are those of the *Questiones*. He began with 1 on folio 88^r and ended with 97 on folio 136^r, which goes beyond the last entry of the *Questiones* on page 95 of folio 135^r. Newton did not number every page, though the sequence is always retained. In what follows, our references will be to both the page number assigned by Newton and the folio number; in every case, the 'Newton number' is followed by the folio number. Thus, the title page of the *Questiones* is on folio 1 88^r. With respect to folios that have no Newton number, reference will be to the folio number alone.

2. Chronology of the *Questiones*

The dating of the earliest entries of the *Questiones* must rest on conjecture concerning Newton's developing interests and our knowledge of the dates on which he acquired certain texts. Abraham De Moivre's report on Newton's acquisitions at the Sturbridge fair of 1663, and his developing interest in Euclid, appear to make 1663 the earliest possible date for the start of *Questiones*. More decisive is the publication of *De vitis dogmatis* in 1664 (Old Style).[2] The entries based on

have retained Questiones. A problem also arises over whether Newton wrote "philosophiæ" or "philosophcæ" (failing to place an *i* between the final *h* and *c*). We are convinced that the latter is the case and have used "philosophicæ." The former is genitive and would have yielded the somewhat odd title *Certain Questions of Philosophy*, whereas the latter is adjectival and gives the more applicable *Certain Philosophical Questions*. Our subtitle, *Newton's Trinity Notebook*, also requires some explanation. Our aim in using this subtitle is to aid the reader by supplying a ready, albeit indefinite, description of a notebook Newton kept during his undergraduate years at Trinity College, Cambridge. It is not our intention to name the manuscript, which has been called, and we hope will continue to be called, by its library designation, Add. 3996.

[2] Until the year 1752, the new year commenced in England on March 25. In addition, whereas much of Europe had adopted the Gregorian calendar by Newton's time, England persisted in using the Julian calendar until 1752. There is a ten-day difference between the two calendars. Thus, June 15 on the Old Style (O.S.) Julian calendar becomes June 25 in the New Style (N.S.) Gregorian calendar. We have retained Newton's Old Style dates unless otherwise noted; however, we have regularized the changing of the year to January 1 in our Commentary. Thus, February 2, 1662/3 (O.S.) has been avoided in our Commentary in favor of February 2, 1663 (O.S.).

Newton's reading of this text come soon after the earliest entries in *Questiones*, and they can be no earlier than late March of 1664. These entries on matter theory, the earliest of which are taken from Charleton's *Physiologia Epicuro-Gassendo-Charltoniana*, provoked Newton's interest in Epicurus and ancient atomism. This, given Newton's inclination to go back to original sources, no doubt led him to the edition of Diogenes Laertius that we know he read.

Thus, the *Questiones* was most likely begun in early 1664, but not earlier than the long vacation (summer) of 1663. The motiviation for starting this section of his notebook was probably the large number of interesting ideas he found in Walter Charleton's *Physiologia*. Charleton's book contains eighteen of the original thirty-seven headings Newton was to use in the *Questiones*. It has pride of place, being the source of the first entries to be made in this new section of Newton's notebook.

Initially, Newton wrote thirty-seven headings[3] at what appears to have been a single sitting, as judged by hand, ink, and pen. They are the following:

Of the First Matter (1 88^r)

Of Atoms (3 89^r)

Of a Vacuum (4 89^v)

Of Quantity (5 90^r)

Of Time and Eternity (9 92^r)

Of the Celestial Matter and Orbs (11 93^r)

Of the Sun, Stars, Planets, and Comets (12 93^v)

Of Rarity, Density, Rarefaction, and Condensation (13 94^r)

Of Perspicuity and Opacity (14 94^v)

Of Fluidity, Stability, Humidity, and Siccity (15 95^r)

---

[3] The *Questiones* were written at a time when spelling had not been regularized and when large numbers of shorthand devices and conventions were used. This makes reading of the manuscript difficult for those unfamiliar with these devices and conventions. We decided, therefore, to include both a 'diplomatic' Transcription of the manuscript that exhibits all of those devices, as well as the vagaries of seventeenth century spelling, and an Expansion, which treats the Transcription almost as if it were in another language (i.e., expanding the shorthand notations and devices, regularizing the spellings and punctuation, etc., while keeping some of the flavor of the original). In our Commentary we shall draw our quotations from the Expansion for ease of reading. However, the reader can refer to the Transcription to check on any point of interpretation.

Of Softness, Hardness, Flexibility, Ductility, and Tractility (16 95^v)

Of Figure, Subtility, Hebetude, Smoothness, and Asperity (17 96^r)

Of Heat and Cold (18 96^v)

Of Gravity and Levity (19 97^r)

Of Reflection, Undulation, and Refraction (23 99^r)

Of Fire (24 99^v)

Of Air (25 100^r)

Of Water (26 100^v)

Of Earth (27 101^r)

Magnetic Attraction (29 102^r)

Electrical Attraction (31 103^r)

Of Light (32 103^v)

Of Sensation (33 104^r)

Of Species Visible (34 104^v)

Of Vision (35 105^r)

Of Colors (36 105^v)

Of Sounds (37 106^r)

Of Odors and Sapors (38 106^v)

Of Touching (39 107^r)

Of Generation and Corruption (40 107^v)

Of Memory (41 108^r)

Sympathy and Antipathy (44 109^v)

Of Oily Bodies (45 110^r)

Of Meteors (46 110^v)

Of Minerals (48 111^v)

Of the Creation (83 129^r)

Of the Soul (85 130^r)

The various spacings between these initial headings probably reflect Newton's views as to how much he would have to write about each. The large gap left between the last headings, "Of the Creation" and "Of the Soul," and the earlier headings is a likely indication that Newton saw this as a scientific work about the natural world and that at the outset he distinguished the scientific topics from the theological considerations, which he placed at the back, thus leaving what must have seemed ample space for expanding his physical and mathematical entries.

After making entries under only eight of these headings, further reading, as well as widening interests, prompted him to add still more headings, to cancel at least one original heading as his jottings demanded more space, and to preface the whole with "A Table of the Things Following." As still more headings and entries were added and the table of contents expanded, the *Questiones* took on its present form, in which a single topic, such as colors or motion, might involve reading entries at the beginning, middle, and end of the notebook.

We have attempted to date Newton's entries on the basis of handwriting, ink, internal consistency, and the few dates he gives. The variety of hands, inks, and pens used over the short period of time in which the *Questiones* was written is astonishing, until one realizes that Newton often experimented with new inks, that pens were usually cut by their users and lasted only a short time before having to be trimmed, and that handwriting is affected by emotion, health, and stamina, to mention only a few variables. Nonetheless, we believe that we can hazard a rough chronology of the *Questiones*.

We have already suggested that the *Questiones* was begun in early 1664. The latest dated entry is April 5, 1665, on folio 58 116^v, where Newton recorded the position of a comet he had observed on that date. There is a later entry on folio 28^r that consists of a note clearly drawn from the *Philosophical Transactions* of the Royal Society for November 1665; it is among the Astronomical Notes and is not a part of the *Questiones* proper.[4] On the basis of these dates it seems sound to place a closing date on the *Questiones* of middle or late 1665. Thus, the *Questiones* were composed in a short period of less than two years in 1664–5. As will become apparent as the chronology unfolds, we believe that by far the larger part of the *Questiones* was written in 1664, when Newton was twenty-one years of age.

This period of two years can be divided into five parts, into which the various entries can be placed as follows: early

[4] The present authors have prepared an article on the contents of these Astronomical Notes that further traces the history of Newton's astronomical education.

1664, when entries were made under ten different head-ings, including eight of the original thirty-seven; middle 1664, when the largest number of entries were made, using thirty-four different headings; late 1664, when entries were made under thirteen headings; very late 1664/early 1665, the period of Newton's comet observations in December of 1664 and January of 1665 (Old Style), when entries were made under eleven headings; early to late 1665, containing five entries made after January 23, 1665 (Old Style), includ-ing a dated entry for February 19, 1665 (Old Style), on folio 46 110^v, as well as the one for April 5, 1665, mentioned earlier.

The Descartes Notes on folio 83^r and 83^v belong to the middle 1664 period, as do most of the Cartesian notes of the *Questiones* proper. The Cartesian notes are more strictly phil-osophical, whereas those of the *Questiones* pertain more to nat-ural philosophy. This probably accounts for their separation.

The reader can examine the differences in handwriting that help to differentiate the earlier and later entries of the *Questiones* by comparing photographic plates 1 and 2. The hand in folio 75 125^r (plate 2) is more vertical, with de-scenders more nearly perpendicular to the bottom of the page. The ascenders on *d*'s in 75 125^r have, in general, less curvature and extend less far to the left than they do in folio 1 88^r (plate 1). Although the letters in the two hands are about the same size, the more vertical character of the hand in 75 125^r allows the letters to crowd closer together, though the space between words remains about the same. All of these features mark a characteristic development in Newton's hand, though the change is far less evident here than it is between samples taken many years apart.

Our placement of the entries of the *Questiones* and the Descartes Notes into the five periods of time is as follows:

Early 1664

Of the First Matter, 1 88^r, 2 88^v, 4 89^v
Of Quantity, 5 90^r (from "As finite" to "of infinite points")
Conjunction of Bodies, 6 90^v, 7 91^r
Of Time and Eternity, 9 92^r (from "The representation" to "sand")

amicus Plato amicus Aristoteles magis amica veritas.

Questiones quædam Philosophicæ

Off y^e first matter

Whither it be mathematicall points: or Mathe
maticall points & parts: or a simple entity
before division. indistinct: or individualls vz. Atoms

1 Not of Mathematicall points since y^t wants di=
mentions cannot constitute a body in their
conjunction Because they will sinke into y^e
same point. An infinite number of mathema=
ticall points sink into one being added together
& y^t being still a mathematicall point is indivis=
ible but a body is divisible. bifins a Mathe=
maticall point is nothing since it is but an
imaginary entity.

Not of parts & Math: points; for such a point is
either something or nothing. if something tis a p^t
& so added betweene 2 p^ts will make a line
of 3 p^ts. if nothing, then added betwixt two
parts there is still nothing betwixt y^e 2 p^ts &
consequently ye line consists of nothing still but
2 p^ts.

Not of simple entity before division indistinct. for this
must be an union of ye parts into wch a body is
divisible since these parts may againe bee united &
become one body as they were before at the creation
now y^e nature
of union (being but a modall ens) is to depend on
its p^ts (wch are absolute entities) therefore it
cannot be y^e terme of creation, or first matter
for tis a contradiction to say y^e first matter de=
pends on some other subject [crossed out] since
y^t implyis some former matter on wch it must
depend.

I gather yt my Phantasie os yt ☉ had yt same operation
upon ye spirits in my optick nerve or yt yt same
motions are caused in my braines by both.

4 Opening my eye os looking in ye darke upon ye like
imaginations there appeared ye like pantasme as when I
shut it

5 Looking uppon white paper there appeared (by meanes
of a strong phantasie) first a spot something darker yn
ye paper wch grew blacker os blacker until there seemed
to be a dusky red spreading almost over all ye pap
sometime this spot would be red os or sometime blew

6 looking on a bright cloude there appeared ye same
phantasme as when I looked on ye white paper
untill at last I was able to make this spot glitter
amidst ye dusky red
whither I looke on ye paper or cloude
like ye ☉ in a cloude so bright yt my eye watered.

7 Imploying my selfe in other exercises for two or 3
houres, when I thought my eye was pretty well recov
ered I repeated all ye former experiment adding this
to ym yt though I shut ye distempered eye os opened
yt wth wch I looked not on ye ☉ yt I could see ye
☉ pictured on ye cloudes or other white objects almost
as plaine as if I had looked wth my distempered eye
ye other being shut os wth doeing this I made such
impress on ye optick nerve yt let me looke wth wch
eye I would ☉ offered it selfe to my vew vnless I set
my fantasie to worke on other things wch wth much
difficulty I could doe

8 If after I had thus seene ☉s image wth my left will
eye I shut it os opened my right eye all objects would
appeare coloured as when I had new seene ☉ But I
could not perceive any such motion in ye spirits of my left
eye for all objects appeared in their right colours to it vnless

Of Motion, 10 92^v, 59 116^r (from "joining" to "no parts")

Of the Celestial Matter and Orbs, 11 93^r

Of the Sun, Stars, Planets, and Comets, 12 93^v (from "Whether" to "comet's tail")

Of Rarity, Density, Rarefaction, and Condensation, 13 94^r (from "Cork" to "in water")

Of Sounds, 37 106^r (from "A man" to "pulse")

Of Touching, 39 107^r

Middle 1664

Of Atoms, 3 89^r, 63 119^r, 64 119^v, 65 120^r (from "other" to "same place")

Of Quantity, 5 90^r (from "It is indefinite" to "a greater"), 87 131^r

Of Time and Eternity, 9 92^r (from "Problem" to "water C")

Of the Sun, Stars, Planets, and Comets, 12 93^v (from "Hebrews" to "rainbow")

Of Rarity, Density, Rarefaction, and Condensation, 13 94^r (from "By" to "times")

Of Perspicuity and Opacity, 14 94^v

Of Softness, Hardness, Flexibility, Ductility, and Tractility, 16 95^v (from "Why flints" to "Descartes")

Of Heat and Cold, 18 96^v (from "Whether things" to "summer?")

Of Gravity and Levity, 19 97^r, 67 121^r (from "another's" to "liberty")

Of Fire, 24 99^v

Of Air, 25 110^r

Of Water and Salt, 26 100^v, 47 111^r (from "insomuch" to "new moons")

Of Earth, 27 101^r

Philosophy, 28 101^v

Magnetic Attraction, 29 102^r

Vegetables, 30 102^v

Filtration and Electrical Attraction, 31 103^r

Of Light, 32 103^v (from "Why light" to "mill sail")

Of Sensation, 33 104^r

Of Species Visible, 34 104^v

Of Vision, 35 105^r

Of Colors, 36 105^v, 69 122^r (from "Try" to "a white")

Of Rarity, Density, Rarefaction, and Condensation, 13 94ʳ
(from "Given two" to "body *qr*")

Of Softness, Hardness, Flexibility, Ductility, and Tractility,
16 95ᵛ (from "Why diamond dust" to "diamond")

Of Violent Motion, 21 98ʳ, 22 98ᵛ, 51 113ʳ, 53 113ᵛ, 53 114ʳ

Of Water and Salt, 47 111ʳ (from "Whether water" to
"other not")

Of Minerals, 48 111ᵛ (from "Four ounces" to "a link")

Of Motion, 53 114ʳ, 59 117ʳ (from "How much" to "pendulum")

Of Comets, 54 114ᵛ through 57 116ʳ

Of Gravity and Levity, 67 121ʳ (from "The gravity" to "than
water")

Of Colors, 69 122ʳ (drawing and table)

Of Sleep and Dreams, 89 132ʳ

Early to Late 1665

Of Place, 8 91ᵛ

Imagination and Fantasy and Invention, 43 109ʳ (from
"When I had" to "Whence"), 75 125ʳ, 76 125ᵛ

Of Meteors, 46 110ᵛ (from "Feb." to "yellow")

Of Comets, 58 116ᵛ

Of the Soul, 86 130ᵛ (from "Perhaps she" to "visible objects")

3. Newton's reading in earlier traditions and some principal sources of the *Questiones*

Newton's notebook also contains notes in Greek from Aristotle and Latin notes drawn from various Renaissance com-
pilations of ancient and Scholastic thought. These notes
were made earlier than the *Questiones* – perhaps in part as
early as 1661 – but almost certainly during the years 1662–3.
Newton either copied or paraphrased from the text that he
was reading, and he seldom provided an interpretation of
the material that he chose to record. For this reason, we
have decided not to reproduce these notes in full. But because they are Newton's notes, we shall give an account of
them in order to evaluate their significance for understanding the origins of his thought. In what follows, these notes

will be considered in the order in which they occur in the notebook.

3.1. The Greek and Latin notes

On folios 3^r–10^v Newton made notes based on Aristotle's *Organon* and Porphyry's *Isagoge*. These are probably the earliest in the notebook. In a youthful hand, they are from a four-volume edition of Aristotle's works entitled *Aristotelis Opera omnia, quae extant, Graece et Latine,* third edition (Paris, 1654), which was under the general editorship of Guillaume Du Val. There is a copy of this work in the Cambridge University Library and two in the British Library. Newton's notes are drawn from Volume I, Book I, of the edition. The contents of the initial notes match Chapters 2–5 of Part I; these are concerned with the five predicables after the *Isagoge* (pp. 2–10), and there are no notes on the remaining Chapters 6–17 of this section. Although the notes are in Greek, Newton maintained running headings in Latin in the margins. Again from Part II he has made notes on Aristotle's *Categories,* including the *Antipraedicamenta* (pp. 20–32). Of Aristotle's complete list of categories, Newton covers substance, quantity, relation, quality, and passion. Folio 8^r begins notes on the *Postpraedicamenta,* which are found on pages 42–51 of Part III, Book I, of the *Opera omnia.* Lastly, folios 10^r and 10^v contain notes entitled *De interpretatione* that begin on page 52 of Book I. There is nothing surprising in Newton's notes. They merely reproduce essential information from the Aristotelian *Organon* as it is presented in a fine edition of Renaissance scholarship.

Beginning on folio 16^r and continuing to folio 26^r, Newton compiled a series of notes entitled *"Johannis Magiri Phisiologiae Peripateticae Contractio."* These are based on the *Physiologiae peripateticae, Libri sex cum commentariis* of Joannes Magirus, a popular Scholastic compendium that went through many editions during the first half of the seventeenth century.[5] In customary style, Magirus gave the data

---

[5] The copy of the work that Newton used probably was the 1642 Cambridge edition. There was a 1619 London edition, but no other English editions existed before mid-century.

relevant to each problem to be discussed and then proceeded to adjudicate the issues in his commentary. From
Book I, Newton took notes on general topics such as motion, rest, the finite and the infinite, place, the vacuum,
and time. On the whole, the notes are brief, and they
reduce Magirus's discussion to what Newton considered
the essential points. There is, however, a distinction drawn
by Magirus (one found in other commentators) that is
present in Newton's mature thought. In Chapter 4, "Of
Motion," Magirus distinguished between two different sorts
of affections, namely, the internal in contrast to the external.[6] Newton recorded the distinction thus: "The accidents or affections of principles are either internal or external. Of these Chapters 8, 9, 10. Internal are motion (to
which the opposite is rest) and the infinite, of which
Chapter Seven."[7] According to Magirus, an internal affection is one that characterizes an individual as such (e.g.,
the specific motion or the limiting boundaries of *this* thing
as opposed to *that* thing).[8] On the other hand, an external
affection is something that any number of things have in
common (e.g., things can endure through the same stretch
of time and can exist together at a place). Thus, place and
time are external affections, because they do not, in themselves, specify the sort or kind to which a thing belongs;
for an external affection can be considered either as a
second-level or supervenient characterization or as a feature of the world that is not strictly constitutive of anything's defining nature. A distinction between internal and
external affections that invokes the second interpretation
of the latter term finds a home in the articulation of Newton's theories of absolute space and time in *De gravitatione*
(ca. 1668) and the *Principia*.[9]

From Book II of Magirus, Newton took rather extensive
notes on the Aristotelian cosmology. This is in keeping with

---

[6] Magirus, *Physiologiae peripateticae*, Lib. I, Cap. IV, p. 26.

[7] "Accidentia vel affectiones principiorum sunt vel internae, vel externae. De
quibus Cap. 8. 9. 10. Internae sunt motus (cui contraria quios) & finitum. De quo
cap. septimus." (Folio 17^r)

[8] Magirus, *Physiologiae peripateticae*, Lib. I, Cap. IV, p. 26.

[9] For a discussion of this point, see J. E. McGuire, "Existence, Actuality and
Necessity: Newton on Space and Time," *Annals of Science*, 35(1978):466–75.

an interest that is evident elsewhere in the *Questiones*. It is Descartes's cosmology that is at the center of Newton's attention, together with the cosmological views of the ancients. Apart from taking notes on the general structure of the cosmos, Newton recorded Magirus's account of the stars, the planets, and the phenomena of eclipses. From Book III he took notes on the elements, the qualities, mixtures, the temperaments, and generation and putrefaction. From Book IV there are notes on meteors, comets, and such meteorological phenomena as lightning, thunder, the rainbow, and vapors and exhalations. Magirus's commentary was no doubt part of the prescribed curriculum that Newton's tutor had him follow. It is a lengthy text, but Newton took notes from only four of its books.

The next set of notes (folios 34^r–36^r) are entitled "*Aristotelis Stagiritae Peripateticorum principis Ethicae.*" They are in Greek and in the same youthful hand as the notes from the *Organon*. As before, Newton based himself on the Du Val *Opera omnia*, and his notes are drawn from Books I and II of Volume III. Again he has supplied Latin headings in the margin. The notes have the flavor of an exercise, merely recording definitions of the various virtues and their interrelations. Folios 38^r–40^r contain another set of notes based on the *Ethica* of Eustachius of St. Paul, a popular seventeenth-century compendium.[10] Judging from the hand, these are probably the last of the notes that he entered from a Scholastic source. They are more substantial than those on Aristotle's ethics, and they deal with topics such as the appetites, free will, and the consequences of actions.

The next set of notes stretches from folio 43^r to folio 71^v, and these are the most extensive notes that Newton entered in his notebook. They are drawn from Daniel Stahl's *Axiomata philosophica, sub titulis XX*, another popular compendium of Aristotelian philosophy.[11] The hand suggests that they are from the same period as his notes from Magirus, and certainly both are later than the ethical notes based on Eustachius. Stahl's text is more advanced than that of Magi-

---

[10] Eustachius, *Ethica, sive summa moralis disciplinae in tres partes divisa* (Cambridge, 1654).

[11] Newton was probably using the second edition (Cambridge, 1645).

rus, and it is thoroughly metaphysical in its orientation. Newton took full notes on the following topics: the nature of essence; the doctrine of actuality and potentiality; the theory of causes, including the concept of final cause (exceptionally lengthy notes); the theory of the appetites and the doctrine of the will; the conception of agent and patient (lengthy notes); the doctrine of matter and form, including the view that prime matter is *"pura potentia"*; the theory of predication, including abstract and concrete terms; the doctrine of genus, species, and difference; the conception of definition, including division and distribution; the distinction between subject and accident; the theory of distinctions; the problem of truth and falsity.

Here it is interesting to note that an important concept (later to become a constituent part of Newton's conception of inertia) is present in Stahl's discussion of the causes. Newton wrote the following based on Stahl:

The same thing, insofar as it is the same thing, always does the same thing, insofar as the same means holding itself in the same manner. This axiom extends to causes that act by nature or necessity, but not to free causes (such as God is). The same [thing] always does the same, i.e., inasmuch as in it lies [*quantum in se est*], but because of diversity of objects, it would do different things. . . . [12]

Much has been written regarding the origin of Newton's use of the conception that each thing acts in accordance with its nature "inasmuch as in it lies."[13] It seems that Newton's notes from Stahl on the causes answer this question, at least in part. For it must be remembered that Descartes also used the expression in his formulation of the principle of inertia, a source that clearly influenced Newton's own formulation of the principle. Nonetheless, when Newton says in the *Principia* that "Materiae vis insita est potentia resistendi, qua corpus unumquodque, quantum in se est, perseverat in statu suo vel quiescendi vel movendi uniformiter in direc-

---

[12] "Idem quatenus idem semper facit idem quatenus idem, i.e., eodem modo se habens. Hoc axioma ad causas natura vel necessario agentes, non ad liberas; (cujusmodi est deus &c.) extendit. Semper facit idem, i.e., quantum in se est, sed ob diversitatem objectorum faciat diversa." (folio 46^v)

[13] I. Bernard Cohen, " 'Quantum in se est': Newton's Concept of Inertia in Relation to Descartes and Lucretius," *Notes and Records of the Royal Society of London*, 19(1964):131–55.

tum,"[14] he is not in a world that is conceptually different in any significant way from Stahl's text on the causes.

The last set of notes (folios 77ʳ–81ᵛ) are in Latin and are based on Gerardus Vossius's popular *Rhetorices contractae, sive partitionum oratoriarum, libri V* (Oxford, 1631). Again, the evidence of the hand places these later than either the Magirus or Stahl notes, and perhaps later than the Greek notes from Aristotle. Newton made notes from Book I, Chapters IV through XII, and from Book II, Chapters I through XV. For the most part, the notes are brief, dealing with such topics as demonstration, deliberation, conjectural reasoning, the nature of legal reasoning, and the various states of mind, such as love and anger.

It has recently been observed that the traditional curriculum at Cambridge during Newton's undergraduate days was "rapidly approaching a state of crisis."[15] Certainly, it may be the case that some of the vigor had gone out of the curriculum and that the exercises it ordained had become to some extent a matter of performance by rote. But this interpretation is not immune to challenge. Moreover, it remains the case that it points primarily to a fact about the institutionalized pedagogy of the University; it does not impugn the intellectual potential of the studies at Cambridge that it organized. As a matter of fact, Aristotelianism, in its variety of forms, continued to provide a coherent and powerful account of human experience. As a systematic metaphysical system, it had few serious pedagogic rivals throughout the seventeenth century. And as already indicated, it would be a mistake to think that Newton's intellectual pilgrimage began only after he rejected the traditional curriculum in favor of the 'mechanical philosophy.' Moreover, it will be made abundantly clear in the Commentary that Newton had definite metaphysical and epistemological interests. In many instances these can be traced to the Scholastic curriculum of his youth. All things considered, they constitute rather more

[14] Alexandre Koyré and I. Bernard Cohen, *Isaac Newton's Philosophiae Naturalis Principia Mathematica* (Cambridge, Mass.: Harvard University Press, 1972), 2 vols., Vol. I, Definitio III, p. 40. In pointing out this connection, we are not implying that Newton formulated the principle of inertia in the *Questiones*.

[15] Richard S. Westfall, *Never At Rest: A Biography of Isaac Newton* (Cambridge University Press, 1980), Chap. 3, p. 85.

than the fact that Newton derived certain doctrines and distinctions from his reading; in particular, his exposure to traditional thought (especially that originating in Greek culture) made a permanent contribution to his mental makeup.

3.2. Some principal sources of the *Questiones*

One of Newton's main sources was Walter Charleton's *Physiologia Epicuro-Gassendo-Charltoniana*. First published in 1654 (London), Charleton's book presents ancient atomism in the context of mid-seventeenth-century science and philosophy. A vast compendium of lore, the *Physiologia* makes copious use of a wide range of ancient, Medieval, and Renaissance sources in mathematics, science, atomism, and metaphysics. But it is more than an eclectic intellectual repository. Charleton presented, within a Christian context, a philosophical perspective of his own that was Epicurean in spirit.[16] In carrying out this task, he introduced into the English vernacular the work of Gassendi on the natural philosophy of Epicurus. Although there is no positive evidence to show that Newton knew Gassendi's *Syntagma philosophicum* (1658), a familiarity with the latter's edition of Diogenes Laertius, Book X, cannot be ruled out.[17]

From ancient sources on atomism, Newton was in fact familiar with Diogenes Laertius's account of the philosophers' lives. Present in his library is a copy of a seventeenth-century edition of Diogenes' *The Lives of Philosophers*. Published in London (1664) under the title *De vitis dogmatis et apophthegmatis eorum qui philosophia claruerunt, Libri X,* this distinguished work of Renaissance scholarship includes notes and annotations by Thomas Aldobrandini, Henri Estienne (Stephanus), Isaac Casaubon, and Gilles Ménage. As was common in the sixteenth and seventeenth centuries, the text is presented with Greek and Latin in facing columns. As a Greek text of the *Letter to Herodotus, Epistle I, De vitis dogmatis* is well up to modern textual standards, when com-

[16] It is often difficult to take Charleton seriously, given his prolix language, bad arguments, and changes of position.

[17] Gassendi's *Animadversiones in decimum librum Diogenis Laertii* (Lyon, 1649) was used by Charleton and may have been available to Newton. But, again, there is no positive evidence that he used it, though a copy of the Lyon edition is in the Trinity College Library (T.7.11).

pared with the text established by Herman Usener in 1887.[18] It also stands up well to the work of Furley, Giussani, and Mau.[19] Aldobrandini, Estienne, Casaubon, and Ménage were scholars of considerable ability, and each made positive contributions to the text, as well as to our understanding of it. Certainly, their collective treatment of the Epicurean fragments of Diogenes is clearly superior to Gassendi's emendations to the text in his edition of 1649.[20] Gassendi's Latin translation is also seriously misleading.[21]

Newton's copy of Diogenes is extensively dog-eared, but it has no marginal annotations. As Newton used the technique of dog-earing, a dog-ear was used not only to mark a page but often to mark a place on a page. The dog-earing corresponds in most cases with topics that are known (on independent grounds) to have been of interest to Newton during the period of the *Questiones*. Besides this, the topics on some of the folded pages express ancient views on the Divine creation of the world, a topic that interested Newton in the *Questiones*. Also significant is the fact that the dog-earing in the sections on Melissus, Leucippus, Democritus, and Epicurus is extensive; in each case the pages in question contain statements about infinity, the nature of the cosmos, the void, and minima.

Both the Huggins and the Musgrave catalogs of Newton's library list the 1621 Paris edition of Sextus Empiricus's *Opera quae extant . . . Phyrrhoniarum hypotyposeon libri III . . . Henrico Stephano interprete; adversus mathematicos . . . libri X, G. Herveto Aurelio interprete.*[22] In Latin and Greek it includes, among

[18] Herman Usener, *Epicurea* (Rome, 1963) (a reprint of the 1887 edition), pp. 3–32. Usener favorably though critically, discussed Aldobrandini, Estienne, and Ménage as editors of Epicurus, pp. xv–xviii. Also see Usener's discussion of the six manuscripts of Diogenes Laertius (especially those that are inferior) in his Preface, pp. v–lxxvi.

[19] David J. Furley, *Two Studies in the Greek Atomists* (Princeton University Press, 1967), Part I; C. Giussani, *Studi lucreziani* (Turin, 1896), Vol. I; Jürgen Mau, *Zum Problem des Infinitesimalen bie den atiken Atomisten* (Berlin: Akademie-Verlag, 1954).

[20] See Usener's Preface to *Epicurea*, pp. xv–xvii.

[21] For example, Gassendi, *Diogenis Laertii*, interpreted Epicurus's text as not only making a well-formed distinction between proportional and aliquot parts but also using these terms to mark that distinction, p. 43.

[22] See John Harrison, *The Library of Isaac Newton* (Cambridge University Press, 1978), Item 1503, p. 237. Subsequent references to books in Newton's library will contain a Harrison number.

other Pyrrhonic works, critical annotations by Henri Estienne on the "Outlines of Pyrrhonism," and observations by
the classicist Gentien Hervet on "Against the Mathemati-
cans." There is thus a good likelihood that Newton was familiar with another fine work of Renaissance scholarship concerned with an important tradition of ancient thought. Apart
from the Du Val edition of Aristotle and *De vitis dogmatis* of
Diogenes—both of which we know he read—Sextus's *Opera*
provides further grounds for supposing that Newton's
knowledge of the ancients was first hand and not restricted
to indirect sources. The availability of these works to Newton
is far from surprising when the vast groundswell of editions,
commentaries, and translations of the Greek philosophers
during the sixteenth century is considered. A great many of
these critical editions (including the individual works of Aristotle) were available in England to the British scholarly com-
munity during the seventeenth century.

The 1621 *Opera* of Sextus that Newton probably owned is
not at present in Trinity College Library's collection of his
books, though one copy is possessed by the Cambridge University Library, and another by the British Library. To establish that Newton read or consulted Sextus is important,
for one can then show that he knew the views of Diodorus
Cronus concerning atoms of time, a position that Newton
himself argues for in the *Questiones* (10 92^v and 59 117^r).
The need to posit time atoms for an indivisibilist account of
motion may well have impressed itself on Newton through
an acquaintance with Aristotle's discussion. As we have already indicated, he made notes on the *Organon* and ethical
writings from the Du Val edition of the *Opera omnia*. Volume I (Books III and IV) contains Aristotle's views on contact, continuity, and motion. Here Newton probably
encountered Aristotelian views on topics such as points,
minima, the infinite, and divisibility. Given his indivisibilism, and his practice of turning to an original source once
his interest was aroused, Newton may well have encountered Aristotle's arguments against time atoms in *Physics* (VI, 1, 231b18–30 and 232a1ff.). Aristotle, Diodorus,
and Epicurus are the primary sources for this opinion—the
first by posing the challenge to indivisibilism, the others by
consciously responding to it.

There is no evidence in Charleton for the sort of indivisibilism for which Newton argued. Nor is Henry More's *The
Immortality of the Soul* a source of this view, an opinion commonly expressed.[23] Newton certainly knew More's work,
which discussed many of the topics that interested him in
the *Questiones*. But the nature of More's influence is different from what is generally understood.

Another source that Newton made extensive use of is
René Descartes's *Opera philosophica*, third edition (Amsterdam, 1656). This work was part of Newton's library and is
now in the library of Trinity College. It contains extensive
dog-earings that match exactly the page references to Descartes's works that Newton made in the *Questiones*. There
are two further copies of this third edition in the British
Library. One is similar in every respect to Newton's copy,
except that the title page of the *Meditations* bears the date
1654 and the text is fuller. The other copy does not contain
the *Meditations*, but is otherwise the same. Here, in one volume, Newton had access to the major scientific and philosophical writings of Descartes, as well as selected pieces of
correspondence.[24] It has become fashionable to think that
Newton had a deep antipathy toward Descartes that caused
him to be excessively critical of all that the latter wrote. That
was true for certain doctrinal beliefs of the Frenchman,
such as those that Newton perceived to clash with the pre-
suppositions of his atomism. On matters of metaphysics and
epistemology, however, the early influence of Descartes is
considerable, though in some cases it is of a negative nature.
As with the *Opera*, so with certain of Newton's other texts we
have used his dog-earings as a guide through his early reading. When used in conjunction with direct or implied reference in the text, the dog-earings indicate another dimension

[23] (London, 1659). See Harrison 1113.

[24] The *Opera* does not contain Descartes's *Geometrica*. The *Opera* bears Trinity
shelf number NQ. 9.116 and is incorrectly listed by Harrison in entry 509 as an
edition of the *Principia philosophiae* (the error probably resulted because the title
page of the *Opera* is missing and the first work is the *Principia*). The confusion is
compounded by the fact that there is a 1656 Amsterdam edition of the *Principia*;
however, it was never in Newton's library. For the editions of Descartes's mathematical works that Newton used and annotated, see D. T. Whiteside, *The Mathematical Papers of Isaac Newton* (Cambridge University Press, 1967–), 8 vols., Vol. I,
pp. 20–1. The Schooten compendium of Descartes's works (Amsterdam edition
of 1659–61) is the one that Newton annotated during the middle of 1664.

of the developing thought of the young scholar. Newton made copious references to Robert Boyle. He took notes from the following works: *New Experiments Physico-Mechanicall, touching the Spring of the Air* (Oxford, 1660); *Experiments and Considerations Touching Colours* (London, 1664); *New Experiments, and observations touching Cold, or An experimental history of cold* (London, 1665). In almost every case, Newton simply took information from Boyle, often using it as a basis for posing a question or for proposing an experiment.

During the year 1664, Newton began to master an impressive range of mathematical texts, including Euclid's *Elements*. By late 1664 he had made a repeated study of Euclid's text (Isaac Barrow's 1655 edition), as his annotated copy testifies, and had begun to take the measure of Descartes's *Geometrica*.[25] Also evident in the *Questiones* is the influence of John Wallis's mathematical works, especially the *Arithmetica infinitorum* and the *Mathesis universalis*.[26] From these works Newton would learn of the definitions of number, point, line, surface, and solid, would learn of discrete and continuous magnitudes, would delve into the wonders of Euclidean proportion theory, would discover the geometry of infinity and the infinitesimal, and would confront the problems of geometrical construction. In this regard, we should not forget the inaugural mathematical lectures of Isaac Barrow, some of which Newton attended. On matters pertaining to the traditional categories, Newton also owned and read Robert Sanderson's *Logicae artis compendium* (Oxford, 1631). In the section on Aristotle's conception of continuous and discrete quantities, Newton entered the Greek terms for some key concepts in the margin of the text.[27] Sanderson's work was prescribed by the curriculum as a first-year text, but its importance for Newton's

[25] See Whiteside, *Mathematical Papers*, Vol. I, Introduction, pp. 3–15.
[26] See footnote 130 in Chapter 1.
[27] *Logicae artis, liber primus,* Cap. 10, "De Praedicamento Quantitatis," pp. 34–5. Newton wrote "Isaac Newton Trin Coll Cant 1661" on the verso of the title page. See Harrison 1442. This is identical with the inscription on the flyleaf of the notebook containing the *Questiones*. It is also interesting to observe that Newton wrote the Greek term συνεχής (continuity) in the margin beside Sanderson's discussion of continuous divisibility of quantity. This indicates that he was already aware of the Greek source for the concept, almost certainly from Aristotle, who used it as a term of art to refer to the unlimited divisibility of magnitude.

introduction to these Aristotelian topics should not be underestimated.

One surprising source is Thomas Hobbes's *Elements of Philosophy. The First Section Concerning Bodies* (London, 1656). From this work, Newton probably became familiar with the method of analysis and synthesis in a philosophical context. Moreover, Hobbes had an important influence on the development of Newton's epistemological thought in the period of the *Questiones*. These sources by no means exhaust the texts with which Newton was familiar, but they are among the more significant, either because they are cited frequently or because they contributed directly to his basic conceptual framework. Most are significant on both counts.

1

INFINITY, INDIVISIBILISM,
AND THE VOID

These topics have intrigued thinkers at least since the time of their appearance in Greek philosophy. It would be difficult to imagine the character of cosmology or metaphysics or mathematics in the absence of their persistent influence. Needless to say, they embody some of the most difficult, though fascinating, puzzles that have engaged speculative thinkers. With some simplification, it is fair to say that the puzzle of Zeno of Elea concerning infinite divisibility continues to whirl inquiring minds around in its vortex. And it is notorious that Zeno's disarmingly simple paradox can be conceived from many different perspectives. For this reason it is important to realize the various assumptions and motivations that inform approaches to the infinite. Accordingly, we have endeavored to place Newton's speculations within the broad context of traditional responses to Zeno.

1. The debt to Charleton

The influence of Charleton on Newton is seen to be direct at 1 88^r, 2 88^v, and 3 89^r. Indeed, 1 88^r consists partly of passages quoted from Charleton and partly of a simplified paraphrase of his exposition. The topic that Charleton discusses in Book II, Chapter III, Section II, of his *Physiologia* is the perplexing problem of the composition of the continuum. Charleton considers four possibilities, which Newton quotes directly on 1 88^r, with minor changes: that the continuum "must consist either (1) *of Mathematical Points;* or (2) *of Parts and Mathematical points, united;* or (3) *of a*

26

Simple Entity, before actual division, indistinct; or (4) *of Individuals,* i.e., *Atoms.*"[1]

The chapter in which Charleton's discussion is embedded is entitled "Atoms, the First and Universal Matter." It is clear from his analysis that he is concerned with the fundamental physical principles of the world. For Charleton, these are embodied in the Greek doctrine of the full and the empty (i.e., atoms and the void). So when he considers the continuous in Section II, it is the 'physical' continuum, not the 'mathematical' continuum, that is his concern. His only attempt to state and to justify this distinction comes in Chapter II of Book II, which is entitled "No Physical Continuum, infinitely Divisible." There he argues that it is impossible that the physically continuous be infinitely divisible.[2] On the distinction between the 'physically' continuous and the 'mathematically' continuous, Charleton has little to say, his main point being that the mathematical continuum is an idealization 'abstracted' from physical quantity to facilitate demonstration and proof in geometry.[3]

What should be noted at this point is the intrinsic connection in Charleton's mind between 'atoms' being the "First and Universal Matter" and his claim that the 'physical' continuum is ultimately composed of indivisibles. Although he nowhere says so explicitly, from his discussion it is clear that he supposes his atoms are physically, but not conceptually,

---

[1] *Physiologia*, p. 107. The italics are Charleton's own. It is virtually certain that Charleton derives his position on 'the continuum' from Johannes Magnenus's *Democritus reviviscens sive de atomis* (Pavia, 1646), a source he cites frequently. In Disputatio II, Caput I, Magnenus states and rejects (in favor of the atomist position) precisely the opinions on the continuum that Charleton discusses and rejects: Proposition XIV, "Continuum non constat ex punctis mathematicis"; Proposition XV, "Continuum non componitur ex partibus, & punctis mathematicis simul" (Magnenus attributes this conception to Suarez); and Proposition XVI, "Continuum constat ex simplici entitate, ante divisionem indistincta." Magnenus's discussion of this position is fully reproduced by Charleton in English. The Hague 1658 edition is cited as once having been in Newton's possession (Harrison 1014), though it is not now among the collection of his books in Trinity College Library. There is, however, a copy in Cambridge University Library. This edition is also in the library of the British Museum, along with the 1658 London edition. Although Newton undoubtedly follows Charleton's presentation of the composition of the continuum, his having direct knowledge of Magnenus cannot be overlooked.

[2] *Ibid.*, Book II, Chap. II, Sect. II, pp. 95–8.

[3] *Ibid.*, pp. 95–7; and Book III, Chap. X, Sect. I, pp. 264–5.

indivisible (as indeed he should), and hence the ultimate reason why physical things are not *in fact* infinitely divisible.[4] Given this background from his reading of Charleton, it is not surprising that Newton should begin his *Questiones* with an inquiry into the nature of 'First Matter' and conclude that it is composed of individual minima called atoms.

Let us turn to those arguments that Newton derives from Charleton. He begins by stating three alternative views, which (following Charleton) he rejects: that first matter is composed entirely of mathematical points, that it is a combination of points and parts, and, lastly, that it is a simple and unitary entity without intrinsic differentiation.

Newton wastes little time disposing of the first alternative. Taking as his premise the notion that mathematical points are dimensionless indivisibles, he argues (1 88ʳ) as follows: If mathematical points are dimensionless indivisibles, they are by nature indistinguishable, because they lack parts and therefore shape (this is implied). If they lack parts, they lack extremities, and so cannot touch part to part; but if they touch whole to whole they coincide. In this case, any number of such points will become a single point, and so even an infinitude cannot constitute the dimension of a body. Because the first alternative allows that bodies are composed *only* of mathematical points, they all become but one point – and that is indivisible. But this is contradictory to the nature of body, which is divisible.

Newton's argument avoids the prolixities of Charleton's discussion. Indeed, the commonplace principle he takes from Charleton and Henry More (though he would find it in Epicurus, Euclid, and Aristotle, as well as in other traditional discussions) is that a point added to a point cannot augment quantity.[5] And in the spirit of the Zenonian tradition, New-

---

[4] *Ibid.*

[5] *Ibid.*, Chap. III, Sect. II, p. 107, and Book III, Chap. X, Sect. I, p. 265. The original basis for the notion is Zeno's fragment 29B1. See H. Diels and W. Kranz, *Die Fragmente der Vorsokratiker* (Dublin: Weidmann, 1968). This fragment is discussed in Section 3 of the present chaper. For Epicurus's use of the principle, see David J. Furley, *Two Studies in the Greek Atomists* (Princeton University Press, 1967), Chap. 1, pp. 13–14, and Chap. 5, pp. 63–77. The principle is embodied in the Epicurean argument in Diogenes' *De vitis dogmatis* (the text that Newton owned, Harrison 519); see Section B, lines 9–11 of the *Letter to Herodotus*, p. 280. The commonplace is not explicit in Euclid, but it is certainly compatible with

ton holds that a sizeless entity cannot contribute to the size of a whole, because something without size is dimensionless. Henry More, Newton's contemporary, claims that it is "the very essence of whatsoever is, to have Parts or Extension. . . . For to take away all Extension is to reduce a thing onely to a Mathematical point, which is nothing else but pure Negation or Non-entity."[6] More states clearly a positive assumption that informs the argument (i.e., that the *having* of magnitude is identical with *being* a physical existent).

But Newton's version of the argument should also be seen in relation to another principle he holds. As just noted, he implies that points are indistinguishable because they lack parts and thus shape and size. This much he could extract directly from Charleton's and More's accounts, both of whom base their arguments on this conception. However, throughout his entire discussion of indivisibles and minima, Newton employs a principle that holds that parts are distinguishable just in case they satisfy the possibility of separation. So mathematical points are distinguishable just in case they are not in one place, but "if they touch they will touch all over and be in one place" (3 89^r). As we shall see, Newton proceeds *secundum imaginationem mathematicorum* concerning the 'places' of points. In a model designed to determine a least distance, he supposes that points can be conceived in the imagination as if they occupy 'places' separate from one another or from another thing (3 89^r). In reality, the 'parts' of bodies in fact satisfy this criterion, because bodies are divisible wholes and so are constituted of separable parts. This principle of separation, as it pertains to points considered as entities of the mathematical imagination, will be examined

Definition 1 of Book I: "A point is that which has no parts." See Thomas L. Heath, *The Thirteen Books of Euclid's Elements* (New York: Dover Publications, 1956), Vol. I, p. 153. See also Heath's discussion of the Definition on pp. 155–6. Aristotle employs the principle in *Physics*, IV. 8. 215b12–13, and in *De gen. et corr.*, A. 2. 316a11–14, 316a24–34, and A. 2. 316b18–27. All references to Aristotle are based on the Oxford editions of his works.

[6] Preface to *The Immortality*, p. 3, in *A Collection of Several Philosophical Writings* (London, 1662). In "Axiome XIV" of Chapter VI, he argues that "Magnitude cannot arise out of mere Non-Magnitude," p. 27. More's equation of the having of magnitude with being an existent is present in Zeno's fragment 29B1. This is certainly an essential part of Newton's conception of natural objects. In what follows, we shall refer to both the first edition of *The Immortality* (1659), Harrison 1113, and to the *Collection*, because Newton may well have known the latter.

carefully, especially because Newton's use of it is based on a version of the real distinction. Although the principle is found in Charleton's discussion of the parts of extension, it has Aristotelian antecedents. Aristotle describes points as unities having position relative to one another or to another thing, whereas the monad, the principle of number, is an indivisible unity that lacks position.[7] Elsewhere he is less definite, saying that mathematical objects such as points lack a real place, except "in respect of their position relative to us."[8] Here Aristotle may have Plato in mind, who holds that an indivisible (as such) cannot have a place, because it lacks a size.[9] Newton probably knew of Aristotle's discussion. Moreover, Barrow discussed the Aristotelian view in his *Lectiones* of 1665.[10] Newton attended these, but they were probably too late to influence his own conception of the 'place' of points in the *Questiones*. At 8 91^v, Newton tells us that "place is the *principium individuationis*" of quantities. He makes two claims: First, he states that the term 'place' functions as an individuating device for describing the positions of the parts of an extension, as "days, years, etc" are said to individuate the parts of duration. In this sense, the 'place' of a straight line is its position in relation to other lines, and two figures can be said to be congruent if they have the same 'place.' Second, he states that bodies become one when they are continuous, "because but in one place." Here he echoes one of Aristotle's criteria for saying that two things are continuous and one; that is, if they have common extremities they are continuous, because their extremities have the same 'place.'[11] Because points can only strictly touch whole to whole, they occupy the same 'place' and are thus indistinguishable.

[7] *Meta.*, Δ. 6. 1016b23–27; *De caelo*, II. 13. 296a17; *An. Post.*, I. 27. 87a36–37. At *Meta.*, Δ. 6. 1016b24–27 he says: ". . . that which is indivisible in quantity is called a unit if it is not divisible in any dimension and is without position, a point if it is not divisible in any dimension, and has position."

[8] *Phys.*, IV. 1. 208b22–25.

[9] *Parmenides*, Part II, 138B. In the *Physics*, Aristotle makes the related remark that there is no distinction between a point *per se* and the place where it is (*Phys.*, IV. 1. 209a11–12). The influence of the *Parmenides* on Aristotle's ontology is considerable.

[10] Isaac Barrow, *Lectiones habitae in scholis publicis academiae Cantabrigiensis* (London, 1684), Lecture VIII, 8. These lectures were delivered in 1664, 1665, and 1666, but were not published until 1684.

[11] *Phys.*, VI. 1. 231a21–231b18. See note 26.

There is nothing in Newton's position that goes beyond an assumption common to both the Aristotelians and the Epicureans (i.e., that only that which has magnitude can exist *in re,* and so points cannot). But his style and sensibility are different from those of the ancients, as well as from those of his contemporaries, though his characterization of points as imaginary (1 88^r) comes from Charleton.[12] At this stage in his thought, Newton also seems to be aware of the Aristotelian distinction between the division of quantity into 'nothings' that are merely illusory and into points that are 'somethings,' in the sense of having position, though lacking magnitude.[13] After all, he cancelled his initial claim that "a Mathematical point is Nothing, since it is but an imaginary entity" in favor of the latter view (1 88^r). This conception of points was to be important for Newton's account of nonarbitrary units of measurement with respect to continuous magnitude (3 89^{r}ff.).

But the view that points cannot be 'somethings' in the sense of having magnitude is crucial to Newton's denial that matter is composed of separable parts *and* mathematical points. He again denies that a dimensionless mathematical point can be an element in the ontology of extended matter. So his argument is this: If a point is an element in extension, it cannot, as such, contribute to matter's extension. But if it does contribute, it is a part, not a point. This, however, contradicts the hypothesis. So much is stated by Charleton in his version of the argument.[14] Here it is clear that Newton wishes to take a consistent actualist position regarding points; he wants to claim that they lack existence *simpliciter,* because they are without positive magnitude. If he had decided to talk of points as *entia modalia* in the manner of Charleton, his line of argument would have been considerably obscured.

In considering Newton's rebuttal of the third alternative that matter is "a simple entity before division indistinct," we are confronted with the fundamental assumptions of his

---

[12] *Physiologia,* Book III, Chap. X, Sect. I, p. 265.

[13] *De gen. et corr.,* A. 2. 316a24–34. This is not to claim that Newton knew Aristotle's discussion at first hand.

[14] *Physiologia,* Book II, Chap. III, Sect. II, p. 108.

position. As before, Charleton provides the framework for the argument, which in this case, though reduced to its essentials, is one that Newton follows closely. There are two important principles at stake here. But first let us analyze Newton's summary of Charleton's position, which conceives the nature of first matter in a creationalist context.

First matter, though characterized by the third alternative as a simple unity at creation, is in fact an entity composed of actually distinct parts. Were this not so, it would be impossible to speak of it as a unity, because that notion presupposes the unity of parts that compose the entity. But these parts must be its ultimate parts; otherwise the problem of unity arises again.

Newton gives two reasons to support this view. The first is adopted from Charleton, and the second is characteristic of Newton's treatment of minima throughout the *Questiones*. Following Charleton, he holds that the very "nature of union" is "but a modal ens" that depends on "its parts (which are absolute entities)."[15] As used in the Schools, a 'modal being' is one that depends on a thing's basic ontology. For example, a thing's determinate shape presupposes its extension, but being extended does not entail that something have a particular shape, merely that it be so determinable. In this case, Newton's use of the notion 'modal being' (1 88^r) specifically characterizes unity as a supervenient feature of a thing, a feature present just in case its least parts combine to form a spatially distinct individual. Indeed, this is the sense of Charleton's discussion, that is, "that a *Modal Ens* cannot subsist without conjunction to an *Absolute*." Thus, a union cannot "be conceived without *Parts,* though on the contrary, Parts may be without Union."[16] Given this line of reasoning, Charleton (and following him, Newton) concludes that only that which is capable of existing independently of other things can be the "Term of Creation." But because in this view unity depends on the elements it unifies, "the Parts of the First Matter are the Subject from whence Union is derived."[17]

Newton's second and related reason holds that the *funda-*

[15] 1 88^r. The language is Charleton's. Book II, Chap. III, Sect. II, p. 108.
[16] *Physiologia*, Book II, Chap. III, Sect. II, p. 109.   [17] *Ibid.*

mentum in re of unity must be *least* parts. If first matter is a unified being at creation, this can only be because it is constituted of parts. But these must be least parts; otherwise their unity presupposes further parts. However, this contradicts the notion of a least and indivisible part. Therefore, the notion of a simple entity without parts before division "cannot be the terms of creation or first matter." And, like Charleton, Newton concludes that first matter cannot be a "simple entity before division indistinct."

An appraisal of this reasoning must focus on two related but unexamined principles in the reasoning of Charleton and Newton: the nature of unity, and the ground of distinctions. Thus far, Newton's arguments are far from compelling as they presuppose what he will independently attempt to establish later, that is, that minima alone are truly indivisible and undifferentiated unities. The view that the unity of being reduces to the unity of parts presupposes, at this stage in the argument, that the basic notion of unity is a coming-together through aggregation. Newton's line of reasoning at this level does nothing to disenfranchise the 'Eleatic' conception that ultimate being, far from being aggregational, is a unity that involves an indissoluble connection between individuality and indivisibility.[18] To be sure, both Charleton and Newton hold that only atomic individuals satisfy the criteria necessary for the unity of ultimate being. But the claim ought to be stated here and argued for, that is, that atomic individuals that are constitutive of coming-to-be-by-aggregation can indeed exemplify 'Eleatic' being. For Newton this involves more than the claim that they are unsplittable; it means showing that minima are conceptually indivisible in the appropriate sense, namely, by virtue of their smallness and partlessness. Only if these sorts of premises are supplied and established can Newton's arguments go through with plausibility. Given these considerations, it is clear that he is not saying that all individuals are merely collective unities. This designation applies only to those en-

[18] In speaking of 'Eleatic being' we do not intend to attribute any specific doctrine to the Eleatic philosophers. Rather, the term is used to connote the view that the unity in being of a thing depends on its being identically one and the same individual everywhere. The conception is clearly in contrast to the notion of unity-through-aggregation.

tities that are capable of differentiation into actually separable parts.

But Charleton does nothing to sort out levels of unity either. He in fact confounds the issue in answering a self-imposed objection: When we grasp first matter, do we not in fact grasp a type of unity not reducible to its parts? It simply begs the question to say that because the unity of the whole is not essential to the understanding's conception of the parts of first matter that the mind can therefore have no conception of first matter's generic unity independent of its parts.[19] This is to confuse the independent ground necessary for a 'modal' conception of unity with the concept through which we make an identifying reference to an individual thing.

The second unexamined principle is perhaps more basic for Newton and Charleton. Indeed, the nature of unity as they understood it presupposes a view concerning the ground of distinctions. In his argument to show that the composition of the continuum cannot be "a Simple Entity, before actual Division, Indistinct," Charleton denies emphatically that the characteristics that are attributable to a magnitude can come into being by the mind's acts of differentiation. In other words, he holds that to say that a magnitude has the potentiality to be conceived thus and so implies that those features necessarily preexist in the thing as such. Charleton puts his view thus: "Those things which can exist being actually separate; are really distinct: but Parts can exist being actually separate; therefore are they really distinct, even before division. For Division doth not give them their peculiar Entity and Individuation, which is essential to them and the root of Distinction."[20] Newton puts the position succinctly on 2 88^v: "Those things which can exist being actually separate are really distinct, but such are the parts of matter."

These passages indicate that in the minds of Charleton and Newton there is a close relationship among individuation,

---

[19] *Physiologia*, Book II, Chap. III, Sect. II, p. 109.

[20] *Ibid.*, p. 108. Part of Charleton's argument, and his use of the real distinction, comes directly from *Disputatio II* of Magnenus's *Democritus reviviscens*, p. 168. "Quae possunt existere actu separata, distinguntur realiter, sed partes possunt existere acta separatae, ergo distinguntur realiter."

divisibility, and unity. Their particular sensibility concerning this relationship derives from an acceptance of a strong version of the real distinction. This says that the mind can individuate distinct parts in a magnitude just in case those parts are in principle capable of actual separation. But if they are capable of existing separately, they are distinct even before actual division. Thus, the ultimate basis for making spatial distinctions pertains to the intrinsic separability of the parts that constitute a given whole. Not only is a whole dependent on the priority of its parts; in addition, the mind's capacity to conceive of distinct parts *in* a thing presupposes the intrinsic separability of those parts themselves. Each of Newton's arguments thus far turns on this principle of individuation that he derives from Charleton's discussion: Only those items that satisfy the possibility of separation are *real* parts. But points in themselves cannot satisfy this principle. Not only is it inconceivable that a point could have independent existence in the manner of a physical object; in addition, it cannot be said to exist *in re*. Therefore, points cannot form the basis of divisible wholes, the unity of which is supervenient on the combination of separable parts whose distinctness the mind is able to apprehend.

The position that Newton and Charleton take raises some fundamental issues concerning the conception of quantity. Here it will be useful to compare their assumptions with the views of the Aristotelians whom they reject. In the first place, it is clear that Newton is an actualist. Like the Epicureans, he holds that in the nature of things, reality must be constituted of ultimate parts. According to this way of thinking, it is absurd to say that a quantity can be infinitely divided 'through and through.'[21] The Epicurean asks: What if an infinite process of division is completed? Surely the quantity will have been decomposed into indivisibles of zero magnitude — mere nothings. Furthermore, if the Aristotelian says that a finite quantity is infinitely divisible, is he not stating a genuine possibility? Accordingly, to say that the quantity is infinitely divisible is to say that it *can* be so di-

[21] Furley, *Two Studies*, pp. 13–14. In *De vitis dogmatis* (the text that Newton owned), the argument is at Section C, lines 1–9, p. 280. It derives ultimately from Zeno's fragment 29B1.

vided. But this is contradictory, because a finite quanitity cannot be constituted of infinite parts. The assumptions of the actualist are clear. The quantity of objects that exist extramentally is determinate and nonarbitrary. Each quantity is a whole constituted by a determinate number of parts consistent with its size. So if a quantity is mentally divided, ultimately the indivisibles of which it is constituted are reached. According to the Epicurean attitude of mind, the parts of a quantity are prior to its whole ontologically. Wholes depend on their parts. After all, we can imagine that any whole can be decomposed into its ultimate parts and then reassembled.

Aristotelianism is far removed from these attitudes of mind. Reality is not uniquely constituted of immutable and indivisible constituents. Any magnitude is infinitely divisible in the sense that the principle of division can proceed endlessly without any ultimate parts being reached.[22] Thus, in place of the actualist's need for ultimate constituents that can serve as units of measure, for the Aristotelian, measure is a matter of comparing one divisible quantity with another in terms of an indivisible unit without position.[23] Both geometry and Aristotle hold that all continua are divisible into continua. The geometer depends on such divisions for his constructions, for example, those involved in the comparison of lines, in the employment of ratios for establishing congruences between figures, in giving quadratures, and in treating problems of incommensurability. In Aristotle's view, continuous quantities are infinitely divisible potentially. For example, a line is infinitely divisible potentially, but this does not entail that it can be divided into an actual infinity of parts.[24] Lines are not composed of points. Nor can one point of a line be immediately next to another of its points.[25] Between any two points there is a line and hence a point. In effect, Aristotle denies that continua have ultimate

[22] *Phys.*, III. 6. 206b7–25. Aristotle's treatment of the epistemic and ontological priorities between parts and wholes is complex indeed, as witness his discussion in *Meta.*, Z. 10 and 11. However, his treatment of part–whole relations in geometrical contexts is relatively straightforward.
[23] *Meta.*, A. 1. 1052b15–37; 1053a9–30; *Phys.*, IV. 12. 220b1–30.
[24] *Phys.*, III. 6. 206a14–17.
[25] *De gen. et corr.*, I. 2. 317a1–12; *Phys.*, IV. 11. 220a1–25; VI. 1. 231b6ff.

parts in any actualist sense. Geometry, far from requiring them, is positive evidence for their nonexistence.

It is clear that the Aristotelian sensibility regarding unity, divisibility, and individuation is opposed to the Epicurean. For Aristotle himself, the notions of continuity and divisibility are linked with the individuation of wholes. If two things are contiguous and share a common boundary, they are one thing. But the sharing of common and unified boundaries makes things continuous; thus, there is a close link between being a whole and being continuous.[26] But "everything continuous is divisible into divisibles that are infinitely divisible."[27] This means that wholes are prior to their differentiation. For example, to say that a line is infinitely divisible is to say that the line *as such* is brought under a principle of division. There can be no question of the whole supervening on ultimate parts. For the Aristotelian, continuous wholes are ontologically and epistemologically prior to their 'parts,' whereas for the actualist, parts are prior to quantitative wholes, and the ultimate basis in terms of which the mind understands the whole. Thus, the Aristotelian not only denies that quantitative wholes are unities of their parts but also denies the actualist doctrine that the parts of such wholes are distinct just in case they are separable. Given that the Aristotelian conceives a close link between the continuity and divisibility of a magnitude, his conclusion could scarcely be otherwise. And within this framework, it need not be asked whether the 'parts' taken successively are the same kind as, or different from, the whole; for the Aristotelian, wholes are not constituted out of proper parts in the actualist sense. The actualist (Newton included) must deny Aristotelian continuity. Thus, he faces the difficult task of showing how his discrete indivisibles can be juxtaposed to each other without the advantage of Aristotle's theory of contiguity.

If the Epicurean takes his minima to be the ultimate basis of reality, Aristotle emphatically affirms that physical objects are continuous and are not composed of indivisibles. There

---

[26] *Phys.*, VI. 1. 231a21–b18; V. 3. 227a10–17. Note, however, that Aristotle distinguishes natural unity from artificial unity in this context: ". . . if there is continuity then there is necessarily contact, but if there is contact, that alone does not imply continuity: for the extremes of things may be 'together' without necessarily being one." 227a20–25.   [27] *Ibid.*, VI. 1. 231b15–16.

would seem to be a problem. In the *Physics,* Aristotle tells us that "Number and mathematical magnitudes and what is beyond the heavens seem to be unlimited due to that which is in the thought not ceasing."[28] In the *Metaphysics,* he says that the geometer brings the geometricals of his constructions (e.g., lines) to actualization by means of his thought, itself an actuality of the mind.[29] These assertions could be read as advocating a subjective interpretation of the objects of geometry.[30] Is there not a hiatus between what can be thought mathematically and what is the case in the world of nature? According to the actualist attitude of mind (one that Newton accepts), an understanding of quantity must rest on the objects of the external world. If conceptual tensions occur between the techniques of mathematics and what is the case in the world, then either the foundations of mathematics must be reformulated to reflect how the world is or mathematics must be treated as though it does not provide fundamental knowledge of natural objects.[31]

We must now ask: Whom do Charleton and Newton have in mind when they reject the notion that first matter is "a single entity before division indistinct"? A traditional candidate is the conception of *materia prima* that has frequently been attributed to Aristotle. As often understood in the seventeenth century, prime matter is bare 'stuff' devoid of all determinations, a potentiality that exists only as actualized in some determinate matter.[32]

---

[28] *Phys.,* III. 4. 203b25–26.   [29] *Meta.,* Θ. 9. 1051a21–34.

[30] For a discussion of the realist and conceptualist poles in traditional mathematical thought, see A. G. Molland, "Mathematics in the Thought of Albertus Magnus," in *Albertus Magnus and the Sciences: Commemorative Essays,* edited by James A. Weisheipl (Toronto: Pontifical Institute of Medieval Studies, 1980), pp. 464–78.

[31] For a discussion of this view among the Epicureans, see Furley, *Two Studies,* Part II, Chap. II, pp. 155–6.

[32] For seventeenth century discussions, see W. T. Costello, *The Scholastic Curriculum at Early Seventeenth Century Cambridge* (Cambridge, Mass.: Harvard University Press, 1958), Chapter III. Costello provides evidence from student notebooks at Cambridge and from writers like Keckermann that ὕλη πρώτη (prime matter) was considered to be *nec quale nec quantum, nec quid* (Chapter III, p. 74). It is interesting to note that Newton uses these characterizations (in an early metaphysical treatise now entitled *De gravitatione et aequipondio fluidorum*) in order to describe prime matter and to distinguish it from generic extension: ". . . quod extensio (cum sit et quid, et quale, et quantum) habet plus realitatis quam materia prima." See A. Rupert Hall and Marie Boas Hall, *Unpublished Scientific Papers of Isaac Newton* (Cambridge University Press, 1962), p. 107.

However, according to the Scholastic manuals of Newton's education, prime matter cannot have quantity as such.
So a more plausible target is the notion that matter is pure
extension that, though of itself indivisible, is capable of divisibility. Interestingly enough, Descartes's conception of generic extension affords an example. In his *Principia* (Part II,
Article XXII), Descartes defines extension itself as a homogeneous entity, the being of which is everywhere one and
the same. Article XXIII goes on to make his conception of
divisibility explicit:

Accordingly, all the matter in the whole universe is one and the same,
which is known by this one fact that it is extended. All the properties which
we clearly perceive in it, reduce to this one, that it is divisible (*partibilis*), and
movable (*mobilis*), according to parts, and thus is capable of all those affections, which we perceive to follow from the motion of its parts. Thus
partition, arising from thinking alone, changes nothing; but all the variety
of matter, or all the diversity of its forms, depends on motion.[33]

It is clear from this that Descartes conceives extension as
lacking intrinsic differences, because it is everywhere and
qualitatively the same. But consistent with its nature is its
capacity for divisibility. This can be realized in reality only if
the 'parts' of extension that are movable are in fact moved.
Although Descartes takes his principle of individuation to be
that which moves, this presupposes that the 'part' that is movable is already in some manner 'picked out.' How a 'part' of
truly homogeneous extension can be distinguished from extension itself is difficult to understand on Descartes's principles. He cannot appeal to a form of occasionalism, which
allows that God simply occasions mental discriminations of
extension, because he is committed to holding that the human mind is disposed to believe that such discriminations are
founded in the extramental reality of a physical world. To
deny this would be to deny that God is not a deceiver. But be
this as it may, Descartes's conception does exemplify the notion of a 'simple entity' that, though said to be distinguishable, in itself is 'indistinct.' In view of his great interest in
Descartes's philosophy, Newton (and probably also Charle-

---

[33] *Opera philosophica, Principia*, Part II, Art. XXIII, p. 32. See also Charles
Adam and Paul Tannery, *Oeuvres de Descartes* (Paris, 1973), 12 vols., Vol. VIII,
pp. 52–3. Henceforth, citations to this work will be to AT, with a reference to
the volume and page numbers.

ton) may well have had Descartes's conception of first matter in mind. After all, Descartes is vehemently opposed to atomism.[34]

At 2 88ᵛ, beginning with paragraph 2, Newton considers the homogeneity of first matter (the full) in relation to the vacuum (the empty). Although this is an important principle for an atomist, it is not surprising that Newton has deleted his remarks at this place. His intention here is to show that atomism has superior explanatory power (if compared with rival accounts) in accounting for phenomena such as rarefaction and condensation. But this puts the cart before the horse. It assumes the explanatory capacity of atomism with respect to any given phenomenon, whereas at this stage the ontological first principles of atomism are still in need of further justification. This realization probably caused Newton to cancel his remarks, intending no doubt to return to them. Because they help to amplify his views on atomism, we shall now comment on them.

Once again Newton is indebted to Charleton, in this case for his discussion of quantity, vacuum, rarefaction, and condensation.[35] Newton begins in a characteristic manner. Matter, considered in itself, is homogeneous whether it is hard or soft or "of a middle temper." So, given that it is hard, "all the parts into which it is divisible" are also hard. In this case, the intrinsic nature of matter provides no basis by which to explain change, because it is everywhere and uniformly the same. Rarity and density are therefore to be explained by one of two hypotheses: by allowing them to arise either "from vacuities interspersed or from several proportions that quantity has to its substance." At the end of Article 5 of Chapter X, Book III, Charleton concludes his discussion of rarefaction and condensation by rejecting the views of those who "imagine the Quantity of a thing to be absolutely distinct from the matter, or substance of it: and thereupon to conclude, that Rarity and Density doe consist only in the several proportions, which substance hath to Quantity."[36] There can be no question that Charle-

[34] *Opera philosophica, Principia,* Part IV, Art. CCII, p. 219. AT.VIII.325.
[35] *Physiologia,* Book III, Chap. X, Sect. I, pp. 261–5.
[36] *Ibid.,* Art. V, p. 263.

ton is a source from which Newton derives the doctrine he rejects: Even their language is similar. Like Charleton, he leaves no doubt that he supports the first hypothesis. On the assumption that it is the case, Newton observes that bodily things are not strictly continuous "as to be without distinct parts," because they are "everywhere divided by interspersed inanities." This is what we would expect, given that Newton accepts the view that parts are distinct just in case they are separable and, if in fact divided, are divided by the presence of voids between them. So, in place of the Aristotelian notion of substantial change, Newton reduces all change to the alteration of parts (specifically, to the alteration in the ratio of the full to empty parts that compose a composite thing). On traditional grounds, he takes it as evident that what is real is not identical with the full or the bodily. Whereas it is contradictory to deny that there is no part of what is that is not, there is no contradiction in asserting that some part of what is is empty: What exists need not be coextensive with what is full. Of course, it does not follow from this commitment that Newton thinks that things contain voids in the manner in which they are composed of atomic parts. This is the view of Gassendi, who holds that although the void is a genuine first principle, it is not an element, because unlike atomic parts, it does not enter into the composition of things.[37] Folio 2 88^{v} also provides definite evidence that Newton had read Kenelm Digby. Newton characterizes the position he rejects as holding that a change in density is to be explained by "the several proportions that quantity hath to its substance." Digby uses the same words to characterize the same view in his discussion of theories of rarity and density in his *Two Treaties*. Section 8 of Chapter 3 is entitled "Rarity and Density consist in the Several proportions which Quantity hath to its Substance."[38] It is no coincidence that Newton uses the same language as Charleton and Digby to state the same opinion.

[37] P. Gassendi, *Syntagma philosophicum* (London, 1658); also found in *Opera omnia* (Stuttgart-Bad: Cannstatt, 1964), 6 vols., Vol. I, Sectio I, Liber III, Cap. VIII, p. 281.

[38] *Two Treatises: in the one of which, The nature of bodies; in the other, The nature of man's soul, is looked into* (London, 1658), p. 17. See Harrison 516.

Because Newton never returns to his claim that the position he rejects "will in its due place be proved impossible," a discussion of it is warranted. It will help to clarify his own position, because the principles of the rejected account oppose the atomist explanation that Newton develops. According to Newton, this account of individual change holds that "a harder nature" is to be explained "by less quantity and a softer by more," and by implication the same explanatory principle (i.e., increase or decrease of quantity) applies to the phenomena of rarefaction and condensation. Newton's observation on this conception of change is rooted in Charleton's text. There we are told that the quantity or extension of anything is one with its matter and thus "the proper and inseparable Affection of matter or Substance."[39] This being so, there is no question, according to Charleton, that the quantity of anything can be distinguished absolutely from its matter or its substance (though the two can be distinguished by reason). It is therefore senseless to suppose, as do the Aristotelians and many of the Schoolmen, that a quantity can stand in different proportions to one and the same substance or matter.[40] If Charleton holds that quantity is not a thing distinct from substance, this does not mean that it is identical with substance, or that it is simply nothing at all. It is merely a condition of a thing's being material, namely, that its parts are capable of moving to and from one another; and this is a condition that does not necessarily exhaust the nature of individual things. So a substance is not quantified by the addition of a separate entity named quantity; rather, it is quantified by virtue of an intrinsic disposition or arrangement of its parts. Apart from Charleton's text, there are other sources known or available to Newton from which his knowledge of quantitative change could have been derived. There is, of course, the classic discussion in Aristotle's *Physics,* as well as discussions in the texts of Scholastic writers like Magirus.[41] Of the moderns, there are Descartes and Gassendi, though there is no positive evidence that Newton had read the

---

[39] *Physiologia,* Chap. X, Sect. I, Art. IV, p. 262.   [40] *Ibid.*
[41] *Phys.,* IV. 7. 214a26–35, 214b1–9; 9. 216b22–35; 217a1ff. Magirus's discussion is found in his *Physiologiae peripateticae,* Lib. I, Cap. IX, pp. 48–51.

latter at first hand.[42] In fact, we shall have occasion to note that Newton's views on the legitimization of certain atomist doctrines differ from Gassendi's.

2. Least distance and the vacuum

At paragraph 4, 2 88^v, Newton returns to his main line of argument. He makes explicit his support for the vacuum as the principle necessary (along with least parts) to explain change and to ground the divisibility of differentiable wholes. It is at this point that Newton's direct indebtedness to Charleton ends, and the place at which he begins to develop his own reasoning. Folio 2 88^v initiates a series of arguments that have as their ultimate aim the establishment of atoms of matter, distance, time, and motion. Again and again, Newton's arguments either exemplify, or are compatible with, the following questionable pattern of reasoning: If one agrees to certain assumptions that show that something has least parts, one cannot deny that they are the least. Furthermore, he appears throughout to accept the assumption that there are intrinsically small indivisibles (or absolute minima) in contrast to leasts that are comparatively small in relation to a given finite quantity. This, of course, leads to a further assumption: that if a least distance is established, it follows that it is conceptually indivisible. In other words, Newton wishes to show that Aristotle's principle – that to have magnitude and to be indivisible are incompatible notions – does not apply to indivisibles of matter, distance, time, and motion.[43] It is difficult to suppose that Newton's assumptions are tenable or even possible. But they are the sorts of assumptions that a thinker who accepts an Epicurean framework must either suppose or argue for. In mitigation, it must be said that Newton cancelled the most theoretical part of his argument in "Of Atoms" (63 119^r through 65 120^r) – probably as soon as he had finished composing it. The argument also shows signs of having been composed

[42] For Descartes, see the *Opera philosophica, Principia,* Part II, Arts. V–IX, pp. 25–7. Also see AT.VIII.42–45. For Gassendi, see *Opera,* Vol. III, *Exercitationes paradoxicae adversus Aristoteleos,* Exercitatio, III, 10, p. 171.

[43] *Phys.,* VI. 2. 232a23–27.

rapidly, directed by a compulsion of thought so characteristic of the man. In the analysis that follows, we shall treat Newton's arguments in some detail. They are often truncated and sometimes not immediately apparent in their purpose. But more than this, in the course of arguing Newton comes to realize the untenability of the principles he is employing. The chief consequence of this realization is the perception that his arguments employ three different sorts of indivisibles—the point, the indivisible of quantity, and the unit of number—that cannot function together in the numerical model he devises for measuring quantity. Although he abandons the doctrine of conceptually indivisible minima, it is from this rejection that Newton's commitment to physically unsplittable atoms is born. So, too, is a commitment to infinitesimals and indivisibles in mathematical analysis. With his invention of a 'fluxional' method for treating continuously changing quantities, this commitment begins to fade, but it is never entirely abandoned.

In reverting to his main argument, Newton invokes the principle (paragraph 3, 2 88$^\text{v}$) that all distinctions in magnitude demand the possibility of separable and hence real parts. He assumes, as well, that real parts presuppose leasts that are by nature indivisible (a principle not justified here). He argues for the following conclusion: Given that first matter is uniformly without parts, as the partisans of continuous wholes maintain, its division in fact is possible only on the supposition of a vacuum. The argument is this. If first matter is in fact divided, matter no longer exists between the two halves, because it is entirely in each of these halves: Thus a vacuum. But in order to outflank the claim (i.e., Descartes's) that first matter is indefinitely divisible, so that parts smaller than the two larger parts first divided can fill the gap between the larger, Newton asks how these smaller parts are divided without supposing a vacuum. So either a vacuum is admitted or there is an actual regress of smaller and smaller parts *ad infinitum,* but no account of how first matter is in fact divisible.

To illustrate his argument, Newton supposes that first matter is divided as small as the grains of sand. Should one of those grains be divided into parts, a third grain cannot, as such, interpose itself between them, unless they are at the right distance apart. So, on the assumption that first matter is in fact entirely divided into grains, nothing remains to fill the least distance between the divided parts of the first grain, except vacuum. To suppose that there is "smaller matter to run in and keep out vacuum" is to assume that matter is already divided into smaller parts before it is in fact divided: Thus, a vacuum exists, or else there is an infinity of smaller and smaller parts. But the latter assumption begs the question of how a homogeneous matter can actually get divided.

Next, Newton applies the same reasoning to matter that he supposes is in fact divided into *least* parts. Assuming that two of these leasts move away from one another, he states that they will not touch when the distance between them is "half their diameter." If they do, "their semidiameter will be but a mathematical point, and their diameter as two mathematical points together, i.e., as nothing, for two nothings put together make a third nothing." But this is absurd, because "the least parts of matter would be mathematical points." So, given that the leasts are indivisible minima of nonzero magnitude, only "Vacuum will then come between" their respective intervals, "and no matter will come between, since the diameter of the least particle will be as big again as that space." Newton's argument assumes what is not here established—that there are indivisible minima with dimension. But this is an assumption at odds with the Aristotelian view that the relation between the terms 'indivisible' and 'without magnitude' expresses a conceptual truth.[44] Thus, Newton appears to assume that there is a viable distinction between a mathematical indivisible that lacks parts and an ultimate unit of quantity that has size but is indivisible.

After formulating this argument, Newton deleted it, perceiving that "what is said of so little bodies [i.e., the least parts] may be said of greater," so that two arguments can be given as one. He asks us, therefore, to consider two globes, each moving from a distance toward the other. If they are

[44] Newton does not directly argue for indivisible minima until the beginning of 63 119^r: "now that this distance is indivisible (therefore the matter contained in it) is thus made plain. . . ." (see 3 89^r).

at a distance equivalent to "half the breadth of the least particle of matter," Newton states as before that no other kind of matter can "interpose itself" between them. If they did actually touch, that would imply "that the semidiameter of the least atom has no breadth, but had it not breadth the diameter could have none, and so the least particles of matter would be mathematical points." Once again, the conclusion is that "a vacuum must interpose, unless you say those atoms are as far divided as they are divisible." In other words, one must admit the existence of least parts of matter along with the principle of division – the vacuum – or simply hold that atoms are directly divided at creation by an unknown principle.

The globe argument applies the same principle as the initial argument that Newton deleted, namely, that an ontology of points results from denying that a vacuum alone can separate the small distances between the least parts. And both arguments assume the rebuttal of the claim that first matter is a combination of parts and mathematical points. It is likely that Newton's use of the globe example is adopted from More, who argues that if a globe touches a plane at a "purely indivisible" point, this is tantamount to saying that the plane consists of mathematical points. But this is absurd, because "Magnitude cannot arise out of a mere Non-Magnitude."[45] We shall return to Newton's globes in Section 5.

There is yet another reason why Newton cancelled the first argument. Thus far he had done nothing to secure the notions of indivisible magnitude and least distance. So the first argument was cancelled, and Newton embarked on his lengthy essay "Of Atoms," to which he refers us at the end of the section we have just examined. When he broached the second version of his argument (i.e., "what is said of so little bodies," etc.), he took for granted that an imaginary reader would already know what he had to say concerning the indivisibility of minima and the nature of least distance. These principles could now be assumed, though a restating of them would have helped the arguments. As we shall shortly see, Newton was far from satisfied with what he had

---

[45] *The Immortality* (1659), Axiome XIV, Book I, Chap. VI, p. 31.

written on indivisibility and least distance, for he later cancelled his entire discussion.

Before turning to the entry "Of Atoms," there is another
argument that must be considered. On 4 89ᵛ Newton offers
a rebuttal – also cancelled – to another objection that can be
made against the necessity of a vacuum. This is an objection
based on the principle of antiperistasis. Given his keen interest in Cartesian cosmology, it would appear that Newton
is referring to Descartes's theory of the vortical motion of
extension, and only secondarily to the Aristotelians.

The context for the argument is the denial, on the part of
the Aristotelians and Cartesians, that a vacuum is necessary
in order to explain motion.[46] The conception of motion that
Newton wishes to reject is the claim "that matter may move
over so little space in an instant, and other matter succeed
in an instant and so there need be no vacuum." Newton's
argument, as we shall attempt to reconstruct it, is in the
form of a *reductio ad absurdum,* and it has two separate but
related parts.

Given that the supporters of antiperistasis are committed
to the claim that motion takes place in an instant, where 'an
instant' is conceived as having atomic duration, Newton's
first argument appears to derive from the principle that
"instantaneous motion is infinitely swift." But then it can be
said that in one and the same instant it is possible that
matter (i.e., Cartesian extension) can move a minimal distance as well as through the distance of "an infinite space."
This is absurd, because moving an infinite distance is in-
commensurate with moving a finite distance in the same
unit of time.

The second part of the argument 'falls out' from the general principle that Newton appears to employ. If anything
can be said to move with infinite swiftness through an infinite distance, the speed is infinite. But this would need an
infinite impulse whose action would produce a "violent effect," the counteraction of which would demand something
with infinite power. But how can properties of this sort be

---

[46] For Aristotle, see *Phys.,* IV. 8. 215a1–30, 215b1–30, 216a1–35. Descartes's
arguments are in the *Principia,* Part II, Arts. XVI–XVIII. AT.VII.49–51, and
Opera philosophica, pp. 29–31.

attributed to "so little an agent," that is, to a small and passive bit of extension that, though capable of being moved, in itself lacks a natural principle of motion.

If this reading of Newton's argument is correct, why does he not simply reject the physical impossibility of infinite motion and all that it entails? Instead, he stresses the lack of causal powers "in so little an agent." In the first place, Newton will later give a detailed analysis of the difficulties he supposes are involved in antiperistasis in his entry "Of Violent Motion." There he argues for the need to posit real causes to account for real effects in nature, and for the claim that there can be natural motion in a vacuum. So the answer to the question may well be that Newton concentrates on the deficiencies that he perceives in the Cartesian ontology of antiperistasis. Consequently, he stresses his belief that mere extension cannot provide the resources necessary to speak of real motion and real agency, let alone prevent the absurd consequence of infinite speed.

But if this is so, Newton's argument misfires. He is imposing his own principles (at least in connection with the second part of his argument) on the antiperistasis assumptions he is subjecting to a *reductio*. After all, in the *Principles,* Descartes conceives motion as a mere mode of matter, not as something real that can exist apart from moving things. Furthermore, it is possible to read Descartes on antiperistasis in such a way that the term 'instant' denies only the temporal priority of parts, but not the sort of dependence on preceding parts involved in specifying the last part in a chain of successive motions (e.g., a Cartesian vortex).[47] Although he may not have been aware of all that is involved, in some sense Newton probably became aware of these difficulties and cancelled his argument. After all, he had decided at this point that antiperistasis had to be discussed in the context of violent motion. And he no doubt recalled the fact that here his argument properly concerns itself with indivisibility, least distance, and the vacuum as a principle of division, and not directly with the doctrine of antiperistasis.

[47] To Mersenne, May 17, 1638. See Charles Adam and Gérard Milhaud, *Descartes' Correspondance publiée avec une introduction et des notes* (Paris, 1936–63), Vol. II, pp. 261–74.

3. Finite quantity and infinite divisibility:
an Epicurean argument

"Of Atoms" is intended to give a justification of the notions of 'least distance' and 'indivisible quantity,' "the chapter" to which Newton makes reference on 4 89^v. There he characterizes "the least parts of matter" as being so little that "there cannot be a place too little for them to creep into, and then you will grant what I plead for, namely, indivisible particles."

Newton begins his account by noting with approval what Henry More has to say about the indiscerpibility of matter.[48] Consequently, he assumes that a physical minimum of matter exists, or at least supposes that More has given reasons for believing that matter is in fact unsplittable *ad infinitum*. Newton nevertheless provides another argument to show that matter "cannot be divisible *in infinitum*." And it becomes clear that he is speaking of the quantity of things in general.

Newton presents his case against the infinite divisibility of finite quantity in a Scholastic mode. It is an argument that has its *locus classicus* in Epicurus's *Letter to Herodotus,* and its probable origins in the fragment 29B1 of Zeno of Elea. Newton's version employs two premises: a major premise, that "nothing can be divided into more parts than it can possibly be constituted of," and a minor premise, that "finite matter cannot be constituted of infinite parts." Because he has already disposed of the view that a real quantity's ontology is composed of indivisible mathematical points, he now wants to show that a *finite* quantity is not divisible into an *actual* infinity of least parts. Suppose it were so divisible; then "those infinite parts added would make the same finite quantity they were before, which is against the minor." In other words, an actual infinity of extended least parts is incompatible with the intrinsic extension of a finite quantity. From this it follows "that an infinite number of extended parts (and the least parts of quantity must be extended) make a thing infinitely extended," not finitely extended.

---

[48] *The Immortality* (1659), Book I, Chap. X, pp. 68–74.

The most probable source for Newton's direct knowledge of the argument is his copy of *De vitis dogmatis,* which includes Epicurus's *Letter to Herodotus.*[49] This section of the work contains a large number of dog-earings, including the page on which the argument is found. The Greek text of the *Letter* is reproduced on the inner column of each page together with a Latin translation in the adjacent column. It provides a fair picture of Diogenes Laertius's account of Epicurus on divisibility and infinity.

After declaring for physically indivisible atoms, Epicurus argues as follows:

Moreover, in addition one must not consider that in the limited body there are an unlimited number of parts (ὄγκους), nor parts of any size you please. Therefore, we must not only do away with infinite division into smaller and smaller parts, so that we may not make everything weak and in our conceptions of the wholes be forced to let things waste into non-being by crushing them, but also we must not suppose that in limited bodies there can be a passing on (μετάβασιν) that continues to infinity, or to continually smaller parts. For it is impossible to see, when someone says that there are unlimited parts in something, however small you please, how this can still be limited in size; for it is evident that the unlimited parts must be of some size, and whatever size from which (ἐξ ὧν) they may happen to be, the size [of the whole] would be infinite.[50]

[49] *Epistle I,* Section B, lines 7–9; Section C, lines 1–9; and Section D, lines 1–2, p. 280; see Usener, *Epicurea,* pp. 3–32 for the complete text. That Newton formulates the argument in a Scholastic manner is no doubt an artifact of his Cambridge education.

[50] The translation is the responsibility of the present authors, and the Greek text from *Epistle I* is as follows:

πρὸς δὲ τούτοις οὐ δεῖ νομίζειν ἐν τῷ ὡρισμένῳ σώματι ἀπείρους ὄγκους εἶναι,[(1)] οὐδ' ὁπηλίκουσουν. ὥστε οὐ μόνον τὴν εἰς ἄπειρον τομὴν ἐπὶ τοὔλαττον ἀναιρετέον, ἵνα μὴ πάντα ἀσθενῆ ποιῶμεν. καὶ[(2)] ταῖς περιλήψεσι τῶν ἀθρόων εἰς τὸ μὴ ὂν ἀναγκαζώμεθα, τὰ ὄντα θλίβοντες, καταναλίσκειν. ἀλλὰ καὶ τὴν μετάβασιν[(3)] μὴ νομιστέον γίνεσθαι ἐν τοῖς ὡρισμένοις εἰς ἄπειρον, μηδ' [ἐπὶ] τοὔλαττον. οὔτε γὰρ ὅπως ἐπειδὰν ἅπαξ τις εἴπῃ ὅτι ἄπειροι ὄγκοι ἔν τινι ὑπάρχουσιν ἢ[(4)] ὁπηλίκοιουν. ἔστι νοῆσαι πῶς τ' ἂν ἔτι πεπερασμένων τοῦτο εἴη τὸ μέγεθος πηλίκοι γάρ τινες, δῆλον ὡς οἱ ἄπειροί εἰσιν ὄγκοι, καὶ οὗτοι ἐξ ὧν[(5)] ὁπηλίκοι ἄν ποτε ὦσιν, ἄπειρον ἂν ἦν καὶ τὸ μέγεθος.

De vitis dogmatis, Section B, lines 7–9; Section C, lines 1–9; and Section D, lines 1–2. See Usener, *Epicurea,* § 56, p. 16, for a modern setting of the text. In the transcription of the text, the Renaissance ligatures have been expanded, and the punctuation has been changed. Apart from a few minor differences (to be noted), the text is similar to the best modern editions (see Furley, *Two Studies,* Introduction, pp. 10–14, for the differences in the texts established by scholars

Epicurus's main conclusion is clear: There cannot be an un-limited number of parts in a finite magnitude. Each part must be of some size; so the whole would be infinite. But this contradicts the supposition that the body is finite.[51] Notice also the claim that an infinity of division would reduce things to nothing, and thus weaken them. And a few lines later Epicurus argues that in the infinite divisibility of a quantity the mind would be forced to traverse an infinity of parts in grasping the totality of a finite magnitude. This particular claim is not to be found in the Zenonian fragment.

This passage clearly provides the intellectual context from which Newton derives his position. But what motivates this sort of argument? In the ancient world the Zenonian puzzles concerning the nature of plurality and infinite divis-ibility exercised both the Aristotelians and the Epicureans. In effect, Zeno's argument in fragment 29B1 is this: If a process of infinite divisibility is completed, the parts either have magnitude or do not. If they do, the whole is infinitely large; if not, they cannot make up the divisible whole. On one and the same conception – infinite divisibility – it ap-pears that a magnitude is both reducible to nothing and infinitely large.[52] Newton joins the Epicurean tradition in

like Arrighetti, Brieger, Giussani, Usener, and Schneider). We have used Her-man Usener, *Epicurea* (Rome, 1963), a reprint of the 1887 Leipzig edition, as a basis of comparison for the text that Newton used. As to the significance of the in-text numbers, the following remarks are relevant: (1) Furley (pp. 11–12) takes considerable pains to establish the sense of οὐδέ in the phrase οὐδ' ὁπηλίκους οὖν, concluding that it means 'not even.' But it appears quite natural to take the sense as 'nor.' Epicurus is making two separate points: (a) that the parts of a finite magnitude cannot be infinite in number and (b) nor can they be of any size that is desired (e.g., to make them infinitely small in size does not allow an infini-tude of them to compose a finite magnitude). (2) In Usener's edition the text reads κἄν, rather than καί. (3) Concerning the sense of μετάβασιν, a term of art with Epicurus, see Furley, p. 14. (4) In some editions ὑπάρχουσιν ἤ is given as ὑπάρχουσι καί. (5) ἐξ ὧν, 'out of which,' is usually deleted from modern texts because it is redundant. We are grateful to Professor Alexander Nehamas for a discussion of the Greek text. The Latin translation of the text is often wordy and inaccurate in places, it succeeds to some extent in capturing the sense of the original. For example, Section B, lines 11–12, p. 280, reads: 'Praeterea in finito corpore, hec infinita, nec quantavis corpuscula, inesse existimare oportet." (More-over, it is necessary to reckon that in a finite body, neither infinite bodies, nor those however small [you please], are present.) Newton has made a large fold at the top of p. 281, probably to mark the text of both p. 280 and p. 281.

[51] *Ibid.*, Sections C and D, lines 1–9 and 1–3, p. 280.

[52] Zeno ends his arguments thus: "Hence if there are many things present, they must be large and small – so small as to have no magnitude, and so large as

response to this challenge: He posits indivisible minima—
units of positive yet minimum extension. Like the Epicurean actualists, Newton abandons a conception of division
that rejects ultimate parts. To say a finite quantity is infinitely divisible is to say that it consists of an infinitely large
number of parts, which is contradictory. As Newton puts
the point: "nothing can be divided into more parts than it
can possibly be constituted of, but finite matter cannot be
constituted of infinite parts." If to divide a quantity is to
divide into parts, divisibility must be constrained by the
parts that obtain. These must be proper parts, parts similar
to the whole itself, not dissimilar dimensionless points. Newton's observation that a whole of quantity must consist solely
of the parts into which it *can* be divided is related to his
argument that first matter cannot be *composed* of points, or
of points and parts. But it is also anchored in Epicurus's
specific claim that infinite division 'through and through'
would reduce the existent into the nonexistent.[53] It is probably the case that Newton knew of Zeno's difficulties with
infinite divisibility, either through Aristotle's report of them
or in the course of his undergraduate training. In any
event, the message is there in Epicurus's text.

Aristotle's strategy against Zeno uses his distinction
between potential and actual infinity. He agrees with the
Epicureans that 'through and through' division to infinity
cannot take place in actuality. In fact, in *De generatione et
corruptione* he gives a reconstruction of the reasoning that he
thinks induced the early atomist to posit indivisible minima
of various sorts. The gist is this. If a quantity is actually
divided, the result is division into real and separate parts.
This means that the quantity is everywhere divided into its
proper parts. But actual division in this sense cannot be
completed if the procedure is infinite divisibility. So, in actual-

---

to be infinite." ["οὕτως εἰ πολλά ἐστιν, ἀνάγκη αὐτά μικρά τε εἶναι καὶ μεγάλα,
μικρὰ μὲν ὥστε μὴ ἔχειν μέγεθος, μεγάλα δὲ ὥστε ἄπειρα εἶναι."]. See Diels and
Krantz, pp. 255–6, and Furley, *Two Studies,* Chap. 5, p. 64. There were many
editions of Simplicius that contained Zeno's argument available during the sixteenth and seventeenth centuries. See *Simplicii commentarii in octo Aristotelis physicae
auscultationis libros* . . . (Venice, 1526). Separate editions appeared in 1532 (Paris),
1565 (Venice), 1587 (Venice), and 1646 (Venice).

[53] *Epistle I,* Section C, lines 1–7.

ity, a quantity cannot be everywhere divisible into an infinitude of separate parts, nor, says Aristotle, can it be infinitely divided potentially at every point.[54] This would entail the possibility that the quantity could be divided everywhere simultaneously. So, although *any* given point can be actualized, *every* point cannot be actualized; there must remain divisions that can be made but that have not yet been made. And if all possible divisions of a quantity were actually made, no quantities would remain; for quantities are divisible, which contradicts the assumption that division has been made through and through. Nor is it possible to say that points without magnitude remain, because these cannot be the appropriate constituents of a quantity.[55] Thus far, Aristotle's reasoning agrees with the Epicureans and Newton. Moreover, his view of through and through division into infinity is compatible with the criteria of distinctness that Newton takes from Charleton. To say, as Newton does, that the parts of a quantity are distinct and thus distinguishable just in case they are separable is reasonable enough if the parts can in fact be separated, but to apply the criteria to actual division into infinity is an inconceivable state of affairs, as Newton's argument implies.

Let us consider two 'thought experiments' that Newton takes to illustrate further his central argument. He states that it cannot be denied that an infinite number of extended parts (when added) make an infinite extension, "if I can prove that things infinitely extended have fine [small] parts." Newton supposes that the cosmic vacuum and matter are both infinitely extended. If one of the infinitely numerous worlds that exist in the infinity of the void were to be annihilated, the void region that remained, that "very vacuum," would not be infinite. So the infinity of void extension can be said to contain finite regions and subregions. The same reasoning applies to the infinite extension of matter if we suppose that it contains "interspersed vacuities." They are not infinite, but finite, "though an infinite number of them would be so [i.e., infinite]." It follows that the parts of infinitely extended matter are themselves finite, and also that "an infinite number of finite units cannot be finite."

[54] *De gen. et corr.*, A. 2. 316b18–27.   [55] *Ibid.*, A. 2. 316a24–34.

Given the Epicurean structure of Newton's argument, the cosmic or 'separate' vacuum is fundamental, but the interstitial vacuum is present only if matter is present. Newton's source for this distinction is no doubt Charleton.[56] An evaluation of the argument depends on establishing what Newton means by infinite extension. This question will be taken up in Section 7.

It is clear that Newton's argument against the infinite divisibility of a finite quantity draws on the Epicurean tradition. Although his first direct acquaintance with the Epicurean perspective probably came from Diogenes Laertius, Charleton's interpretation of the argument, though vague, was doubtless an influence.[57]

But his treatment of the issues involved is loose and partisan for atomism. He does, however, draw an obvious consequence for the actualist: If a quantity is infinitely divisible, there are as many parts in the smallest quantity as in the largest. Thus, any proper subset of an infinite totality will be equinumerous with the whole.[58]

Whereas it is certain that Newton knew the discussions of Charleton and More, another important figure is Isaac Barrow, who became Lucasian Professor of Mathematics in 1663. Newton twice implies that he attended Barrow's inaugural series of lectures, which began on March 14, 1664. In one of his references, Newton indicates that he attended the series given in 1665.[59] Barrow deals in systematic fashion with the foundations of mathematical thinking, especially in the lectures of 1665. Among the topics treated in that year are extension, number, continued quantity, divisibility, and infinity, all issues that concern Newton in the *Questiones*.[60] Whereas it is probably the case that Newton wrote his essay "Of Atoms" during the first part of 1664, and thus too late to be directly influenced by the 1665 lectures, the series in

---

[56] *Physiologia*, Book I, Chap. IV, Sect. I, pp. 21–5.

[57] *Ibid.*, Book II, Chap. II, Sect. I, p. 91.

[58] *Ibid.*, Book II, Chap. IV, Sect. I, pp. 91–2.

[59] Univeristy Library Cambridge, Add. 3968.41, folio 84. See Westfall, *Never At Rest*, note 55, p. 131, for a discussion of Barrow's influence on Newton, and his attendance at the lectures.

[60] Isaac Barrow, *Lectiones habitae in scholis publicis academiae Cantabridgiensis 1664, 1665, 1666* (London, 1684), Lectures II and III of 1664, pp. 23–47; Lecture I of 1665, pp. 10–15.

1664 would certainly have given him a good indication of Barrow's views. Barrow is basically an Aristotelian. His views on magnitude are formed by Aristotelianism and the Euclidean modes of geometrical reasoning. On the question of continua he agrees with Aristotle and geometry and is against the indivisibilists.[61] But despite his philosophical outlook, and despite the fact that Barrow's views are in many respects antithetical to Newton's, they give a clear indication of the mathematical training in foundations available at Cambridge during Newton's undergraduate career.

Barrow holds that "there is really no quantity different from what is called magnitude or continued quantity, and that this alone ought to be considered the object of mathematics." Magnitude is the "common affection of all physical things" and is inextricably connected with the nature of all things that exist in *rerum natura*.[62] In Barrow's view, arithmetic and geometry "are not about diverse subjects, but rather they equally demonstrate properties common to one and the same subject," namely magnitude. So number (and its demonstrable properties) "really differs nothing from what is called continued quantity, but is only constructed to express and declare it."[63] Barrow reverses the traditional perspective: Geometry is prior to arithmetic.

On the nature of number itself, Barrow is a nominalist: "I say that mathematical number is not a kind of thing which has existence proper to itself and really distinct from the magnitude it denominates [*revera distinctam a magnitudine quam denominat*], but itself is only a kind of note or sign [*nota quaedam vel signum*] of magnitude considered after a certain manner."[64] In Barrow's view, a number is a mere cipher employed by the mind in considering those operations that are appropriate to the nature of magnitude. Of itself it has no mode of existence independent of its representative role.

It is within this framework of concepts that Barrow develops his divisibilist answer to the indivisibilism of the Epicureans. As we might expect, the first lecture of 1665 makes

---

[61] *Ibid.*, Lecture I of 1664, pp. 1–15.
[62] *Ibid.*, Lecture II of 1664, pp. 23–4. Translations are the responsibility of the present authors.
[63] *Ibid.*, Lecture III of 1664, p. 34.   [64] *Ibid.*, Lecture III of 1664, pp. 46–7.

many references to the authority of Aristotle, the Stoics, and Descartes and to the geometrical conception that any magnitude can be divided such that it is always possible to state a proportion of magnitude smaller than a given small one.[65] Throughout his lectures, Barrow's strategy is to argue that the indivisibilist program makes geometrical proofs difficult, if not impossible, should the principle of the perpetual divisibility of magnitude be denied.

Such sentiments do not refute the indivisibilist directly. But Barrow takes up the Epicurean argument against the infinite divisibility of finite quantity. "Epicurus objects for the patronage of his atoms that if the parts of magnitude are infinite it is not understandable how a finite magnitude can consist of such parts."[66] In reply, Barrow considers the distinction between aliquot and proportional parts of magnitude. This is a distinction that Charleton introduces (rather unsatisfactorily) in his discussion of the infinite divisibility of finite quantity, but he does nothing with it.[67] In arithmetic, an aliquot part is an integral fraction of a whole number that when multiplied by an integer reconstitutes that number; for example, 2 is an aliquot part of 10 but not 3. In a geometrical context, an aliquot part is an equal part of a whole that when added to itself is commensurate with that whole. Barrow agrees with Epicurus that "it is against reason for a finite magnitude to have infinite aliquot parts," because that entails that it can be said to have an infinite number of equal parts.[68]

He goes on to argue that it is not contrary to reason to suppose that a magnitude is divisible into proportional parts. For example, a magnitude may be divided into two parts, each of those divided into smaller and smaller parts, there being no last division beyond which another cannot be assigned. After considering various operations in geometry that involve indefinite divisibility, he concludes that it is not contradictory that "something finite contains in it an infinity of parts [*finitum aliquod infinitas in se partes continere*]: especially since nothing agrees with number, which does not

---

[65] *Ibid.*, Lecture I of 1665, pp. 1–15.   [66] *Ibid.*, Lecture I of 1665, p. 19.
[67] *Physiologia*, Book II, Chap. II, Sect. I, p. 91.
[68] Barrow, *Lectiones*, Lecture I of 1665, p. 20.

with more right agree with magnitude, which number represents and denominates [*quam numerus repraesentat ac denominat*]."[69] Although Barrow says that a finite magnitude may 'contain' infinite parts, nothing commits him to a literal interpretation of 'parts.' For Barrow, numbers are ciphers that allow the mind to conceive of the various operations that pertain to magnitude. Because numbers are mere tokens by which the mind signifies the divisions of continuous magnitude, to speak of 'parts' in a magnitude is merely to enumerate the ways in which it can be represented through the constructed properties of numbers. There can be no question of parts actually preexisting in a magnitude before the application of a numerically expressed rule of division.

It is clear that Aristotle, Epicurus, Charleton, Barrow, and Newton agree that a finite quantity cannot be reduced to an infinitude of aliquot parts. It is clear, too, that Charleton and Newton hold the doctrine of composition in a strong sense. They suppose a quantity to be divided so that no part can be divided into further parts. Consequently, the composition is such that another account of its structure will make reference to at least one other part that is divisible. Clearly, composition in this strong sense immediately rules out the infinite divisibility of finite quantity into aliquot parts. But what of infinite divisibility by proportional parts? They do not presuppose uniqueness in an object's structure. Although he does not face the question directly, Charleton's reply would probably rest on the distinction that he makes between what is solely in the mathematical imagination and what exists *in re*. From the mathematical perspective, the geometer can suppose that there is "an Infinitude of points in every the least Continuum,"[70] in order to facilitate the devising of constructions and proofs. In *rerum natura*, however, there are only indivisible physical minima, which offer no strict basis for the geometer's leasts of quantity, namely, the fictive device of dimensionless points. Furthermore, the geometer's lines and surfaces and so forth are simplifying abstractions posited by the need for geometrical exactness. How can we suppose that these ideal and exact notions have

---

[69] *Ibid.*
[70] *Physiologia*, Book II, Chap. II, Sect. I, p. 95.

any foundation in the real world, or that they can be established by laying aside the sensible properties of physical
objects?[71] Given these commitments, Charleton would probably treat a rule of division by proportional parts as a device
necessary to geometrical construction and a creation of the
geometer's imagination constrained solely by the structure
of geometry itself. But a division of this sort seems arbitrary. More important, if it pertains to a perceptible quantity, preexistent and determinate parts must obtain; but this
is absurd, because proportional parts can never exhaust a
quantity. A distinction similar to Charleton's between the
mathematical and the natural was drawn by some Medieval
indivisibilists.[72] Moreover, the view is compatible with the
Epicurean conception that mathematics contributes nothing
to the knowledge of things.[73] If Charleton's position on proportional division can be constructed from his views on
mathematics, Newton's attitude is a matter of speculation. It
is reasonable to assume, however, that he would probably
accept a view similar to Charleton's. After all, the essay "Of
Atoms" was written in the first part of 1664, and Newton
had yet to face the question of how mathematics relates to
its sensible object. It is not difficult to imagine that its full
force did not strike him until he began to construct his own
mathematics some months later. We shall take up this question in Section 6.[74]

Henry More cannot be overlooked as a source of Newton's approach to the problem of divisibility. At the begin-

[71] *Ibid.*, pp. 95–8.
[72] For a general discussion of this distinction, see A. G. Molland, "An Examination of Bradwardine's Geometry," *Archive for History of Exact Sciences,*
19(1978):136–8. The distinction is also drawn by Magnenus in his *Democritus
reviviscens,* which Charleton knew. In Definitions VI and VII of *Disputatio II,* and
in *Lemma II* of Cap. V of the same disputation, he contrasts mathematical and
physical divisibility. The first is "divisibilitas extrinseca," whereby the imagination
can assign 'parts' to a magnitude; in this case the magnitude is said to be 'synca-
tegormatically' divisible to infinity. The second mode of divisibility pertains to the
division of an object into those parts that are intrinsically present in it.
[73] See Furley, *Two Studies,* "Conclusion to Study I," pp. 155–7.
[74] The present authors have prepared a book-length study entitled *Philosophical Themes in the Early Thought of Isaac Newton.* It explores further the pertinent
intellectual background to Newton's theories of indivisibles, continuity, and infinity. It also considers his epistemological commitments in the context of his developing interests in natural philosophy.

ning of his argument for indivisible parts, Newton refers to the Cambridge Platonist's *The Immortality of the Soul,* a work full of references to the physical minima of natural things.[75] It may also be the case that Newton knew More's *A Collection of Several Philosophical Writings,* which in 1662 republished *The Immortality* along with other of his works. A copy of this work was acquired by the Trinity College Library in 1664 and may have been at Newton's disposal.[76]

More's preface to *The Immortality* contains an argument similar to Newton's. He argues that compounded matter is made up of perfect parvitudes, or of "particles that have indeed real extension, but so little, that they cannot have less and be anything at all, and therefore cannot be actually divided."[77] Although a compounded body cannot "physically and really [be] divisible in infinitum," its ultimate parts are intellectually divisible. More puts his position on divisibility in the form of a syllogism: "That which is actually divisible so farre as actual division any way can be made, is divisible into parts indiscerpible. But Matter (I mean that Integral or Compound Matter) is actually divisible as farre as actual division any way can be made."[78] Although More's position resembles Newton's, the purpose of his reasoning is different. More wishes to establish the impossibility of actually splitting a body into infinity. Eventually, physical 'indiscerpibles' are reached, which are divisible only conceptually, "For every Quantity is intellectually divisible,"[79] even an in-

---

[75] Folio 3 89ʳ.

[76] (London, 1662). The acquisition date of 1664 is marked on the title page. The library regulations of 1651 specified that only Masters of Art, Fellows, and Fellowcommoners were permitted to use the Trinity Library; only Fellows were permitted to remove books from the library. Newton became an M.A. and a Minor Fellow of Trinity College in 1667. Thus, the library was not available to Newton during the period of the *Questiones,* though it is possible that Fellows of the college may have acquired books for him from the collection. It is also of interest that in the winter of 1665–6 the roof of the library was destroyed by fire, making the books even less accessible during the period of the writing of Add. 3975 (folios 1–22). See Philip Gaskell and Robert Robson, *The Library of Trinity College Cambridge: A Short History* (Cambridge: University Printing House, 1971).

[77] (1659), Preface. The Preface lacks any consistent pagination; in fact, in places there is none at all. In the 1662 *Collection* this passage is on page 3.

[78] *Ibid.,* ". . . the parts that constitute an indiscerpible particle are real, but divisible onely intellectually" *Collection* (1662), p. 3.

[79] *The Immortality,* Book I, Chap. VI, Axiome XV, p. 31.

discerpible minimum. It is this latter claim that separates Newton from More. Newton will argue (and has already presupposed) that the minima of quantity are conceptually indivisible. This is the characteristic claim of Epicurus in his *Letter to Herodotus,* and a claim that Newton attempts to establish for reasons shortly to be examined. Newton's position differs also from Gassendi's. The latter claims in the *Syntagma* that he is not concerned with mathematical and conceptual indivisibilism, but only with physical atoms.[80]

4. Extension, indivisible quantity, and the metric of least distance

We turn now to the most spectulative aspects of Newton's argument. Having dealt with the three alternative views of first matter to his own satisfaction, and having presented an Epicurean case against the infinite divisibility of finite quantity, Newton now turns to his positive task. This is an attempt to establish that there are indivisible minima and least distances of extension. The reasoning consists of interrelated arguments that will be laid out before considering the principles on which Newton's analysis depends. However, where implicit and explicit intellectual commitments need clarification, this will be done as we proceed.

As his concern throughout has been with least parts, Newton now asks an interesting question (3 89^r): "how they are indivisible, how extended, of what figure, etc." To this end we are invited to consider a way of conceiving or 'modelling' minima that exploits intuitive relationships between number and extension. Newton proposes to compare "mathematical points to ciphers, indivisible extension to units, divisibility or compound quantity to number, i.e., a multitude of atoms to a multitude of units." It is evident that his general strategy is to exploit numerical relationships to illuminate an ontology that constitutes least and indivisible parts of extension. Numbers are collections of units that can be taken to denote least extensions. In terms of his model, Newton will argue

---

[80] Gassendi, *Opera,* Liber III, Sectio I, Cap. V, pp. 256–66.

that mathematical points and ciphers can provide a way of conceiving those least distances and indivisible parts, which form "the basis of all other extensions and the mould of atoms." Moreover, he will attempt to encapsulate, in one and the same model, least and indivisible distances, as well as indivisible minima whose places they are.

Newton begins with the conception of a least distance and asks us to reason hypothetically. Suppose that "a number of mathematical points were imbued with such a power as that they could not touch or be in one place, for if they touch they will touch all over and be in one place." In the latter case they would be ontologically indistinct from one another. But given that they can be conceived as separable and thus distinct, suppose that they are added "as close in a line as they can stand together." On the hypothesis being considered, each successive point will add extension to the length of the line, because each is distinct and separate from all its predecessors. In this case there is "a line which has *partes extra partes*," and one that satisfies the conception of the extension or quantity.

Is the distance that obtains between any two of the successive points that form the line the least that can be conceived of? Newton answers in the affirmative. Another point cannot be placed between any two now in the line without contradicting the original hypothesis, for placing another point assumes that "the former points did not lie so close but that they might lie closer." From these premises, Newton's first conclusion follows: The distance between any two points in the line is "the least that can be, and so little may an atom be and no less." Thus, a least distance is conceived as that which is equal to the smallest distance between any two separate and distinct mathematical points. In terms of the corresponding ontology of extension to which Newton's model refers (i.e., a unit of extension is the "indivisible basis of number," 64 119^v), an atomic indivisible is that which is no larger than that which can occupy the least part of the extension of the line.

Once again there is the characteristic reasoning: If one accepts certain premises, one must also accept the hypothesis based on them. Before following Newton's reasoning further, an anticipation of a major difficulty is necessary. On folios 63 119^r, 64 119^v, and 65 120^r there is incontrovertible evidence that he believes his minima of extension are conceptually indivisible. To complicate matters, his arguments (as we have seen) employ three different sorts of indivisibles: an indivisible minimum of positive extension; the mathematical point; and the indivisible unit of number. As regards the first sort of indivisible, Newton supposes that if an ultimate least distance can be established, then that distance is indivisible in the sense that it is partless. So if a minimum of extension combines its smallness with its partlessness, Newton appears to assume that there is no conceptual purchase for a would-be divider. This is the sort of indivisible that Aristotle rejects: For him, being a least of positive extension entails being divisible.

Something has already been said about Newton's conception of a point. He probably knew Aristotle's definition of a point as an indivisible with position relative to another point or to something else. It is certain that he knew Euclid's definition: "A point is that which has no parts."[81] Nor can it seriously be doubted that he was aware of Aristotle's objection to defining a point as the extremity of a line (i.e., it defines the prior in terms of the posterior).[82] Aristotle's concern is also found in Euclid. Euclid defines the point independently of the line (as does Aristotle), but then connects the two notions in Definition 3 of the *Elements*.[83] In both the Aristotelian and Euclidean traditions, if a line is divided into two parts, their sum remains the same as the original whole. The two points now at the inner ends of the two half-lines make no difference to the ends of the line. So a point is not a positive entity in the sense that it is *added* to the line to terminate it. The point itself lacks existence independent of the line, but it can be distinguished by its position relative to another point, or with respect to the line itself. In Postulates 1–3 of the *Elements*, Euclid presupposes points, lines, and circles; that is, he makes no separate assumption to affirm their existence.[84] His concern is the requirement of geomet-

---

[81] *The Elements*, edited by Heath, Vol. 1, p. 153.
[82] Aristotle, *Topics*, VI. 4. 141b21.
[83] *The Elements*, Vol. I, p. 165.   [84] *Ibid.*, p. 195.

rical constructibility. Consequently, the postulates legitimize certain types of constructions that presuppose the existence of points, lines, and circles. Thus, the postulates are not equivalent to existence claims for points, because to be constructible is a sufficient condition, not a necessary condition, for their existence. In contrast to Euclid, Aristotle supposes, without proof, the definition and the existence of points and lines.[85] His position is similar to Newton's, and the task is to specify and to define the ontological source and nature of points. As to the third indivisible – number – Newton takes over the Greek notion that number is indivisible in the sense that the unit can be neither greater nor lesser than itself. Thus, one unit is not more a unit than any other unit. Given this conception, the unit is the principle of number, and number is a multitude of units. In his attempt to wield his panoply of indivisibles, it should come as no surprise that Newton's arguments come to grief. But it will be instructive to see why. Among other things, the mathematical arguments in "Of Quantity" employ indivisible lines. This assumption raises difficulties for Newton's indivisibilist program[86] (see Section 6).

With this background in mind, notice Newton's language for describing the point. Points are said to be "naked;" to be "embued with . . . a power"; they are such that they "cannot be added into the midst" of a line, and they "cannot be put or conceived in [a] little space to divide it." These expressions indicate that Newton conceives his points as 'free-standing' items that, if separated in the imagination, can be conceived as being separate from one another and from other things.

Now, it is clear from this sort of language, and from the character of his arguments, that Newton proceeds according to the mathematical imagination. Points are pictured as if they exist 'cinematographically' across the imagination's eye. So considered, they are viewed together as demarcating a line *partes extra partes*. Initially, in Newton's model, it is the

[85] Aristotle's position is that the geometer must assume what a thing is (i.e., its definition), but must show that it is. It is only in the case of points and lines that he assumes both their definition and their existence without proof.

[86] As we shall see, Newton fails sufficiently to distinguish the 'point' from the 'unit.'

'power' with which the points are fancifully endowed that prevents them from being indistinct, not just their position in the line. As to their source, there is no need to suppose that Newton 'plucks' his points from some strange meta-physical space. Nothing prevents him from taking a point as the bisector of a line, or as its extremity, the line itself with-out positive reality distinct from a surface, just as the sur-face itself depends ontologically on the solid. Given this procedure, Newton can define a point to be what exists at the intersection of two lines, or as that which terminates a line.

To get to his 'free-standing' points, however, two levels of abstraction are involved. In the first place, according to an abstractionist account, a surface results from mentally 'lay-ing aside' certain sensible features in the object; a line is taken as the boundary of the surface, and the point is the extremity of the line. Again, Newton could agree with this. In fact, he could agree with Aristotle that infinitely many points exist potentially (but not actually on a line) and are brought into being as the mind thinks of divisions of the line.[87] But he must also intentionally 'abstract' points from their grounding in the line and imagine that each has a distinct position in aggregation. For this purpose he adopts the conceit of ascribing to each a 'power' of separation. There is clearly a tension in Newton's treatment of points. His actualist approach to perceptible quantity does not allow him to consider every geometrical object as a pure *ens ratio-nis* of the mind. So it is with points. On the other hand, Newton's demarcation model for leasts of extension de-pends on points being deployed with the sort of freedom explicit in the representative constructions of the mathe-matical imagination.

Why doesn't Newton assume that his points are on the line he wishes to demarcate? He is attempting to construct a model that focuses the mind on the demarcation of least distance and indivisible extension. This requires that points be considered as if they have a mode of existence apart from extension itself. To assume points on a line subject to

---

[87] *De gen. et corr.*, A. 2. 316a24–34; 316b20–25.

mental demarcation of course entails that between any two points another can be taken. So, however small the line segment, further points exist (at least potentially) at which the segment may be divided. In the end, this consequence will clearly defeat Newton's model. Although we cannot know the extent to which he was aware of this as he proceeded, it is impossible to deny its role in his decision to cancel the whole of folios 63 119^r, 64 119^v, and 65 120^r.

There is another reason why Newton treats his points as if they are apart from the extension they demarcate. If mathematical indivisibles cannot increase an extended quantity (see Section 1), it is because they contact whole with whole. As such, they cannot give separate parts. But in order to demarcate the line, the 'elements' themselves must be distinguishable. So the 'powers' fictively ascribed to points are posited to provide a basis for saying that what is separate is distinct and hence distinguishable. It will become apparent that individuation difficulties lurk in Newton's conception.

Newton next states (3 89^r) "that this distance [i.e., a least distance] is indivisible (and therefore the matter contained in it) is thus made plain." Having argued that there are least extended distances, Newton tries to show that these leasts are indivisible. He begins by making a fatal claim. In reference to a least distance, he holds that "a point cannot be put or conceived in this little space to divide it." On the supposition that points are unable to touch, Newton argues as before. If a point is added to a least distance that is the product of two points already as close as possible without touching, does it touch either of those points? If it does not, the original hypothesis is violated; if it does, it is ontologically indistinct from the original points. Then can the least "space be divided into which a mathematical point cannot enter to separate its parts?" Newton concludes that not only is the least distance indivisible, but so too is the atom "which is no larger than to fill up that space." Here he does not see what follows from assuming that if a distance is a least, it is also indivisible.

Newton's argument moves on toward ontological commitments that are far-reaching. His model has two explicit levels. There are mathematical points, for which he substitutes 'ciphers,' and there is the 'real' ontology–least and indivisible parts of distance, and indivisible atoms that occupy them. If points are construed as having position, they are 'somethings.' Newton generalizes this conception by arguing for an ordering relation that pertains to the mutual positions of his points. In terms of the characteristics of the ordering relation, he attempts to establish a metric for the least parts of distance, which are "the basis of all other extensions and the mould of atoms." In some respects the motivation for the argument is similar to a question raised by Grosseteste. If only one line existed, how could it be measured, when there is nothing else to compare it with?[88] Grosseteste says that it is properly measured by the number of point-indivisibles that it potentially contains; Newton, by the number of indivisible lines that actually compose it.

In order to exploit the resources of his model, Newton generalizes by considering the collective properties of a system of ciphers. Whereas a cipher is itself a mere 'nothing,' in conjunction with a system of such characters it provides a basis on which numerical properties can supervene. Newton begins by stating that a close relationship exists between the nature of "number and extension—in so much that nothing can be supposed of one, but may be so of the other." The influence of Barrow cannot be ruled out. He, too, postulates a relationship of dependence between number and continued quantity. But the kinship with Barrow's views must not be pushed too far. Nothing Newton says indicates that (following Barrow) he wishes to construe discrete quantity or multitude as ontologically dependent on continued quantity. Newton's views are compatible with the traditional conception that magnitude and multitude are species in the category of quantity. Nor, indeed, need we suppose that in Barrow's fashion he conceives unit numbers as having the properties of magnitude as their basis of construction. Newton's purpose is simply to "all along draw a similitude from numbers" and to employ a natural concatenation between a numerical series and repeated units of extension. In so doing, he places

---

[88] Robert Grosseteste, *Commentarius in VIII libros physicorum Aristotelis*, IV (231b5–8), edited by R. C. Dales (Boulder: University of Colorado Press, 1963), pp. 90–5. Grosseteste was available to the seventeenth century. See *Summa Linconiensis super octo libris physicorum Expositio Sancti Thome super libros physicorum Aristotelis* (Venice, 1515?).

the model in the arithmetical mode. Indeed, his argument treats the arithmetical mode as prior to the geometrical.

As in the case of points, Newton asks us to entertain a system of ciphers that, hypothetically considered, are "of such a nature and quality that they will resist being the same." Thus, they will not touch and become one when posited together, because their 'nature' keeps them separate and distinct. Given these conditions, Newton proceeds to consider relations between ciphers and the unit "or indivisible basis of number," between the unit and the indivisible extension of least parts, and, as before, between number and "a multitude of atoms." If to a cipher a second is added, their difference is a unit; a third added differs from the first two by "no less than two units," and so on. Thus, each additional cipher differs from the preceding by the "quantity of a unit."

Suppose we have eleven ciphers so arranged. Together they make ten distinct units or numbers. Is it possible to add another "qualified" cipher to this arrangement (e.g., between ciphers five and six)? Using the familiar pattern of reasoning, Newton replies in the negative. On his hypothesis, the 'intruded' cipher must differ individually from five and six; otherwise, "it would be added to them, not between them, and so be the same with that which it is added, which is against its nature." However, because five and six, considered in themselves, are distinguishable from each of the other ciphers, the 'intruded' cipher must differ from each of them individually and *a fortiori* from the whole collectively. This means that the cipher can exist only in the 'space' between any two of the original ciphers, as, for example, between five and six. But given Newton's initial assumption that each of the original eleven ciphers is separated by a least distance, this is impossible. So, assuming that a new or "naked" cipher (one 'unqualified') is added to the system of ten units formed by the eleven ciphers, "there is nothing between five and six (the difference or distance of the number or space between them) in which it can be, therefore it must be either added to five or six, the joint (as one may say of the numbered units)"; and *a fortiori* this applies to any other of the ciphers forming the 'joints' of the system as a whole. The new cipher therefore fails to form a division in any of the units defined by the original ciphers. Once again Newton concludes that the 'space' or unit between each pair of the ciphers is indivisible and reflects the indivisible extension of the least parts of the distance, as well as the dimensions of the atomic quantities whose places they are.

Newton immediately raises a difficulty: How can mathematical points and ciphers be endowed with qualities and powers?[89] His answer is straightforward: "I would not be mistaken as if I thought a point or cipher (which are nothings) were capable of powers or qualities, but because I thought it a supposition easy to conceive of and fit for the purpose, I ventured upon it. And though it be impossible that the thing should be so, yet it is not so to conceive it."

But he does not leave the matter at that. He suggests that ascribing powers and qualities to ciphers is merely a *façon de parler*, a way of specifying their collective relationship considered in aggregation. This is evident when he says of a cipher 'intruded' into the original eleven, "that it is not the cipher's having power to keep separated or different from that sort of being to which it is added, but the actual resistance of being one with them." Thus, he denies that ciphers possess powers that make them capable of separability, and he suggests that it is in their nature, when aggregated, to be separated one from another. When put together in a group, "the first cipher of the multitude thus qualified will be still a plain cipher, because there is no former cipher with which it should be one with. But the second cipher refusing to be what the first is makes the unit, or indivisible basis of number. So a mathematical point is not extended by having power to resist conjunction with another unless there be another point with which it refuses to be joined, and then there is distance between. . . ."

Notice that Newton offers no criteria that distinguish ciphers from any of the three sorts of indivisibles that his argument assumes. There is good reason for this. The

[89] Newton's notion that ciphers and points are endowed with 'powers' or 'qualities' may have been suggested by Magnenus. See *Disputatio II*, Propositio XXII, "Datur symphathia inter atomos," pp. 196–203.

cipher model is introduced as a formal stratagem for handling the three sorts of indivisible items. None of these indivisibles has identity as such. Consequently, the model attempts to specify an abstract ordering relation whereby the
individual character of each indivisible element is derived
from its noninterchangeability with all the other elements in
the totality. Should elements (*per impossibile*) exchange with
one another, they would become numerically indiscernible
and would forfeit the individuality uniquely conferred by
the ordering relation. Notice Newton's claim that any two
ciphers make "the unit or the indivisible basis of number."
Although he is here taking the unit as the principle of number, any unit of itself is indistinguishable from any other.
Thus, the ciphers demarcate the unit, but they also provide
the ordering arrangement whereby a "multitude of units" is
endowed with numerical character. Newton is clearly dealing with the traditional problem of expressing quantity in
terms of number. In this case, he is attempting to 'mark' off
a continuous quantity by the iteration of an indivisible unit
expressed through his ordering relation. The measured
quantity is thus conceived as divided into equal and countable units. Newton's approach is complicated by the fact
that he is marking off indivisible minima of extension,
which he takes to be demarcated by mathematical points
juxtaposed as close as possible. This complication arises
largely because he is attempting to reduce the problem of
the 'measure' of quantity to an abstract ordering relation
that ranges over his three different sorts of indivisibles.[90]

Let us first be clear about Newton's claims. When he tells
us that the collective nature of his ciphers is such that they
do not touch and become one, he invokes the principle that
the character of each cipher is distinguished from all others
according to its noninterchangeability.[91] This principle sub-

[90] It is again well to notice that Newton is attempting to construct a model for
measuring a magnitude as if there were no other in existence. Hence, he cannot
conceive measure as the comparing of one magnitude with another in virtue of
an iterative unit without position.

[91] For an interesting discussion of this principle of individuation, in connection with Newton, see Arnold Koslow, "Ontological and Ideological Issues of the
Classical Theory of Space and Time," in *Motion and Time, Space and Matter*, edited
by Peter K. Machamer and Robert G. Turnbull (Columbus: Ohio State University
Press, 1976), pp. 224–5.

sumes the earlier notion of individuation applied to points (i.e., that they are distinguishable just in case they can be said to have different positions). But the place of a point cannot be distinguished from the point; so, strictly speaking, points cannot have position.[92] Consequently, Newton attempts to avoid the 'spatialization' explicit in his treatment of the individuation of points. The cipher model constitutes his attempt, and it is clear that it involves a high level of formal abstraction. Thus, the individual character of each cipher is to be determined solely through the ordering relation of the whole. For example, if the first pair, 1-2, were to be interchanged with the second, 2-3, they would cease to be the first pair, and conversely. It is clear that Newton's ordering relation must be formal and not based on intuition, if he is to escape the individuation problem evident in the fact that points cannot have individual 'positions' as such.

Newton's model is well formed. His ordering principle is an epistemic device that ranges over *finite* elements – points, extensions, and units of number. There is thus no need to establish the relation to which it applies independently, any more than the individuating device 'taller than,' if applied to a group of men, demands that at least one man can be named separately. Nor does Newton's model suffer from problems arising from counterfactual counterparts. The model says that if a pair of ciphers *A* were to interchange with a pair *B*, *A* would be *B*. This simply unpacks the notion of identity carried by the ordering relation. As we shall see, the principle has less plausibility in Newton's later writings when he uses it to individuate the parts of infinite space that are infinite in number.

First, it is important to see that the cipher model encounters difficulties in connection with the sort of existence claim involved in an ontology of points. Given that points can be conceived in isolation from their source, they aid the mind as a device for demarcating least and indivisible parts of extension and provide a basis for their numerical ordering. So considered, an array of points is a geometrical ana-

---

[92] *Parmenides*, 138B; *Phys.*, IV. 1. 209a11.

logue of the cipher model. And if the points are taken all at once and together, the ordering relation individuates them, because they mark off the units of extension. As we noted previously, this takes place at a second level of abstraction. At the first level, points are taken on a line or surface apart from which they lack positive reality. But this means that in themselves they have no identity. Lacking size, shape, and extremities, any point is indistinguishable from any other. So, if a point can be distinguished only in relation to another or to something else, *a fortiori* Newton's 'detached' points ultimately presuppose the same ground of individuation. This means that the individual character of each point, as defined by the ordering relation, assumes the existence of points, as well as a means of specifying their ontological source and nature. But, in effect, this involves a principle of individuation that mobilizes the notion that a point is distinct just in case it can be assumed on a line, or taken as one of its extremities.

This presents a difficulty for the hypostatized points of Newton's model. If points are said to demarcate the least and indivisible parts of extension, they do so if and only if there is a direct correspondence between themselves and the extension. This renders the individuation of the hypostatized points supernumerary, for it is tantamount to assuming that the points are on the extension itself. It follows immediately that points can be interpolated indefinitely between any two that are assumed on the extension. But this defeats the device for conceiving least and indivisible parts. Ironically enough, Newton raises the issue. He concedes that "the least extension is infinitely larger than a point" and "therefore can contain an infinite number of points." His attempt to deal with this self-imposed objection will be considered shortly. For the present, suffice it to say that the path to the infinite divisibility of extension is left unchecked.

The cipher model for individuation is more than a mere phenomenon of Newton's youth. An analogue occurs in *De gravitatione* and in the *Principia* itself. In the former treatise, Newton argues that the parts of duration and the parts of space are individuated by their positions, apart from which

they have no individual identity.[93] The mind can understand these parts to be what they really are only because of
their order and position, "nor have they any other principle
of individuation apart from that order and position, which
therefore cannot change."[94] It is clear that the principle
annunciated has an intellectual kinship with the ordering
relation of the *Questiones*. There is also an analogue to the
difficulty we have been discussing. In *De gravitatione*, for
example, Newton is concerned with a theory of infinitely
extended space, the parts of which are present altogether
and eternally. If this is so, then so, too, is the order of their
positions. The thrust of Newton's argument is this: If the
mind grasps the principle that orders their positions, does it
not indirectly grasp the identity of the parts themselves?[95]
Newton's individuating principle no longer ranges over
points as demarcators of the parts of extension, but over
positions. But positions are positions of parts, and they depend for their character on the parts themselves. From an
epistemic perspective, the problem of establishing a link between a position and its part would not be serious if it were
not for a unique difficulty. Newton's space and time are
infinite, and their parts and positions are also infinite. If
there is no first or last term in the number of positions over
which the individuating principle ranges, it is necessary that
one position, any one, be nameable independent of the
others. But this cannot be done; therefore, Newton's sec-
ond-order principle cannot get mobilized. It is interesting to
observe that this difficulty is related to assigning individual
constants (names) to infinite domains. This arises in attempting to deal with universally quantified formulas as infinite strings of conjuncts.

In the Scholium on space and time in the *Principia*, the

[93] Hall and Hall, p. 103.

[94] *Ibid.*, "nec habent aliud individuationis principium praeter ordinem et posi-
tiones istas, quas proinde mutare nequeunt."

[95] *Ibid.*, 'For just as the parts of duration are individuated by their order, so
that (for instance), if yesterday were to change its order with today and become
the later, it would lose its individuality and no longer be yesterday but today: so
the parts of space are individuated by their positions, so that if any two were to
change, they would become numerically one. The parts of duration and space
are only understood to be the same as they really are as a result of their mutual
order and position."

same ordering principle appears. The order of the parts of infinite space and time is immutable, and the parts maintain the same unchanging positions "from infinity to infinity."[96] But Newton also argues that the parts of absolute space, as the potential and primary places of things, have an individuality independent of the order of their positions. Indeed, he says that "situations [*situs*] properly and truly speaking have no quantity, nor are they so much the places themselves, as the affections of places [*quam affectiones locorum*]."[97] So to speak of a situation is to conceive a part of space in accordance with its relations to other parts. This indicates that situations derive their character from the parts of space on which they depend, and not the converse as *De gravitatione* states.

Although Newton distinguishes the identity of the absolute parts of space from epistemic criteria used for making identifying references to them through the ordering of their positions, he of course lacks a way of establishing the identity conditions of the former independent of the latter. But his position in the *Principia* is open to yet another individuation difficulty. If at a given time a change of a part's situation with respect to another entails its change of identity, tracing its continuity involves quantifying over places and times. This means that the criterion of identity over time for a part in space requires that its occupant remain numerically one and the same at different times. But, equally, the criterion of identity for things occupying a part of space presupposes that the part itself remains numerically one and the same at different times. Given that Newton holds that the parts of space and time are ontologically independent of their occupants, he lacks a means to break the circle. Once again he has no way to distinguish identity conditions independent of identifying conditions, and understandably enough he stresses the epistemic role of the latter. It is surely this that lies behind his claim that we must not "confound real quantities with their relations and sensi-

[96] Isaac Newton, *Philosophiae Naturalis Principia Mathematica* (Cambridge, Mass.: Harvard University Press, 1972), edited by Alexandre Koyré and I. Bernard Cohen, 2 vols., Vol. I, pp. 48 and 50.
[97] *Ibid.*, p. 47.

ble measure";[98] that is, real places in absolute space are to be distinguished from relative places, the latter being the epistemic means available for talking about the former.

We can now return to another central difficulty in the *Questiones*. If, for Newton, least parts are indivisible parts, how does he answer the objection he puts to himself; that is, if a "least extension is infinitely larger than a point . . . [can it not] contain it and be divided by it." He tells us that atoms are so small that "they are bounded and touch others by mathematical points . . . though held apart by the atom and no power of their own." Given this conception: Newton attempts to argue that although points can be assumed on the atom's boundary, it cannot have a point situated within it.

The incoherence of his strategy will be readily apparent. Nevertheless, it is important to realize that Newton attempts to maintain that his atom is conceptually indivisible. He tells us that the atom has "no inside, no midst, nor center, but is itself all (center, inside, and midst) to the environing superficies, and all it can do is to keep those points on either side of it from touching." Thus, "you cannot put a point within it, because it has no inside. Set a point upon it and it touches its superficies." Points therefore "cannot make out a place for division." Not only that, but they cannot even be "conceived in this little space to divide it" (63 119ʳ).

Newton's claim is clear enough. An atomic minimum combines smallness with partlessness. If it lacks internal differences, it cannot be differentiated, for Newton holds that only parts that are actually separable in a quantity are distinct and hence distinguishable. But in an atom, such parts do not obtain, and so there is no place at which a point can be assumed to divide it. Moreover, "the whole atom is all in the same place." But anything that is all in one place is indistinguishably one, and hence indivisible. Unfortunately, these consequences obliterate the distinction Newton makes between dimensionless indivisibles and indivisibles of minimum extension. Even if an opponent were to grant the latter distinction, nothing Newton says prevents the claim that the mind can *conceive* a half and half again of any given

---

[98] *Ibid.*, p. 52.

minimum. After all, because it is finitely small, it is a positive extension, and so it differs only in size from a visible
extension. Although the imagination may not be able 'to
picture' a minimum quantity, the understanding can conceive that it is divisible. Again, Newton tells us that points
are "held apart by the atom and no power of their own."
Inasmuch as points are assumed on the surface of the atom
(e.g., some at its top, others at the bottom), by Newton's own
criterion of separability they are parts of the atom in that
they depend on its distinguishable 'places.' Once again it
follows that the atom is conceptually divisible, and, *a fortiori*,
so is the minimum place that it occupies. It is no help for
Newton to allow that a least extension "can contain an infinite number of points, but they must be all in the borders or
sides and outward superficies of it, and that cannot make
out a place for division." If an infinitude of points can be
assumed on the "outward superficies" of the atom, on what
grounds can it be denied that its *inside* is divisible? Newton
cannot deny that the atom lacks positive extension. That
expedient would deny that the atom is a unit of minimum
extension. If it lacks inner dimensions, it lacks size. But this
again obliterates the distinction between the point and the
unit indivisible, the very distinction that Newton is attempting to establish.

Where has Newton arrived at in his argument? There is
no doubt that the incoherence of his position forced him to
answer the question Why can a point not be assumed on a
least extension? He immediately cancelled what he had written on folios 63 119^r, 64 119^v, and 65 120^r. It is significant
that the part of the argument that turns on the claim that
the atom "has no inside, no midst, nor center" (64 119^v) is
twice deleted. This indicates that Newton had rejected his
initial assumption that atoms of quantity are conceptually
indivisible. The essays "Of Atoms" and "Of Motion" (10 92^v
and 59 117^r) were written at about the same time. In both
essays he proceeds on the assumption that if a least distance
is established, its indivisibility can also be established. In "Of
Atoms" he argues that leasts can be shown to be indivisible,
and in "Of Motion" he rightly proceeds on the assumption
that an indivisible quantity can be said to move only if ex-

tension, time, and motion are themselves composed of indivisible units. Once he realized that his atom of quantity could not be conceptually indivisible (64 119^v), he cancelled the entire argument of folios 63 119^r, 64 119^v, and 65 120^r, as well as the essay "Of Motion." The cancellation most probably occurred after he had written the whole of his essay on atomic motion and nearly all of the essay "Of Atoms." If the atom cannot be conceptually indivisible, the indivisibilist arguments Newton marshals in its support lose purchase. After all, there is no need to mobilize a framework of atomic times, distances, and motions if the problem of how an indivisible quantity moves no longer must be faced. Here is the birth of Newton's physically unsplittable atom, a commitment that will henceforth characterize his thought on the fine structure of matter. It is interesting to observe that, unlike Charleton and Gassendi, Newton does not concern himself with establishing the parameters of physical atomism (i.e., the various sizes, shapes, arrangements, orders, and positions of the atoms). He is concerned only with the nature of the primordials. All of this raises some prior questions: Why did Newton ever suppose that ultimate atoms and their places are conceptually indivisible? And why did he not see immediately that the deployment of three different sorts of indivisibles within one conceptual framework would lead to disastrous consequences? A consideration of these issues will be postponed until after the arguments in "Of Motion" have been considered, because it will involve features of Epicureanism that have not yet been dealt with.

An important question can now be asked. It raises issues that lead naturally to a consideration of Newton's views on atomic times. In Book VI of the *Physics,* Aristotle makes it clear that continua cannot be composed out of indivisibles because indivisibles cannot be continuous. Two things are continuous "if their extremities are one." But points lack parts. To talk of the extremity of one point touching another implies that they have extremities distinct from the interior. Were a continuous line composed out of points, they could not be in contact. Contact is either whole with whole, part with part, or part with whole. If indivisibles lack parts, their contact must be whole with whole. But a whole touching a whole cannot be continuous, "for the continuous has one part here and another there, and is divided into parts that differ in this way and are separated in place."[99]

It is clear that the indivisibilist must explain how his constituents compose continua (i.e., how they can connect with one another and yet account for the extension of a continuous quantity). Newton never raises this issue. Indeed, he seems to be unaware of the import of Aristotle's challenge, as well as Epicurus's response in the *Letter to Herodotus*. Newton's silence in the face of Epicurus's text is further complicated by the fact that the latter maintains a 'two-tier' theory according to which the atom itself *contains* minimal parts.[100] We agree with the majority of recent commentators on the text that *minimae partes* of the atom are conceptually indivisible, whereas the atom itself is only physically indivisible.[101] Newton's arguments reflect no such distinction, even though the basis for making it is evident in the Epicurean text he used. Indeed, for Newton, it is the atom of quantity itself that is indivisible. Why this should be so will be considered later, together with an answer to the general query already raised concerning Newton's adherence to conceptual indivisibility. But whether it is the *minimae partes* of the atom or the atom itself that is said to be indivisible, the problem of how such indivisibles connect must be answered.

[99] *Phys.*, VI. 1. 231a21–30, 231b1–6.

[100] *De vitis dogmatis, Epistle I*, Section E, lines 5–9; Section F, line 1, p. 280. Of the *minimae partes*, Epicurus says: "We see these minima in succession, starting from the first, neither all in the same place nor touching part to part, but rather in their own particular way providing the measure of magnitudes – more for a larger, fewer for a smaller. This analogy [Epicurus means the analogy between the least perceptible unit and the least conceivable unit] we must consider is followed also by the minimum in the atom. . . ."

ἑξῆς τε θεωροῦμεν ταῦτα ἀπὸ τοῦ πρώτου καταρχόμενοι καὶ οὐκ ἐν τῷ αὐτῷ, οὐδὲ μέρεσι μερῶν ἁπτόμενα, ἀλλ᾽ ἐν τῇ ἰδιότητι τῇ ἑαυτῶν, τὰ μεγέθη καταμετροῦντα, τὰ πλείω πλεῖον, καὶ τὰ ἐλάττω ἔλαττον. ταύτῃ τῇ ἀναλογίᾳ νομιστέον καὶ τὸ ἐν τῇ ἀτόμῳ ἐλάχιστον κεχρῆσθαι,

See Usener, § 58, p. 17, for a modern setting of the text. Again see Furley, *Two Studies*, pp. 17–24, for a discussion of this part of the text.

[101] See Richard Sorabji, "Atoms and Time Atoms," in *Infinity and Continuity in Ancient and Medieval Thought*, edited by Norman Kretzmann (Ithaca: Cornell University Press, 1982), pp. 37–86.

To be sure, Epicurus's argument is far from clear, and scholars are seriously divided on its interpretation. According to Sorabji, Aristotle explicitly denies only that points and instants can be arranged successively, but leaves a loophole for the indivisibles of quantity of Epicurus ánd Diodorus Cronus.[102] If Aristotle has shown that continua cannot be composed of indivisibles that lack size, it does not follow that he has also shown that indivisibles must have no size. In the face of this, Sorabji argues that Epicurus's minima can be 'next to' each other without being in contact, whereas Furley argues that Epicurus arranges his *minimae partes* in edge-to-edge contact.[103] In the end, the question resolves itself into showing the following: (a) whether or not partless indivisibles need boundaries, edges, and extremities in order to be in contact; (b) whether or not extremities are parts; (c) whether or not contact must be defined as a relation between boundaries; (d) whether or not indivisibles can be 'next to' one another without being in contact. We can only guess that Newton did not consider these difficulties. In all probability he had given up his indivisible minima before such consequences could press home.

5. Indivisibles of time and motion

In the last sentence of his essay "Of Motion," Newton inquires about his indivisibles of quantity: "But to explain how these leasts have no parts." There is no follow through in the manuscript. This is probably for the reasons already cited, the chief being that Newton had come to realize in "Of Atoms" that his conceptually indivisible atom was inco-

---

[102] *Ibid.*, p. 58. A text apparently in favor of this position (but not cited by Sorabji) is *Phys.*, V. 3. 227a29–30: ". . . for points can touch, while units can only be in succession." But here Aristotle is not speaking strictly about the relation of points if to touch one another is taken to mean arranging points successively. Thus, what he says does not contradict his opinion that it is not possible to arrange point-indivisibles so that they can constitute the truly continuous. After all, if points are said to touch, in Aristotle's view they can only satisfy the relation of being superimposed.

[103] Furley, *Two Studies*, Chap. 8, pp. 115–16 and 128. Epicurus thought that his minimum was suggested by the experience of the extremity of any perceptible object: If it is divisible, it is not an extremity; but in order to be perceptible, it must have magnitude. *De vitis dogmatis*, Sect. F, lines 1–6, pp. 280–1. See Usener, § 58, p. 17.

herent. But again, why did he ever suppose that individual atoms and their places are conceptually indivisible? It seems clear enough why he deploys the notion throughout both essays. He all along assumes that if a least part can be established, a similar *type* of argument can establish that the least is indivisible. But an argument to show that something is small does not, in itself, show that it is partless. Although partlessness may indicate conceptual indivisibility, smallness does not. If what is small has positive size, a separate argument is needed to show that it lacks parts. Newton realized he could not provide this when he reached the end of his essay "Of Atoms."

But there is a prior question already raised. Why did Newton adhere to conceptual indivisibilism in the first place? The two essays "Of Atoms" and "Of Motion" form a whole in which he deploys an interrelated set of indivisibles. Following Epicurus, Newton accepts Aristotle's supposition that if an indivisible quantity is said to move, then time, distance, and motion must also be composed of indivisible units. He therefore argues that if there are indivisible units of time and motion, then in traversing one unit of motion, an indivisible must also traverse a unit of indivisible distance in one unit of indivisible time. As a result, Newton accepts the consequence that it is never true to say of an indivisible that "it is moving," only that "it has moved."

Newton states his problem immediately: "That it may be known how motion is swifter or slower." His strategy is equally clear: to show that motion can be conceived in terms of least and basic motions that are taken to be uniform in kind. This involves the articulation of a conceptual framework that allows that a motion can be said to be swifter or slower, given that the least parts of motion, time, and distance can be demarcated. As in "Of Atoms," leasts are defined 'operationally' in terms of the relationship of two globes. If they stand together as close as possible without touching, that is a least distance. If one globe moves into contact with the other, that is least motion with respect to a least distance traversed in a least time.

Given an 'operational' definition of his basic units, Newton states his general conception (10 92^v):

There are so many parts in a line as there can stand mathematical points in a row without touching (that is, falling into) one another in it, and so many degrees of motion along the line as there can be stops and stays, and there are so many least parts of time in an hour as there can be τὸ νῦν's.[104]

Clearly, distance, time, and motion are said to be composed of least distances, times, and motions; and each of these basic units is coordinated with the other two. There is nothing in the passage, as such, that demands a commitment to a doctrine of atomic times. The Greek τὸ νῦν, which literally means 'the now,' can refer either to an instant of time or to a stretch of short duration.[105] An instant is not a short period of time, but the beginning or end of a period. Of itself, it is sizeless. A time atom is different. It has size, unlike an instant; but like an instant, it is said, strangely enough, to be indivisible. A controversial temporal entity of this sort needs some arguing for, a task that Newton undertakes in paragraphs 2 and 3 of 10 92ᵛ.

Newton's arguments follow the same pattern of reasoning that he uses to show that there are least parts of extension ("Of Atoms"). In fact, he refers (10 92ᵛ) to that essay when he says that least parts of time and motion are "proved as I proved a least part in matter." He begins in paragraph 2 by stating that "these leasts have no parts, for that implies that they are yet divisible, but they are divisible neither *prius* nor *posterius*." Given this principle, Newton entertains the supposition that the leasts of time, distance, and motion can be considered as having first and last elements that are compatible with their individual natures; but this, he attempts to show, contradicts the principle that they are leasts. The supposition cannot be true of a least distance. This is a distance traversed in an "indivisible part of time." Consequently, "there cannot be a different time ascribed to the entrance of a thing into that part of space and the leaving of it." The same can be said of a least motion; it "too is performed in an indivisible part of time and is no sooner begun than

---

[104] Newton has added an English plural to the Greek τὸ νῦν, the 'now' or 'present.'

[105] For a discussion of this issue and related topics, see Colin Strang and K. W. Mills, "Plato and the Instant," *Proceedings of the Aristotelian Society*, 48(1974): 63–96.

done." A least time is indivisible "because first and last imply several parts of time." But this contradicts the supposition that a least part of time is *the* least part.

Notice that certain conceptual relations between time, distance, and motion begin to emerge implicitly from Newton's reasoning. Both indivisible distance and indivisible motion are conceived on the assumption that time is indivisible. If time is composed of indivisibles (an assumption Newton clearly makes here), motion at an indivisible time entails motion through an indivisible distance; but in this case the motion of an indivisible through an indivisible distance entails having moved without it ever being true that it is moving. These conceptual relations only begin to emerge. The reasoning is not explicit until the arguments of paragraph 3 are developed.

Before turning to paragraph 3, these relations will bear examination. What reason is there for taking Newton's τὸ νῦν to be a time atom, rather than a sizeless instant? If Newton had a sizeless instant in mind he would hardly have bothered to counter the claim—a claim natural enough—that any temporal duration must have an earlier part and a later part. After all, it hardly makes sense to deny that an instant has an earlier and later part when it itself is but the beginning or end of a period of time. The regress is just too obvious. But more than that, Newton's talk of there being least parts of time *in* time shows unmistakably that he has time atoms in mind. An instant, because it lacks size, cannot be said to be a part of time; but a time atom, because it has size, can count as a part *in* time.[106] It can be concluded reasonably that Newton's τὸ νῦν denominates an atom of time.

Granted that motion at an indivisible time is through an indivisible distance, what evidence is there that Newton's atom of distance is a partless place rather than a sizeless point? He says that a least motion "is no sooner begun than done." This implies that least distance is not a sizeless point, but a partless place, because something *can have* traversed it. This is a kind of progress; but the traversing of a sizeless

[106] See G. E. I. Owen, "Aristotle on Time," in *Motion and Time, Space and Matter*, pp. 3–19.

point is hardly to make progress at all. Moreover, it is difficult to attach any sense to something's moving with respect
to a sizeless yet indivisible point. The partless place, then, is
an atom. And the partless time said to be adequate for
anything to traverse it is also an atom.

But can we be confident that Newton's least part of motion, which is "no sooner begun than done," is an indivisible
atom of motion? There seems to be little doubt. If time is
atomic (and it is for Newton), the traversal of a distance
must be all at once. In other words, because "there cannot
be a different time ascribed to the entrance of a thing into
[a least] part of space and the leaving of it," it follows that in
traversing that single and least part of space a moving body
must jerk. Otherwise, it can be said that there is a stage
when part of the thing has entered that least space, while a
part of it simultaneously occupies part of the original space
it was in. But because, on Newton's showing, time and distance are indivisible, this description does not apply. Therefore, the thing can never be said to be moving, but only to
have moved. This is a potential that can be truly ascribed to
it (i.e., the potential to have moved).

In paragraph 3, Newton squarely addresses the question
of how atoms of motion, distance, and time relate. What he
seeks to show, in effect, is that if we accept that an indivisible unit can move, we must also accept, in one and the same
framework, least units of space, time, and motion; whereas
in paragraph 2 he argues that if we grant an atom of time,
we must also grant least distance and least motion.

The arguments in paragraph 3, of course, presuppose the
'operational' definition of distance, time, and motion that
Newton gives in terms of the disposition of the two globes.
But they also presuppose the familiar pattern of reasoning
associated with the various relationships of the globes: If
certain assumptions are accepted that show a unit is a least,
without absurdity it cannot be denied that it is the least. The
proposition that Newton seeks to establish is that "the least
degree of motion is equal to the least distance and time." In
what follows, "is equal to" will be taken to mean 'is proportional to.' Motion is proportional to the least distance "and
not to more, because a thing moves in passing over but one

of them." That is, in order to say that a thing has moved, it need only be true that it has traversed a least distance. It is proportional to the least distance and not to *less*, "because the least motion is over some distance." However, if it were over a lesser distance, and if Newton's assumption is that the globes define a least distance such that they cannot stand closer without touching, the motion over that lesser distance would be tantamount (*per impossibile*) to a distance defined by the globes touching; in this case we would have to say that the thing traverses a dimensionless point, which is absurd. The same reasoning applies to the claim that a least motion is proportional to a least time, and not to more: To say something has moved simply requires that it has traversed a least time. Otherwise, no sense attaches to the notion that a thing can move in a period of time that is compounded of least times. Nor is the motion proportional to a lesser time, because that would imply that the motion "is done in an instant or interval of time" (Newton probably did not intend 'interval' here; the argument seems to make sense only if he meant instant). Again Newton's conception of a lesser time presupposes the structure of the globe argument as it relates to his denial of motion with respect to a distance less than a least distance. The last argument in paragraph 3 illustrates that if there is indivisible motion, there must be indivisible time and distance, so that one unit of motion traverses one unit of distance in one unit of time; otherwise, any one of these units is "liable to divisibility which contradicts the notion of an indivisible part."

Newton's views may well have their source in Sextus Empiricus's account of Diodorus Cronus (*Against the Physicists*), and also in Epicurus's *Letter to Herodotus*. According to the Huggins list of Newton's library, he owned a 1621 Paris edition of Sextus.[107] Consider this passage from Diodorus as given by Sextus:

And if a thing moves in a partless time, it traverses partless places. But if it traverses partless places, it is not moving. For when it is in the first

[107] Sextus Empiricus, *Opera quae extant* (Paris, 1621). Harrison 1503. *Liber nonus, De motu,* Section E, Line 6, p. 400; Section A, Lines 1–6, p. 401. A copy of this edition is at present in the Cambridge University Library, and another is in the British Museum.

partless (ἀμερίστους τόπους) place, it does not move, since it is still in the first partless (ἀμερεῖ) place. And when it is in the second partless place, still it does not move, but [rather] *has* moved. Thus a thing never *is* moving.[108]

This is speculation, but it seems reasonable to suggest that Newton is to some extent basing himself on this text. After all, the sources for this sort of reasoning are few indeed. As in Newton's argument, motion is construed in terms of a partless time that provides the framework for the traversing of an atom of distance. Also, like Newton's exposition, the argument assumes that a partless magnitude cannot be partly in one place and time and partly in another; this would contradict the assumption that it is partless. From this conclusion, Diodorus states explicitly that it can be said only that the magnitude *has* moved; for the magnitude is now in one place, now in another, and it cannot move *in* a place it is not yet in. In any event, the structure of an argument for atomic times is more readily available to Newton here than it is from Epicurus's references to "the smallest continuous period of time," or to "the times which are distinguished only in thought."[109] There is further evidence

---

[108] *Against the Physicists*, Book II, Chap. II, "Does Motion Exist?" lines 120–6, pp. 270–2 (Loeb Classical Library). *Sextus Empiricus*, 4 vols., Vol. III (Cambridge, Mass.: Harvard University Press, 1936).

εἰ δ᾽ ἐν ἀμερεῖ χρόνῳ τι κινεῖται, ἀμερίστους τόπους διέρχεται. εἰ δὲ ἀμερίστους τόπους διέρχεται, οὐ κινεῖται· ὅτε γὰρ ἔστιν ἐν τῷ πρώτῳ ἀμερεῖ τόπῳ, οὐ κινεῖται. ἔτι γὰρ ἔστιν ἐν τῷ πρώτῳ ἀμερεῖ τόπῳ. ὅτε δὲ ἔστιν ἐν τῷ δευτέρῳ ἀμερεῖ τόπῳ, πάλιν οὐ κινεῖται ἀλλὰ κεκίνηται. οὐκ ἄρα κινεῖταί τι.

The Greek in Harrison 1503 is the same in every respect as that in the Loeb.

[109] Usener, *Epicurea, Epistle I*, § 62, p. 19. *De vitis dogmatis*, Section D, lines 6–8; Section E, lines 1–4, p. 281.

ἀλλὰ μὴν καὶ κατὰ τὰς συγκρίσεις θάττων ἑτέρα[(1)] ἑτέρας ῥηθήσεται, τῶν ἀτόμων ἰσοταχῶν οὐσῶν, τῷ ἐφ᾽ ἕνα τόπον φέρεσθαι τὰς ἐν τοῖς ἀθροίσμασιν ἀτόμους καὶ κατὰ τὸν ἐλάχιστον[(2)] συνεχῆ χρόνον, εἰ μὴ ἐφ᾽ ἕνα κατὰ τοὺς λόγῳ θεωρητοὺς χρόνους, ἀλλὰ πυκνὸν ἀντικόπτουσιν, ἕως ἂν ὑπὸ τὴν αἴσθησιν τὸ συνεχὲς τῆς φορᾶς γίνηται.

(1) ἑτέρα is not in the text of *De vitis dogmatis*. (2) Usener agrees with the accusative after κατὰ rather than the genitive form in the text of *De vitis dogmatis*. Furley translates as follows (p. 123): "But of course, in the case of compounds, one will be said to be faster than another, although atoms all move with equal speed, by virtue of the atoms in the compounds moving in one and the same direction even over the smallest *continuous* period of time, though in the times which are distinguishable only in thought they do not move in the same direction but have frequent collisions, until the continuous tendency of their motion comes within the scope of perception."

that Newton may have gone directly to the text of Sextus. Ironically enough, Charleton also quotes from the chapter in *Against the Physicists* from which the preceding quotation is taken. Charleton is concerned to argue against those who hold that motion is impossible, and he cites the positions of Parmenides, Melissus, Zeno, and Diodorus.[110] In the course of his discussion he quotes Sextus's report of Diodorus's argument to show that a thing can move neither where it is nor where it is not, and "therefore nothing is moved."[111] Charleton does not make any mention of Sextus's report on Diodorus's arguments for various sorts of indivisibles (including time atoms) that appear in the same chapter. But given his habits of mind, it is reasonable to assume that Newton turned to these arguments once his interest in Sextus was aroused on reading Charleton's account of the Eleatics on motion.

Notice that Newton says that his leasts "are divisible neither *prius* nor *posterius*." The use of these terms argues for the influence of Boyle. In summarizing Francis Hall's views on indivisibles, Boyle says that "he [Hall] supposes Time to consist of a determinate number of indivisibles, (that is, such as have neither *prius* nor *posterius* included in them) which he calls instants."[112] These descriptive terms are Boyle's, and although appropriate to Hall's arguments, they are not found in the latter's *Tractatus*. They fit the use that Newton makes of them in his argument perfectly; so it seems highly likely that he took them over from Boyle's account of Hall. This is not to say that the conception originates with Boyle. An indivisibilist who posits atomic times must deny Aristotle's claim that every duration has a beginning and end in time; otherwise, he contracts the claim that an atom of time is partless.[113] On the whole, Boyle is rightly critical of Hall's argument, indicating where he thinks that contradictions lie. On the question of time atoms, Boyle offers a criticism motivated from the side of Aristotle: "As for his indivisible parts of Time, those also must necessarily be in *quotvis partes divisibles;* for else the same body or indivi-

---

[110] *Physiologia*, Book IV, Chap. II, Art. 7, p. 443.   [111] *Ibid.*
[112] Boyle, *A Defence of the Doctrine Touching the Spring . . . of the Air* (London, 1662), p. 102.   [113] *Phys.*, IV. 2. 219a20–29.

sible must necessarily be in divers places at the same instant."[114] But if this is so, Hall's indivisible of quantity is no longer indivisible; and so time itself must be divisible as the Aristotelians claim. Here Boyle misses the thrust of Hall's indivisibilism. In Chapter XXVI of his *Tractatus,* Hall attempts to show that there must be indivisible units of time, distance, and motion if there are indivisible units of any of these.[115] To be sure, his arguments are full of obscurity (e.g., true indivisibles are 'virtually extended' and divisible on account of that).[116] Moreover, it is not clear what motivates them. Yet they present an account of the sort of interrelations among indivisibles that may well have been suggestive to Newton if he knew of the treatise directly. There could be few contemporary sources available to the young undergraduate that, like Hall, would argue that indivisibles of quantity lack intrinsic parts.

In "Of Atoms," Newton attempts to argue for a partless atom by drawing on the interconnections among his indivisibles. On 65 120^r he supposes that if a point is put at "one extreme" of the atom and let "move toward the other extreme, then it is no sooner from one extreme but it is at the other." This suggests that if the point traverses the distance from one extreme to the other, there is a stage at which it is neither at rest nor at either of the extremes, but between them. Newton asks: "Can it then be where it is impossible for it to rest from motion?" He answers in the negative, seeming to claim that the point cannot be described as being *in* motion. Rather, it can be said to have traversed "the least distance and the least distance is from one side to another in the atom." To describe the point as moving across the distance entails that it can be partly in one place and partly in another. But this implies that the point, its motion, and the distance traversed are divisible, which contradicts the supposition that the point and the atom of distance are indivisible entities. There can be no doubt that Newton makes the same assumption as in "Of Motion," namely, that if an in-

---

[114] Boyle, *A Defence,* p. 103.
[115] Francis Hall, *Tractatus de corporum inseparabilitate* (London, 1661), pp. 175–84. Hall is, of course, Francis Linus, who objected so strenuously to Newton's early paper on light. [116] *Ibid.,* Cap. XIV, pp. 163–4.

divisible is considered to move, this requires indivisible units of distance, time, and motion. In this case it is the partless distance across the atom that Newton is trying to establish. He rejected this argument at the same time as he rejected his attempt to show that the atom is without parts.

This brings us to the two-tier structure of Epicurus's atoms. According to Epicurus, his conceptual indivisibles are *parts* of physically unsplittable atoms.[117] But for Newton it is the atom itself that is conceptually indivisible. Why is there this difference, if Newton has based himself on Epicurus's text? That text is not a model of clarity on this point; consequently, its import could be easily missed. But apart from the difficulty of the text, there is a good reason that arises from the character of Newton's thought. The model that he devises for demarcating least parts is designed to mark off the places that his atoms can occupy. Now, it is clear from his arguments that Newton's atoms of extension are all of the *same* indivisible size. This means that the units of time, distance, and motion must all have the *same* indivisible sizes if Newton's partless atoms can be said to move. Otherwise, if they move they will be partly in one time and place and partly in another. Given this picture, Newton also accepts (as indeed he must) that his partless atoms never move over a whole distance as such, but over each of the indivisibles of which the whole is composed in a series of jerks. That is, it is never true that they are moving, but only that they have moved all at once across each successive unit. Epicurus is also concerned with the places successively occupied by his moving atoms. According to Furley, Epicurus's atoms can be of *different* sizes, which means that some can "occupy more than one unit of spatial extension."[118] Accordingly, his atoms are not partless, but contain *minimae partes*. But then all of the *minimae partes* of the atoms, as well as those of space and time, must be equal; otherwise, an indivisible unit of time could be too large or too small for an indivisible unit of the atom to occupy exactly.

Apart from the problem of accounting for indivisible mo-

---

[117] *Epicurea* § 58 and § 59, pp. 17–18. *De vitis dogmatis*, Section E, p. 280, lines 6–8. See footnote 100 for passage and translation.

[118] Furley, *Two Studies*, Chap. 8, p. 129.

tion, Epicurus needs minima to forestall the problems of infinite divisibility. He argues that we have experience of such minima whenever we perceive the extremity of a perceptible quantity. If in perception there is a smallest perceptible unit that lacks discernible parts, so there is a smallest conceivable unit that lacks conceivable parts. On the basis of this analogy, Epicurus argues that because there is a partless limit in a perceptible object, there is a partless limit to the conceivability of the atom.[119] He argues further that such minima can be conceived as arranged successively across the size of the atom itself.[120] Because Newton's atoms are all of the same size, he avoids Epicurus's two-tier theory of the atom. Whether or not he was fully aware of the theory is difficult to determine, though certainly the logic of his position inclines one toward a negative answer.

Although Newton's position obviates a consideration of the number, shape, and arrangement of the *minimae partes,* he needs a surrogate for the role they play. He needs a purchase on observable states of affairs that can motivate a belief in his sort of indivisible minima. This is the function of the globes. If they are perceived to come closer and closer together, a least distance is the last that occurs before they touch. But the logic of Newton's position *seems* to imply more; he seems to say that the mind can extrapolate from the last perceptible distance between the globes to the least *conceivable* distance before they touch, a theoretical distance that is beneath the threshold of the senses. Newton is not explicit about this, but surely some such notion lies behind his reasoning. After all, as a partisan of atomism, he naturally enough holds that the sensible features of things are to be explained by the imperceptible properties of the indivisible realm beyond the senses. The choice of the globe is not accidental; it is an artifact of Newton's analysis. His model for marking off the units of extension concentrates on the notion of a least distance. Perceptible globes, rather than Epicurus's minimal perceptible units, provide the purchase

---

[119] Usener, *Epicurea,* § 58 and § 59, pp. 17–18. *De vitis dogmatis,* Section E, lines 5–9; Section F, lines 1–3, p. 280; Section A, lines 1–5, p. 281. This last section, which states the analogy, is clear and non-corrupt in Newton's edition of the *Letter to Herodotus.*

[120] *Ibid.,* § 58, p. 17, Section F, lines 1–2, p. 280.

on experience that Newton requires. In the end, he is prevented by the logic of his reasoning from showing that his
minima are partless, though for a time his critical awareness
of the problem is forestalled. And it is forestalled because
Newton attempts to encapsulate in one abstract model a
means of determining the nature of all minima at once,
whether they are least parts of extension, distance, time, or
motion. It is one thing to hold that a minimum, by virtue of
its smallness, can be said to lack integral parts in the sense
that is has no parts that can be spatially separated; it is quite
another to hold that this provides a basis for saying that it is
conceptually indivisible. Ironically, it is the falseness of this
entailment that More's use of the example of a globe touching the plane is intended to illustrate (see Section 2).[121]

We can now consider Newton's initial question: how to
conceive whether motion is swift or slow. Unfortunately,
he does not give an answer to this, because the entry "Of
Motion" was never completed. But it can be conjectured
that he may well have had some Epicurean arguments in
mind. Epicurus distinguishes atoms that all move at the
same speed "as quick as thought" and in "the times which
are distinguishable only in thought" from entities compounded of atoms that exhibit different speeds over continuous periods of time.[122] Thus, Epicurus envisages his
atoms as having traversed atomic distances in atomic times,
thus providing a uniform basis for conceiving the different
speeds of bodies compounded from them. Epicurus seems
to have the following picture in mind. Suppose we compare the motions of two atoms with respect to ten indivisible units of time. Furley presents the situation thus: Suppose that "an atom moves upward for 4 of them and
downward for 6; then it will have moved 2 space units
downward in 10 time units. It will then *appear* . . . to move
more slowly than an atom which moves upward for 3 units
and downward for 7, and therefore moves 4 space units
downward in 10 time units."[123] Although a difference in

---

[121] More, *The Immortality*, Book I, Chap. VI, Axiome XIV, p. 31 (1659), p. 27
(1662).

[122] *De vitis dogmatis*, Section D, lines 5–7; Section E, lines 1–4, p. 281. See
footnote 109.

[123] Furley, *Two Studies*, Chap. 8, p. 124.

speed appears to occur over a continuous period of time, there is no real difference within each indivisible unit of time. Such a picture may well have motivated Newton's arguments for atomic times and motions. After all, least times, distances, and motions provide a uniform basis for comparing speeds of compounds, especially when the 'apparent' motions of the latter are conceived as multiples of the former. They also provide a ready 'measure' for uniform rectilinear motion. But all of this is conjectural. And, in any event, Newton had probably given up his program in the face of the difficulties he faced.

Aristotle has another formidable argument for his theory of continuity that may well have contributed to Newton's decision. In order to combat indivisibilism, Aristotle employs a distinction between two things, one of which traverses a given distance more quickly than the other.[124] Given this condition, he shows that time and distance must be divisible together with motion. The argument is this. If a magnitude A traverses D_1 in time T_1, another can traverse D_1 in T_2 less than T_1; this divides T_1. But it also follows that during the T_2 of the faster thing, the slower can traverse a D_2 less than D_1, so that D_1 is now divided. This procedure, Aristotle argues, can be iterated endlessly, with the consequence that the faster will continuously divide the time, and the slower, the distance. Thus, whatever magnitude is initially taken – time, distance, or motion – the continuous divisibility of the other two can be shown. Epicurus himself accepted Aristotle's claim that faster and slower motions entail the divisibility of time and distance. His response is the theory that we just sketched, that is, that there really are no differences in atomic speeds, and only apparent differences in visible motions. It is generally agreed that the texts in which Epicurus explains his theory are probably corrupt and certainly difficult to understand. The text at Newton's disposal is no exception. It is difficult to see that the young Newton could have easily interpreted Epicurus's obscure argument as an attempt to respond to Aristotle's argument. However, if the indivisibilist does not want to give up the

---

[124] *Phys.*, VI. 2. 232b26–233a21.

possiblity of faster and slower motions, he must devise a way either to refute or to assimilate Aristotle's position.[125] If the first alternative is his course, the atomist can simply deny Aristotle's theory of continuity. Thus, in the case of a motion that is slower over a given distance, he can postulate a greater number of rests (i.e., stays) than in the case of a motion that is faster over the same distance. In other words, the atomist can adjust the motions of different things so that they 'rest and jerk' according to different patterns, and cover different numbers of spaces at the same time.[126] The result is a strange ontology in which things disappear and reappear at various intervals, but it is not incoherent.

6. Mathematics and indivisibilism

The entry "Of Quantity" (5 90^r) reveals some interesting and controversial themes in mathematical reasoning. It was almost certainly written in early 1664, though the closing entry on 5 90^r was composed after the main body of the argument. It should come as no surprise that Newton is concerned with the 'method of indivisibles,' a technique well represented in the mathematical culture of his period.[127] For reasons that will soon be apparent, "Of Quantity" is part of the same intellectural framework as "Of Atoms" and "Of Motion." Newton's strategy is straightforward. He wants to provide an account of geometrical indivisibles that is compatible with the indivisibilism of his Epicurean atomism. Clearly, the strategy raises queries concerning Newton's view of incommensurability, as well as his attitude toward Euclid's general theory of proportion. And it also brings us back to his implicit attitude toward the status of the mathematicals in relation to the structure of the world.

Newton's essay exhibits a consistent pattern of reasoning.

---

[125] For an interesting discussion of the response of a Medieval indivisibilist to Aristotle's argument, see Murdoch, "Henry of Harclay and the Infinite," to appear in *Studi sul XIV secolo in memoria di Anneliese Maier*, edited by A. Maieru and A. Paravicini-Bagliani.

[126] See Moses Maimonides, *The Guide for the Perplexed* (New York: Dover Press, 1956), Chap. LXXIII, p. 122.

[127] D. T. Whiteside, "Patterns of Mathematical Thought in the Later Seventeenth Century," *Archive for History of Exact Sciences*, I(1961):199–388.

The analysis that follows falls into three parts. First, the pattern of Newton's main argument will be laid out; second, its sources will be discussed; third, the argument will be put into critical perspective together with an appraisal of its import for understanding Newton's mathematical development. The entry added later to the bottom of 5 90^r will be considered separately, because it confuses absolute and relative infinities.

Newton begins with an analogy that suggests that points are indivisible or atomic lines. This is to be expected, because he intends to establish a parallel with the Epicurean reasoning of the first statement of the entry. Notice the claim that "points added between points infinitely are equivalent to a finite line." Newton is not saying that a finite line is composed out of an infinite number of atomic lines; rather, he is claiming that a finite whole can be analyzed mathematically as if it were composed of an infinitude of units consistent with its nature. In other words, the phrase "are equivalent to" suggests that Newton wishes to assume an infinity of atomic lines in order to consider the measure of the line in terms of units appropriate to itself. His perspective has shifted from a denial that dimensionless points with position can compose a magnitude to the claim that finite wholes can be mathematically 'decomposed' into an infinity of appropriate units.

The argument continues by treating a surface as comprising an infinity of lines, and each body (i.e., a solid) as comprising an infinity of surface laminae. Thus, a point is an indivisible of a line, and a surface atom is an indivisible of a surface, as a surface lamina is an indivisible of a solid. But surfaces and solids differ in extension only if they are finite. For example, although surfaces taken individually have the same proportion to an appropriate finite quantity, a given surface may be greater than another. Thus, Newton states that "though all infinite extensions bear the same proportion to a finite one, yet one infinite extension may be greater than another." So if a surface is bisected by a line, each of the halves comprising an infinitude of lines is infinitely greater than the bisecting line, and one half may be greater than the other. Newton's principle is motivated by the strategy of analyzing a finite whole as the sum of an infinitude of appropriate units: The whole and its 'parts' have the same reference, but differ conceptually. Thus, Newton can talk of unequal infinities, though referring to the same finite whole. Moreover, the principle is general in scope and goes beyond the cases that Newton considers. It says that infinite quantities of the same sort may vary in extent if compared with one another, but that they are the same relative to the appropriate finite quantities. It is clear that the following relations are encapsulated in Newton's reasoning. There is the relation of the infinitely large to a finite extension, and the relation of the infinitely small to a finite extension. As a matter of fact, Newton's examples deal only with the infinitely small – the angle of contact (horn angle) as an indivisible of a rectilinear angle, and the indivisible line itself. As we shall see, there is a good reason for this. On the infinitely small side of his principle, it is clear that he conceives points as infinitely small lines, lines as infinitely small surfaces, and surfaces as infinitely small solids, whereas in regard to the relation of the infinitely large to finite extension, there is solid to surface and surface to line. Newton is not committed to points of different sizes; he says only that one infinite extension *may* be greater than another. This is the case with angles of contact, the first example he gives of his principle. He regards them as atomic angles that may be of different sizes. Viewed as a whole, Newton's pattern of reasoning questions whether or not the ratio $\frac{a}{c} = \frac{b}{c}$ implies that $a = b$ is a valid inference for infinite quantities.[128] He is, consequently, close to committing himself to non-Archimedian quantities.

Newton gives a second example. He tells us that "$\frac{2}{0}$ is double to $\frac{1}{0}$, and $\frac{0}{1}$ is double to $\frac{0}{2}$, for multiply the first two by 0 and divide the second two by 0, and there results $\frac{2}{1} : \frac{1}{1}$ and $\frac{1}{1} : \frac{1}{2}$." Thus, he states:

(a) $\frac{2}{0} : \frac{1}{0} :: 0 \times \frac{2}{0} : 0 \times \frac{1}{0} :: \frac{2}{1} : \frac{1}{1}$

(b) $\frac{0}{1} : \frac{0}{2} :: \frac{0}{1} \div 0 : \frac{0}{2} \div 0 :: \frac{1}{1} : \frac{1}{2}$

But if $\frac{2}{0}$ and $\frac{1}{0}$ are compared to a unit, they are both infinite and "ought therefore to be considered equal in respect of a

---

[128] Heath, *Euclid's Elements*, Vol. II, Book V, Prop. 9, pp. 153–4.

unit." In other words, Newton relates the case he is considering to his general thesis and observes (c) that $1 : \frac{2}{0} : : 1 : \frac{1}{0}$. This conclusion is generalizable for any finite number. He then reverts to his analogy between infinitely small angles of contact and points. But this requires that points be indivisible lines, in that angles of contact are treated as indivisibles of quantity whose sum is a larger finite angle. The last set of analogies between the point, the line, the parallelogram, and the parallelepiped simply reiterate the reasoning of the first analogies ("as finite lines . . . in respect of a line").[129]

So much for the structure of the main argument. It raises a number of critical issues that will now be considered, together with their sources. The first important issue is Newton's calculus for treating infinite numerical quantities, especially the concept of number that it employs. In the numerical analogy he uses, 0 is not numerical 0. If Newton's reasoning is considered as a piece, 0 is the apparant result of dividing the unit an infinite number of times. Thus, 0 is not a number, but a nonnumerical first principle of number: It is not a number because it is not measured by the unit. In fact, 0 bears to the numerical unit, 1, the same relation that the point bears to the line. Consequently, $\frac{2}{0}$ and $\frac{0}{2}$ are not numbers, strictly considered. The first is a numerical quantity that "exceeds all number"; the second is a numerical quantity that is less than all number. As Newton employs 0, it seems to be *defined* by dividing the unit an infinite number of times.

It appears that Newton takes 0 as the principle of number and not the unit as is the case in Greek mathematical thought. What is the source for his conception? It is almost certainly Wallis's *Mathesis universalis* (1657). Following Stevin, Wallis argues that the Greeks overlooked the fact that the analogy is not between the 'point' and the 'unit,' but

---

[129] For a discussion of these mathematical entries from the *Questiones*, see D. T. Whiteside, *Mathematical Papers*, Vol. I, pp. 89–91. Whiteside's edition is authoritative reading for anyone who wishes to understand the development and nature of Newton's mathematical reasoning. We are grateful to Dr. Alan Bowen for his insightful and critical discussions of Newton's arguments. Also, we wish to thank Dr. George Molland for a general discussion of Newton's position. Neither is responsible for the interpretative views here put forward.

between the 'point' and 'naught.'[130] Thus, the true principle
of number is naught, which is the sole numerical analogue
to the geometrical point, as the instant is its temporal
analogue.[131] Ever willing to reconcile his views with tradition
(in this case Aristotle), Wallis proceeds to argue that there
are two senses of 'principle' involved. It can mean (a) the
first that is such (*primum quod sic*) as to be of the same nature
as the thing itself or (b) that which is the *last* that is not
(*ultimum quod non*) such as to be of the same nature as the
thing itself.[132] In the first sense, the unit can be the principle of number; in the second sense, nought is the principle.
So considered, if 0 is to have a parallel with the geometrical
point, the latter can be considered a principle of quantity
only in the second sense. In other words, the point must be
treated as a different species from such quantities as a line.
There can be little doubt that Newton's numerical calculus
appears to employ 0 as the principle of number and the
point as its geometrical analogue. Equally, there can be little
doubt that the source is Wallis. But given the nature of his
analogies, Newton's interpretation of the geometrical point,
in relation to his numerical calculus, differs from Wallis's
conception of 0. This will become apparent shortly.

Newton seems to define 0 as the result of dividing the
unit an infinite number of times. Again, Wallis's conception of the unit appears to be the source. For Greek
thought, the indivisible unit is the principle of number.
Although he does not deny that the unit so conceived has
a numerical character, Wallis argues that the unit can be
treated as something continuous: "When arithmetic wishes
to imitate in some way the infinite divisibility of geometry,
it supposes a unit or a one which is something whole, as it
were, but divisible into as many parts as you please [*in
quotvis Partes divisibile*]."[133] Thus, Wallis conceives the unit

[130] John Wallis, *Opera mathematica* (Oxford, 1657), *Parte prima, Mathesis universalis*, Cap. IV, pp. 17–18.   [131] *Ibid.*
[132] *Ibid.*, pp. 17–18. For an interesting discussion of this distinction, see Jacob
Klein, *Greek Mathematical Thought and the Origins of Algebra* (Cambridge, Mass.:
M.I.T. Press, 1968), Part Two, Sect. C, pp. 212–15.
[133] Wallis, *Mathesis universalis*, Cap. XLI, p. 364, ". . . Infinitam Geometriae
divisibilitatem, cum quodammodo imitari velit Arithmatica; supponit Unitatem
sive Unum, quasi jam quid integrum, in quotvis Partes divisibile."

as *internally* continuous, as capable of divisibility into 'parts'
such as fractions. But in what sense is the unit continuous?
For Wallis, numbers are the product of abstraction and are
understood as symbols apart from things countable. He
argues not only that numbers compose a system of symbolic representation but also that the representations are
themselves mathematical objects.[134] So when he tells us
that the unit is internally divisible into numerical 'parts,'
this, it seems, is to be understood within his system of
symbolic representation. Furthermore, Wallis initially vacil-
lates on whether or not fractions are true numbers. He
then proposes to treat them as "an index of the ratio numbers have to one another." But this is to claim, as he
readily admits, that fractions can be considered as numbers, that is, as "indices of all possible ratios whose common consequent [i.e., reference] is 1, the unit."[135]

Wallis is an Aristotelian about continuity. To say that
quantity is continuous is to say that it is infinitely divisible.
Neither in geometry nor in arithmetic does division stop
with the unit, which is always indivisible. But Wallis is also a
conceptualist about mathematical objects. Thus, he treats
ratios as continuous quantities that are infinitely divisible.
So when the unit is taken as the unique and defined quantum, "all the rest of the numbers (whether whole, or broken
[*fracti*], or even irrational [*surdi*]) are the indices or exponents [*indices sive exponentes*] of all the ratios possible with
respect to the defined quantum."[136] Here, of course, number is no longer tied to what is directly countable, but indicates a 'ratio' or *lógos*. It is a 'whole' or a 'broken' or an
'irrational' number only with respect to a ratio determined
in reference to the unit. Ratios result from division when
the quotient indicates how many times one magnitude is
contained in another. But in the case of incommensurable
magnitudes of the same type, the proportions are construed
in terms of Definition 4, Book V, of Euclid, in which reference to 'parts' and divisors is avoided by the operation of

[134] For a discussion of this interpretation, see Klein, *Greek Mathematical Thought*, pp. 215–21.
[135] Wallis, *Mathesis*, Cap. XXXV, p. 315.   [136] *Ibid.*, pp. 315–16.

multiplication.[137] Thus, numbers in Wallis's schema are conceived as homogeneous and are identified with their indexical role as indices of ratios. Throughout his analysis, Wallis has Euclid, Book V, Definitions 3–5, and Book X, Propositions 1 and 2, uppermost in mind.[138]

It is difficult to determine at this early stage in his mathematical development the extent to which Newton was familiar with Wallis's views, especially those concerning the theory of ratios. But there seems to be little doubt that Wallis's *Mathesis universalis* is the source that occasioned Newton's numerical calculus. Wallis nowhere says that the unit is infinitely divisible by 0, such that $\frac{2}{0}$ can be considered an infinitely large number. In fact, he is wary of the term 'infinite number.'[139] But he does provide a model of a unit that is internally divisible, such as Newton employs in his calculus, though it is not clear in what sense Newton himself regards his unit as internally continuous.

Given the character of Newton's numerical calculus, the following commitments seem clear enough. He wants to say that the point bears the same relation to the line as 0 bears to the numerical unit, 1; that 0 is a nonnumerical first principle of number; and that 0 appears to be an element of the unit for all possible divisions (i.e., the unit is internally continuous just because it is everywhere divisible). Wallis construes the analogy between 0 and the point as requiring the latter to be a dimensionless indivisible with position.[140] Accordingly, it is of a different species from the line. The same argument applies to atoms of the surface and the

[137] Heath, *Euclid's Elements*, Vol. II, p. 120. Later in this section there is a discussion of Definition 4. Heath's comment is worth quoting to the effect "that the ratio, as defined in the preceding definition [viz., "A ratio is a sort of relation in respect of size between two magnitudes of the same kind"], and about to be used throughout the book, includes the relation between any two incommensurables as well as between any two commensurable finite magnitudes of the same kind," p. 120.

[138] Wallis, *Mathesis*, Cap. XXV, pp. 221–30. These propositions will be discussed later.

[139] *Ibid.*, Cap. V, pp. 21–2, ". . . numeros dari actu *infinitos* . . . sit impossibile." In effect, Wallis holds that there cannot be an infinite number in any absolute sense, but "sunt tamen numeri possibiles Infiniti, hoc est, nullus est omnino terminus numeris assignabilis; quo cum accedit numerus, ulterius augeri non possit."

[140] *Ibid.*, Cap. IV, p. 18.

solid. All are principles of analysis in the sense that each is
the last that is not such as to be of the same nature as the
thing itself. But Newton's Epicurean frame of mind and the
geometrical analogies that he develops reinforce his actualist disposition to treat the point as an indivisible line. If the
point is so construed, the numerical analogy breaks down;
in fact, there is no analogy. Strictly interpreted, there is an
analogy between 0 and the point only if the point is a dimensionless indivisible with position as Wallis intends. But if
this is so, Newton loses purchase on any connection between
0 in his calculus and the geometrical point as he interprets it
in his analogies. Moreover, Newton seems to define 0 as the
result of dividing the unit an infinity of times, and he seems
to regard number as a plurality of units so divided. Thus
considered, his conception of number has affinities with the
notion that number is a multitude of units and is measured
by the unit. This is the notion that he employs in his actualist treatment of magnitude (see Section 4). But, equally,
Newton's calculus reflects the structure of the relationship
Wallis sees between the point and 0 as the principle of number, but not Wallis's interpretation of the point. At this stage
Newton appears to have a hybrid conception of number
that he is not fully aware of. Once he perceives that there
cannot be a numerical analogy between 0 and the point
construed as an indivisible line, Newton gives up the mathematical realism motivated by his Epicurean actualism and
turns toward the conceptualism evident in Wallis's treatment of numbers and indivisibles. Moreover, in his use of
the 'method of indivisibles' he also accepts Wallis's interpretation of 0 as designating a vanishing small quantity that
tends toward zero, but never equals it. This is clear from
Newton's "Annotations out of Dr. Wallis, his Arithmetica
infinitorum," whose entries date months later than "Of
Quantity."[141]

The pattern of Newton's argument indicates that he
wishes to minimize the gap between the mathematicals and
the perceptible quantity of physical objects. During the early
part of 1664, Newton's mathematics shifts from an orienta-

[141] Whiteside, *Mathematical Papers*, Vol. I, pp. 96–104.

tion in the Greek tradition to a Wallisian framework. It is as if he moves from the *De lineis insecabilibus* to Wallis's infinitesimal, from the ancients to the moderns.[142]

Sometime in the latter half of 1664 Newton began to interpret indivisibles in a Wallisian fashion. But early in 1664, at the time of writing "Of Quantity," he construed indivisibles within the framework of an Epicurean ontology. There are three characteristics of Newton's reasoning that support this contention. His 'calculus' allows for the possibility that there is an infinitely large quantity and an infinitely small quantity that can be related to a finite quantity of the same sort. Although he gives examples of the infinitely small, Newton never explains how his infinite numerical quantities relate to 0, nor how infinitely large quantities relate to the infinitely small. But the commitment is there. And it is a commitment that distinguishes his position from that of Wallis, because Newton allows the possibility that there are quantities that are infinitely larger or smaller than any assignable quantity. There is further evidence of the depth of this commitment in the *Questiones.* In a short entry also entitled "Of Quantity" (87 131^r), Newton opposes Descartes's view that the extension of the universe can only be characterized as indefinite with respect to human knowledge. For Newton, "all the extension which exists" is infinite, and its inverse is minimally small, not infinitesimally small (see Section 7). Nor is the concern for infinite quantities an aberration found only in the *Questiones.* In 1665 Newton wrote an essay on proportion and ratios that is based on Euclid and is entitled "Of Quantity and Muchness." As part of a long series of definitions, he defines his conception of infinite quantities. He states that "that quantum is infinitely greate or greater y^n any finite quantity wch is not increased or diminished by being joyned to a finite quantum" and "that quantum is infinitely little or less y^n any finite quantum wch doth not increase or diminish a finite quantum by being joyned to it."[143] It is clear that, far from denying the exis-

---

[142] See Furley, *Two Studies*, Chap. 7, pp. 104–10, for an analysis of *On Indivisible Lines*.

[143] ULC, Add. 3995, folios 8^r–10^v; an alternative version is found on folios 17^r–18^v.

tence of infinite quantities, Newton proposes to define them in comparison to a given finite quantity. Nothing he says here indicates how the infinitely great and small relate, nor how he views incommensurability with respect to infinite quantities. This characteristic of his thinking raises the question of his attitude (especially in the *Questiones*) toward Euclid's general theory of proportion. We shall turn to this presently.

In the second place, though $\frac{2}{0}$ and $\frac{0}{2}$ are not strictly numbers – the first exceeds all number, and the latter is less than all number – they appear to represent fractional quantities (whether proper or improper) that are truly infinite. That they are fractions is borne out by the fact that ":" is used to represent ratios, so that "−" appears to signify a fraction. This implies an embarrassing consequence that Newton failed to notice initially. If $\frac{0}{2}$ is a fraction, then $\frac{0}{2}$ is a half of 0. This consequence arise from a failure to perceive that the analogy is between the point and 0, not between the unit and 0. Newton is initially misled in this regard by his interpretation of the point as an indivisible line.

Lastly, Newton tells us in the entry he made later to "Of Quantity" that "$\frac{a}{0}$ exceeds all number and is so great that there can be no greater, but finite number is called indefinite in respect of a greater." Thus, $\frac{a}{0}$ is infinite and is to be distinguished from infinity understood as the indefinite. The expression $\frac{a}{0}$ certainly refers to the numerical quantities in the calculus, and the converse reasoning applies to $\frac{0}{a}$ (i.e., $\frac{0}{1}$, etc.). It is also important to notice that at this stage, Newton thinks that $\frac{2}{0}$ and $\frac{0}{2}$ are not only infinite quantities but also unique. But $\frac{a}{0}$ is a unique and infinite quantity only in comparison to finite quantities like 1 and 2. If it is compared with $2 \times \frac{a}{0}$, it represents a quantity that is infinite, but not unique, in that $2 \times \frac{a}{0}$ is twice as great. Newton has confused absolute infinity with relative infinity.

If the foregoing analysis is correct, 0 in Newton's calculus cannot be interpreted in Wallisian fashion. As Newton interprets the expression, '$\frac{1}{0}$ = a numerical quantity greater than all number,' it does not say that the fraction becomes arbitrarily large as the denominator approaches zero. In other words, Newton's use of the notion cannot bear a Wallisian 'limit' interpretation, which says that 0 is never equal to zero, but signifies a quantity tending to zero. As Newton understands $\frac{2}{0}$ and $\frac{1}{0}$ in the calculus, they are *actual* infinities; equally, $\frac{0}{2}$ and $\frac{0}{1}$ are infinitely small quantities less than any quantity assignable. His contrast between the sense in which he takes a quantity to be infinite and his conception of quantities that are infinite in the sense that they can be extended so far as one pleases puts the interpretation beyond doubt. Wallis nowhere uses expressions like $\frac{a}{0}$ or $\frac{0}{a}$. Moreover, it is clear that his Aristotelianism would cause him to reject quantities of this type, if, indeed, he took them to be meaningful at all. For Wallis, the expression $\frac{2}{0} = \infty$ would simply indicate that 0 had passed into absolute zero, and, for him, that which is less than any assignable quantity is not a quantity. In fact, Wallis contrasts the expressions $\frac{1}{\infty}$ and $\frac{0}{1}$. In reviewing his arguments to show that the angle of contact is a nonquantity, Wallis states: "Nam $\frac{1}{\infty}$ (pars cujuspiam infinitesima) fiet (infinite multiplicatione) toti aequalis: Sed $\frac{0}{1}$ (pars cujuspiam nulla) non potest ulla multiplicatione fieri Aliquid."[144] Thus, the first expression ($\frac{1}{\infty}$) represents an infinitesimal quantity, but the latter designates nothing and cannot by multiplication become something; in Wallis's view it is mathematically meaningless. Newton's calculus treats it otherwise: 0 is not numerical zero, but functions as a non-numerical first principle of number, not itself measured by the unit.

The difference between Newton and Wallis is further illustrated by their treatment of the 'horn-like angle.' In his *De angulo contactus et semicirculi* (1656), Wallis reviews the history of the vexatious controversies concerning this issue in geometry.[145] He defends the views of the French geometer Peletier against Clavius. In Wallis's view, to speak of a 'horn-like angle' is a misnomer, because the area between

[144] This passage is from Wallis's *Defensio tractatus de angulo contactus et semicirculi* of 1685, in which he defends his *De angulo contactus* of 1656. In *Opera mathematica* (Oxford, 1693–9), 3 vols., Vol. 2, p. 657. Although this work was written in 1685, it is a defense of views held in the 1650s. Furthermore, the distinction that Wallis makes is consistent with his views in the *Mathesis* and the *Arithmetica infinitorum*. In translation, the passage reads, "For 1/∞ (an infinitesimal part of anything) will become (by infinite multiplication) equal to the whole: but 0/1 (no part of anything) cannot by any multiplication become something."

[145] *Operum mathematicorum, Pars altera*, Cap. I and Cap. II, pp. 1–5.

the curvature of a circle and a tangent to a point lying on it forms neither a true angle nor a quantity.[146] A straight line that meets a circle does not form an angle with it, because at the point where they meet, the line is not inclined to the circle, but is coincident with it.[147] Wallis argues that circles passing through the same point can have different degrees of curvature and can be compared to the same standard. But it is not the case that the 'angle' that a straight line makes with a curve that it touches at a point is either greater or less than another 'angle' formed by another curve touching the same point.[148] But, according to Newton, the angle of contact (Wallis's designation for it) is a true angle whose sum is an even greater angle. Moreover, he conceives these angles as having quantity. Just as one infinite extension may be greater than another, "so one angle of contact may exceed another, yet they are all equal when compared to a rectilinear angle, which is infinitely greater." Also, the analogy that he draws between these atomic angles and points requires that the latter be construed as indivisible lines. As previously noted, this differs from the view of Wallis, who takes the point as an extensionless magnitude with position. In Wallis's view, anything that is analogous to a point will be a nonquantity, as it is less than any assignable positive quantity.[149]

But the basis for Newton's argument to the effect that "the angle of contact is to another angle as a point to a line" (5 90^r) is to be found in Wallis's *De angulo contactus*. Newton states that the degree of curvature of a circle is as four right angles. Given that an infinite-sided polygon is equal to the perimeter of a circle, he can conclude that "that crookedness [degree of curvature] may be conceived to consist of an infinite number of angles of contact, as a line does of infi-

---

[146] *Ibid.*, Cap. XII, p. 43; Cap. XV, p. 51.

[147] *Ibid.*, Cap. I, p. 2, ". . . *Angulum* qui dicitur *Contactus*, seu contingentiae, non esse revera *Angulum*, nec omnino *Quantitatem;* Sed Rectam, quae circulum tangit, cum Peripheria coincidere, non autem ad illam inclinari; *Angulum* autem Semicirculi omnino Rectum esse, & rectio rectilineo aequalem." Wallis states this as Peletier's opinion, but it is the one he vigorously defends in the treatise.

[148] *Ibid.*, Cap. I, p. 2, and Cap. VI, p. 18.

[149] *Ibid., Pars Prima, Mathesis,* Cap. 4, p. 16. "Ut autem magnitudo, puta Linea, inchoatur a puncto, quod omnino magnitudinis expers est, & quidem (positive) nihil est, sive nullius magnitudinis."

nite points." Basing himself on Euclid, Proposition 32, Book I (the addition of Proclus), and referring to Proposition 16, Book III,[150] Wallis argues that the polygon argument shows that the angle of contact is less than any positive quantity and is therefore a nonquantity, as well as a nonangle.[151] Although the argument differs from the conclusion that Newton draws, it seems nevertheless that he has based himself on Wallis. Like Wallis, he views the perimeter of the circle as an n-sided polygon whose sides increase indefinitely in number but decrease correspondingly in magnitude. As n becomes indefinitely large, in the limit the polygon becomes a circle whose curvature remains equal to four right angles, each exterior angle of which becomes an angle of contact. And, certainly, the two editions of Euclid that Newton used do not contain Proclus's extension of Proposition 32 to polygons.[152] But Newton's reasoning unlike that of Wallis, appears to involve non-Archimedian quantities. This raises the question of his knowledge of, and attitude toward, Euclid's general theory of proportion early in 1664, when the arguments in "Of Quantity" were doubtless written.

It is generally agreed that Newton's first systematic study of Euclid began during the second half of 1664.[153] According to De Moivre, he was acquainted with the *Elements* earlier in 1663, having purchased a copy at Sturbridge fair.[154] It is likely that the copy in question is a 1573 Paris edition that is listed as part of Newton's library. The text simply enunciates the propositions, with diagrams, but without proof or commentary.[155] Newton's copy is unmarked, and he probably read it as a novice, concentrating on the early

[150] For an important discussion of Proposition 32 and Proclus's extension to polygons, see Heath, *Euclid's Elements*, Vol. I, pp. 316–22, and Vol. II, pp. 37–42, for Proposition 16.

[151] Wallis, *Operum mathematicorum, Pars altera*, Cap. XII, pp. 40–5. See Whiteside, *Mathematical Papers*, Vol. I, p. 90, for a discussion of the argument.

[152] Isaac Barrow, *Euclidis Elementorum libri XV, breviter demonstrati* (Cambridge, 1655). Books II, V, VII, and X of Barrow's text are copiously annotated, but there is nothing beside Proposition 16, Book III, on the horn angle. See footnote 155 for details of the other edition. Also see Harrison 580 and 581.

[153] See Whiteside, *Mathematical Papers*, Vol. I, Introduction, Part I, pp. 5–7.

[154] *Ibid.*, pp. 5–7. Whiteside evaluates De Moivre's account and sees no reason to doubt the claim that Newton first became acquainted with Euclid in 1663.

[155] *Ibid.*, p. 6. See Harrison 580, *Euclidis elementorum libri XV, Graece & Latine* (Paris, 1573). See Whiteside, *Mathematical Papers*, Vol. I, p. 6, note 12.

Books. We may conjecture further that Newton's systematic study of Euclid coincided with Barrow's first series of mathematical lectures, which began in the autumn of 1664. As evidence of Newton's involvement with Euclid, we have his well-thumbed copy of Barrow's 1655 edition of the *Elements*, which in 1664 was already an established university text.[156] It is Book II (as well as Books V, VII, and X, which treat of the theory of proportion) that particularly drew Newton's attention.[157] Indeed, it is these Books that contain the bulk of Newton's marginal annotations. All of this raises the presumption that early in 1664 Newton may have had a fuller acquaintance with Wallis than with Euclid. Indeed, it raises the possibility that he had yet to come to terms with the theory of proportion when he wrote "Of Quantity."

But the issue of Newton's attitude toward indivisibles goes beyond establishing when he mastered the resources of proportion theory. For there is no direct correlation between his detailed examination of the relevant Books of the *Elements* and an outright rejection of the existence of indivisibles. Quite the contrary: In his *On Analysis by Infinite Equations* (1669) he tells us that he is not "afraid to talk of a unity in points or infinitely small lines inasmuch as geometers now consider proportions in these while using indivisible methods."[158] Yet this is a work that contains the central ideas of his 'fluxional' method, a method that views quantities as generated by continuous motions 'measured' against the independent variables of time and distance.[159] Even in the *Principia* (1686), Newton does not reject the method of indivisibles. The Scholium to the lemmas of the First Book states that "demonstrations are shorter by the method of indivisibles," though the technique can be used "with greater safety" if his method for taking the limits to which the ratios of quantities converge is presupposed.[160] Nor is

---

[156] *Ibid.* See also Whiteside's "Isaac Newton: Birth of a Mathematician," *Notes and Records of the Royal Society of London*, 19(1964):57–9.   [157] *Ibid.*

[158] Whiteside, *Mathematical Papers*, Vol. II, p. 235.

[159] *Ibid.*, Vol. I, Introduction to Part II, pp. 145–54. Also see Westfall, *Never At Rest*, Chap. 4, pp. 131–2, and A. Rupert Hall, *Philosophers at War* (Cambridge University Press, 1980), Chap. 2, pp. 10–23.

[160] Newton, *Principia*, edited by Koyré and Cohen, Vol. I, Liber I, Sect. I, p. 87.

there an unqualified rejection of indivisibles as such. He tells us that it is in no objection to his method to claim

that if the ultimate ratios of evanescent quantities are given, their ultimate magnitudes will also be given: and so all quantities will consist of indivisibles, which is contrary to what Euclid [has demonstrated] concerning incommensurables, in the tenth Book of his *Elements*.[161]

Newton's main concern is to warn against confusing the notion that magnitudes are composed of ultimate indivisibles with the notion that the *ratio* of two quantities converging on a limit can be evaluated, but not the ultimate values of their numerators and denominators considered individually.[162] After all, $\frac{0}{0}$ is mathematically meaningless. But the warning is not tantamount to a denial of the existence of indivisibles, a claim further warranted by the tenor of the Scholium as a whole.

Let us turn to Euclid. Definition 4 of Book V states that "magnitudes are said to have a ratio one to another if capable, when multiplied, of exceeding one another."[163] This immediately excludes the possibility that there is an infinitely large or small magnitude that can be related to a finite magnitude of the same kind. This definition is closely akin to the so-called axiom of Archimedes, which underlies the method of exhaustion, the principle of which is equivalent to Proposition 1, Book X, of Euclid:

If two unequal magnitudes be set out, and if there be subtracted from the greater a magnitude greater than its half, and from that which is left a magnitude greater than its half, and so on continually, there will be left some magnitude less than the lesser magnitude set out.[164]

It is no doubt this proposition, together with the following, to which Newton refers in his Scholium. In conjunction with Proposition 2, it is a test for incommensurability: "If the lesser of two unequal magnitudes is continually subtracted from the greater, and the remainder never measures that

---

[161] *Ibid.*, p. 88.   [162] *Ibid.*, p. 87.

[163] Heath, *Euclid's Elements*, Vol. 2, p. 114 (see p. 120 for a discussion of Definition 4).

[164] *Ibid.*, Vol. 3, Book X, pp. 14–16. See Koslow, "Ontological and Ideological Issues of the Classical Theory of Space and Time," in *Motion and Time*, edited by Machamer and Turnbull, pp. 250–1, for a discussion of this proposition in relation to Newton's indivisibilism.

which precedes it, the magnitudes will be incommensurable."[165] If indivisibles are construed as ultimate parts of magnitude, Proposition 1 alone provides the basis for their categorical rejection, because it provides a rule for producing arbitrarily small submultiples of any given segment. Furthermore, Definition 5 of Book V provides a tool for handling equimultiples of magnitudes capable of exceeding one another and of the same type that avoids reference to the notion of parts. It thus applies to both commensurable and incommensurable magnitudes by focusing on their inequality through the technique of multiplication, rather than on the division of a greater by a less.

But what about Proposition 2 of Book X? It says that any two magnitudes of the same type are incommensurable if they lack a common measure. If that is so, they lack a *least* common measure. It is therefore impossible that they share a common indivisible. But Euclid's propositions say nothing about commensurable magnitudes; accordingly, Newton can claim that they have indivisibles as a least common measure. In short, if all magnitudes are divisible without limit (Proposition 1), the existence of constituent indivisibles that are least in quantity is impossible. Still, Euclid's position does not deny that an indivisible can be a least common measure of two commensurable magnitudes of the same kind. But a 'Euclidean indivisible' is still a comparative, not an absolute entity: Of itself it is divisible.

If Newton had not mastered Euclidean proportion theory when he wrote "Of Quantity," and if he had also not fully penetrated Wallis's reasoning on "indivisibles," the situation is otherwise some months later. In a pocket book that dates from 1664–5, Newton made a series of entries based on Wallis's *De sectionibus conicis* and *Arithmetica infinitorum*. Newton squared the parabola, and attempted to square the hyperbola by using indivisibles.[166] Following Wallis, he extended the technique of treating indivisibles in terms of arithmetic progressions, using the powers of positive integers to cover fractional, negative, and zero indices (i.e.,

---

[165] *Ibid.*

[166] Whiteside, *Mathematical Papers*, Vol. I, Part I, pp. 91–141.

exponents).[167] In these entries, Newton is employing the 'method of indivisibles,' a technique of reasoning about quadratures by treating the 'indivisible' elements of the line and the surface, and so forth. He says of the "progression $1 \cdot \frac{2}{3} \cdot \frac{8}{15} \cdot \frac{48}{105}$" that it "deduceth its originall from this

$$A \times \frac{0 \times 2 \times 4 \times 6 \times 8}{1 \times 3 \times 5 \times 7 \times 9} \quad \&c$$

in wch A is an [infinite] number $= \frac{1}{0}$."[168] Here 0 is not interpreted as he does in the numerical calculus in "Of Quantity." Rather, he construes it in a manner that (put anachronistically) is analogous to the expression $\lim_{n \to 0} \frac{1}{n} = \infty$.[169] So interpreted, the final ratio of a progression is never reached; it is simply implied that as the denominator approaches zero, the fraction becomes arbitrarily large. This is the manner in which Wallis speaks of infinity in his work on series.

If there is a shift in Newton's treatment of indivisibles in the mathematical context from "Of Quantity" to his later reading notes from Wallis, what status do indivisibles have in the earlier piece of writing? This observation is intimately related to the following questions. Newton is a mathematical infinitist about finite quantity in 5 90^{r}, but about physical quantity he is an Epicurean finitist and compositionist.[170] How are these views to be reconciled? Second, Newton is committed to the possibility of unequal infinities. How is this to be reconciled with the part–whole logic that his actualism demands?

Recall that Wallis treats indivisibles as infinitesimally small elements of the line and surface. So considered, they are indivisibles in the sense that they can be construed as if they are dimensionless magnitudes with position, or as if they are vanishingly small elements tending toward, but never equal to, zero. Conceived thus, they are nothing more than con-

[167] *Ibid.*, pp. 97–115.

[168] *Ibid.*, p. 101. Whiteside suggests that Newton's reasoning is based on Proposition CLXXXVII of Wallis's *Arithmetica infinitorum*, pp. 167–9.

[169] It is misleading to suppose that Newton articulates a limit concept in this context, though, of course, he has an intuitive grasp of the notion.

[170] For Medieval precedents for this view, see Anneliese Maier, *Die Vorlaüfer Galileis Im 14. Jahrhundert* (Rome, 1949), Chap. 7, pp. 181–6.

ceptual tools of mathematical analysis. Thus, wholes can be viewed as being decomposable conceptually into infinitely small elements that are prior in analysis and understanding. And taken as heuristic tools of investigation, to be used in a context of mathematical discovery, indivisibles do not confront Euclid's theory of proportion and can avoid Zeno's strictures in 29B1. Newton also wishes to treat indivisibles as tools of analysis. But initially his actualism causes him to view them as possessing positive quantity; for example, he initially posits indivisible lines. In other words, Newton's mathematical realism inclines him to think of mathematical indivisibles as entities that reflect the nature of physical quantities. Moreover, he wants 'to measure' and compare areas mathematically; this again inclines him toward a non-Wallisian type of indivisible.

That Newton treats indivisibles in certain contexts as tools of analysis can be clarifed in the following way. There is direct evidence that he wants to contrast what is said about the divisibility of physical quantity and the divisibility of a quantity mathematically considered. The entry that concludes the arguments in "Of Atoms" states the following: "Whatever can be objected against indefinite divisibility in bodies may also be objected against the same in quantity and number. But if the fraction $\frac{10}{3}$ be reduced to decimal form it will be 3.33333333 etc. infinitely, and what does every figure signify but a part of the fraction $\frac{10}{3}$ which, therefore, is divisible into infinitely many parts." The first sentence in this quotation refers back to Newton's actualist arguments concerning the divisibility of finite quantity (see Sections 4 and 5). Thus, quantitatively considered, a physical body cannot be divided into an infinite number of separable parts. But a decimal expansion has the property that it can be augmented as far as the mathematician pleases. In this sense, there is no upper limit to the number of decimal places into which the fraction is divisible. The fraction is, of course, a finite number; it can nevertheless be conceived as infinitely divisible according to a rule of iteration. It is clear that Newton is concerned to understand an infinite process in terms of the properties that a numerical series possesses, a commitment also made evident by his exploration of Wallisian progressions some months later. Thus, if there are statable rules, Newton is willing to model a notion of 'potential' infinity on the numerical properties of progressions. In such contexts, he is a conceptualist about infinity and divisibility. And, like Charleton, he would probably construe the division of quantity into proportional parts as a well-founded tool of mathematical analysis. However, with regard to the quantity of objects that exist extramentally, Newton is an Epicurean actualist. According to this attitude of mind, to say that a physical quantity is infinitely divisible states a genuine possibility; it is to say that it *can* be so divided. But this is contradictory; a finite quantity cannot be composed of infinite parts.

This distinction between considering a quantity physically, in contrast to mathematically, is reflected in the status that Newton eventually accords to indivisibles. Whether they are construed as entities of positive quantity (Newton's initial position) or as dimensionless points (Wallis), they are conceptual tools of analysis within the framework of certain mathematical operations. We have already indicated some reasons for Newton's shift to the Wallisian perspective. But there is another consideration. It is the relationship between his mathematical analysis of continua in terms of indivisibles and his positing the possibility of unequal infinities.

Newton wants to hold the possibility of unequal infinities in order to provide a *measure* of finite continuous quantities of varying sizes. Although he does not claim, in the mathematical context, that quantities are composed of indivisibles, he wants to say that a whole line, for example, contains twice the infinitude of indivisibles as its half. This means that Newton initially rejects the claim that continua cannot be measured by indivisibles on the ground that there cannot be more of them in a greater continuum than in a lesser continuum. As is clear from his calculus, however, he is an actualist about the infinitude of his indivisibles. On strict grounds, a Euclidean would question that the axiom that the whole is greater than its parts is applicable to infinite quantities, and this is a criticism not unrelated to the claim that there is no ratio between the infinitely great and small.

Henry of Harclay, an actualist and a compositionist concerning indivisibles, attempted to adumbrate a different part–whole axiom in the face of this sort of objection to infinitism and indivisibilism, but with doubtful success.[171] The difficulty would certainly have occurred to Newton as he became more deeply immersed in the implications of Euclid's proportion theory. Equally unlikely, at this point, is the possibility that Newton failed to appreciate the force of traditional arguments designed to show the inconsistency between indivisibilism and geometry by application of the techniques of radial and parallel projection to simple geometrical figures. But, unlike Harclay, Newton does not attempt to develop a part–whole axiom different from the one standard in Euclidean geometry. As he sees the untenability of conceptual indivisibilism in regard to physical quantity, so he also sees in the mathematical context that the 'fixed' indivisible must give way to the infinitesimal, an indefinitely small unit tending toward, but never equal to, zero. This shift in Newton's mathematical gestalt probably took place at the same time as his rejection of conceptual indivisibilism. And it is a shift away from a mathematical surrogate for his 'Epicurean' indivisible of physical quantity to the Wallisian infinitesimal of analysis. In the Wallisian framework (as we have seen), Newton began to interpret the claim that $\frac{2}{0}$ and $\frac{1}{0}$ are infinite numbers – the first double the second – as follows: They are fractions that become arbitrarily large as the denominator approaches zero. Thus, 0 signifies a quantity tending to zero. In the limit as 0 passes into zero, $\frac{2}{0} = \infty$, and $\frac{1}{0} = \infty$; in this sense they can be said to

---

[171] *Ibid.*, pp. 196–215, for a general account of Harclay's views. See also Murdoch, "Henry of Harclay and the Infinite." According to Murdoch, Harclay attempts to replace "the relation of 'greater than' which occurred in the standard part-whole axiom, with the relation of a whole 'containing something else more than or in addition to' its part." (mss. T83^v–84^r; folio 95^r and folio 95^v). Harclay is an actualist who interprets indivisibles in a fashion akin to Newton's first position. What he fails to specify clearly is a conception of greater numerosity, which compares infinities such that of two equinumerous sets one can be said to contain more elements than the other, but not according to some determinate number more. This distinction is compatible with Wallis's infinitesimal approach to indivisibles in his mathematics, and one that Newton probably grasped (at least implicitly) at this stage. Also, there is nothing in Newton's position that prevents his holding that some infinities are *equal;* for example, the fact that the set of all odd numbers is equal to the set of even numbers.

be the same with respect to a unit. Thus interpreted, Newton's indivisibles allow a comparison of the infinities of lines, surfaces, and solids. But it is clear that he does not give up actual infinities. As we shall see in Section 7 of this chapter, his treatment of actual infinities gets mobilized in an epistemic and theological context.

It is the Wallisian perspective that Newton later urges against Bentley's claim (1693) that all infinities are equal.[172] If infinites are "considered absolutely without any Restriction or Limitation, [they] are neither equal nor unequal, nor have any certain Proportion one to another."[173] However, if Wallis's methods are used, "various Proportions of infinite Magnitude" can be treated "by the various Proportions of infinite Sums."[174] Thus, it can be said that "tho' there be an infinite Number of infinite[ly] little Parts in an Inch, yet there is twelve times that Number of such Parts in a Foot."[175] There is no question here of dealing with ratios of infinite numbers as such; rather, it is a matter of taking the *limits* of the ratios of finite numbers, where the numerator and denominator change determinately in relation to one another. Thus, Newton understands unequal infinities as ratios differing from 1. So the 12 : 1 ratio of parts in a foot to those in an inch is construed as the 'potential' infinity of the indivisibles involved in taking the limit of the ratio of indefinitely increasing finite numbers of the ever decreasing 'parts' of the foot and inch, respectively.[176] In no sense can it be said that the infinitude of parts in the foot is actually twelve times as great as that of the parts in an inch.

If Newton becomes a mathematical conceptualist about indivisibles in 1664, if he continues to use the 'method of indivisibles' throughout his lifetime, it must still be recognized that the mathematics of the *Questiones* predates the beginnings of Newton's mature development as a mathematician. These beginnings can be placed about the autumn of

---

[172] I. Bernard Cohen and Robert E. Schofield, *Isaac Newton's Papers and Letters on Natural Philosophy* (Cambridge, Mass.: Harvard University Press, 1958), p. 299.

[173] *Ibid.*, p. 299.   [174] *Ibid.*, p. 295.   [175] *Ibid.*

[176] See Adolf Grünbaum, "Absolute Relational Theories of Space and Space-Time," in *Foundations of Space-Time Theories*, Minnesota Studies in the Philosophy of Science, Vol. 8 (Minneapolis: University of Minnesota Press, 1977), pp. 312–13, for a discussion of Newton's letter to Bentley.

1665, at which time the elements of the fluxional calculus start to emerge.[177] To be precise, Newton rejects the conception of the indefinitely small increment as a basic tool of analysis. In its place he posits the fluxion, a finite and instantaneous speed, defined in terms of the independent dimension of time, and with respect to the geometry of the line segment.[178] Nevertheless, there remains an important continuity in his attitude toward the well-foundedness of mathematical concepts. In the *Questiones* he attempts to model his mathematical indivisibles on the static structure of physical quantities construed from an actualist perspective. Thus, the initial strategy is intended to show that mathematical indivisibles at least reflect the indivisibilist basis of sensible quantities. This attitude of mind continues to motivate his development of the fluxional calculus. In other words, Newton persists in his attempts to minimize the divide between the mathematical representations of the imagination and the sensible world itself.

About the middle of 1665, he became discontent with the infinitesimal basis on which his method of tangents then rested.[179] Some time prior to that he had begun to employ kinematic models for the generation of curves, defined as the loci of points moving under determinate conditions. The generation of quantities by motion is the key conception of Newton's method of fluxions, though the term 'fluxion' does not appear in the mathematical manuscripts of 1665.[180] The conception is this: Instead of conceiving variable quantities to proceed through infinitely little increments, Newton imagines variables 'to flow' from one value to another according to a determinate rate of speed or motion. Clearly, the generation of figures by motion is comparable to the claim that the mind has a direct intuition concerning the movement of sensible things, an intuition that represents the kinematic character of change in the natural world. The idea is not unique to Newton, nor is it unique to Barrow, whose use of the technique no doubt influenced

[177] Whiteside, *Mathematical Papers*, Vol. I, Introduction to Part II, p. 146.
[178] *Ibid.*, p. 146; see especially Whiteside's note 10, p. 147.
[179] Westfall, *Never At Rest*, Chap. 4, p. 131.
[180] *Ibid.*, p. 134; see also Westfall's note 63.

Newton.[181] Ironically, although one of Newton's powerful tools for handling infinite quantities rests on representations of the kinematics of change, a type of indivisible appears in the model. To be sure, incremental indivisibles are replaced by the concept of instantaneous speed, but instantaneous speed is 'measured' as a ratio between an indivisible unit of independent time and a distance traveled along a line segment.[182]

7. The structure of the universe: infinity and the void

There is no doubt that Newton is committed to the basic principles of physical atomism in the *Questiones*. At 2 88ᵛ he accepts the doctrines of the interstitial and separate vacuum. In the essay "Of Violent Motion" he argues that there is true motion in a vacuum. In his "Of Motion," Newton implicitly indicates that he favors the Epicurean view that compound motions are multiples of invisible atomic motions. Nor is there any doubt that he accepts the atomic conception of the fine structure of perceptible quantities. Throughout his career these commitments remain, transformed, of course, in emphasis and scope. But what is also characteristic of Newton's later thought is a commitment to the notion that the universe is infinitely extended – a conception he usually expresses in the doctrine that spatial extension is infinite. Are these conceptions also present in the *Questiones*?

The difficulties in Newton's mathematical claim that $\frac{a}{0}$ is an actually infinite and unique quantity have been noted.

[181] During his controversy with Leibniz over the invention of the 'calculus,' Newton wrote that "it's probable that Dr. Barrow's Lectures might put me upon generation of figures by motion, tho I not now remember it" (ULC, Add. 3968.41, folio 86ᵛ). The generation of mathematical quantities by motion goes back to the ancient world. See Heath, *Euclid's Elements*, Vol. 2, pp. 156–9, for a discussion of this position. See also Proclus, *A Commentary on the First Book of Euclid's Elements*, translated and introduction by Glenn R. Morrow (Princeton University Press, 1970), pp. 79–82.

[182] Whiteside, *Mathematical Papers*, Vol. I, pp. 385–6. See Westfall, *Never At Rest*, pp. 133–4, for a discussion of Whiteside's views. See also Philip Kitcher, "Fluxions, Limits, and Infinite Littleness. A Study of Newton's Presentation of the Calculus," *Isis*, 64(1973):33–9.

But his views about infinite quantities involve further commitments. He tells us that "all the extension that is, eternity, and $\frac{a}{0}$ are infinite" (5 90^r). Thus, three different sorts of things are said to be determinately infinite, in contrast to things that can be indefinitely extended. Furthermore, we are told that although it is indeterminate how far we can "fancy" (i.e., imagine) the augmenting of number, quantity, time, and extension, this provides no grounds for denying that they are actual infinities. Thus, not only is $\frac{a}{0}$ a quantity that is infinite, but so, too, are extension and eternity. If it is indefinite just how large a quantity can be imagined, then insofar as a quantity is imaginable, it is finite; or, if one imagines it to be infinite, the quantity is thus characterized by saying that it can be extended as far as one pleases, that is, that the mind can formulate an iterative rule for projecting an increase in its size or for dividing it into parts. Nevertheless, Newton implies that although we cannot imagine the infinite, we can conceive of it.

The context of Newton's argument is closely related to the entry entitled "Of Quantity" on 87 131^r. Here, in a 'metaphysical' setting, he states a relation between the infinitely great and infinitely small. To say that extension "is only indefinite in greatness and not infinite" is tantamount to saying that "a point is but indefinitely little; and yet we cannot comprehend anything less." In other words, although it is beyond the capacity of the imagination, if there is an infinitely great extension, then there is an infinitely small extension, one whose smallness and partlessness afford no purchase for conceptual divisibility.[183] The argument indicates that the entry (as well as the whole of 5 90^r) was probably written before Newton finished his essay "Of Atoms," in which he begins to see the difficulties in the view that minima are conceptually indivisible. It is otherwise difficult to account for his positing an 'intrinsic' minimum (such that nothing smaller can be conceived of) as the inverse of an actually infinite extension. The obvious difficulties in making such a commitment mathematically coherent

---

[183] For discussion of this relationship, see Paul Henri Michel, *The Cosmology of Giordano Bruno* (Ithaca: Cornell University Press, 1973), Chap. VI. There is no evidence that Newton knew the writings of Bruno or Cusanus.

have already been noted. But here Newton is making a 'metaphysical' statement about the extensiveness of what exists: It is infinitely, not just indefinitely, extended. His interest in cosmologies that posit infinities is amply attested to; no fewer than twelve pages in his copy of Diogenes Laertius are dog-eared. In many cases the page is turned back so that a reference to the term 'infinity' can be pinpointed by the corner of the page. The cosmological views of Anaximander, Anaximenes, Zeno the Stoic, Empedocles, Philolaus, Melissus, Democritus, and Epicurus are thus marked for further reference.[184] It can be conjectured that Newton's commitment to the infinity of extension (in a cosmological setting) led to his attempt to relate geometrically the infinitely great and its reciprocal, the infinitely small. For it is difficult to suppose that a mathematical impulse could alone propel him in this direction.

The influence of Descartes on Newton's views concerning the nature of extension cannot be overlooked. Newton's vast reading in the Cartesian corpus is amply documented in Chapter 2 of this Commentary, in which evidence of his interest in extension is recorded. On 87 131^r Newton asserts the following:

To say that extension is but indefinite (I mean all the extension which exists and not so much only as we can fancy) because we cannot perceive its limits, is as much as to say, God is but indefinitely perfect because we cannot apprehend his whole perfection.

There is no doubt that Descartes is the target here. It is he who uses the term 'indefinite' when referring to extension in contexts where the 'size' of the universe and the divisibility of matter are at issue.[185] Newton's point is straightforward: The scope of what is imaginable provides no grounds for the conclusion that in reality extension is indefinite, any more than the mind's inability to comprehend the fullness of God's perfection in any way affects the reality of Divine being. Although he offers no indication of how the infinity

[184] *De vitis dogmatis, Liber secundus*, pp. 33–4; *Liber tertius*, p. 197 and p. 199; *Liber octanus*, p. 227 and p. 235; *Liber novem*, p. 243 and p. 248; *Liber decimus*, p. 276 and p. 281.

[185] Descartes, AT.VII.14–15. *Pars prima*, Arts, XXVI and XXVII; *Opera philosophica*, pp. 7–8.

of extension is known, Newton rejects what he takes to be a negative thesis, that is, the view that if something is epistemologically indeterminate, grounds are provided for supposing that it is indeterminate in reality.

A similar anti-Cartesian argument is made in *De gravitatione*:

> Nor is it an objection that space is said to be indefinite in relation to us, that is, that we are simply ignorant of its limits and do not positively know that there are none (Part I, Art. 27 [Descartes's *Principia*]). This is because, though we are ignorant beings, God at least understands that there are no limits, not merely indefinitely, but certainly and positively, and because though we negatively imagine it [space] to transcend all limits, yet we positively and most certainly understand that it does so.[186]

Admittedly, Newton is speaking of the infinite extension *of* space, in contrast to the 'indefiniteness' of Descartes's material extension. Still, the same point is made as in the *Questiones*, that is, a rejection of the view that extension is "indefinitely great" if that nature is expressed in negative terms. Explicit here is an epistemological principle present only implicitly in 5 90^r and 87 131^r of the *Questiones*: To say something is deficient with respect to a kind presupposes a positive conception of the kind itself. If this principle is applied to Newton's conception of the extended, it means that the capacity of the imagination to negate the assigned boundaries of a given extension presupposes a positive understanding of the infinity of extension. Here the distinction between what is properly understood by the intellect and what is grasped by the imagination is made explicitly. As will be made clear in Chapter 2 of this Commentary, it is reasonable to suppose that Newton's knowledge of these distinctions is based in large measure on his reading of Descartes. The use he puts them to is his own.

Unlike Descartes, Newton uses these distinctions to argue for a positive view of the infinity of extension. Present throughout Newton's discussion of infinities in the *Questiones* (as well as in *De gravitatione*) are two principles: (a) though we can only negatively imagine the infinity of something, it is still possible to understand that it is infinite; (b) that any existent "either has limits or not and so is either

[186] Hall and Hall, p. 102.

finite or infinite."[187] So, whether or not we are in a position to justify making true statements about the limits of extension, or able to imagine an obtaining state of affairs of that sort, extension is either limited or not limited. If it is not limited, it is true to say that it is infinite, and true to say that we understand that it is, even though we may not completely grasp its entire and defining nature. Combined together, these principles allow Newton to hold that the capacity of the imagination to negate any limit assigned to extension presupposes an understanding that the nature of extension is infinite. So, to hold that it is indefinite how much extension can be imagined does not deny that extension is infinite just in case its lack of limits cannot be imagined.

It is clear that the parallel Newton draws between saying that extension is indefinite because "we cannot perceive its limits" and the claim that God is "indefinitely perfect" because the mind cannot comprehend his total perfection is meant to point up an asymmetry in Descartes's treatment of infinity. If we cannot comprehend God's infinity, as Descartes claims, but can understand that he is infinite, why does the same pattern of reasoning not apply to the nature of extension? The nature of Divinity differs from the nature of extension. But why should the difference prevent the ascription of positive infinity to different sorts and kinds? In *De gravitatione*, Newton in fact makes this argument.[188] But surely it has its beginnings in the *Questiones*, as well as in his notes on Descartes's ontological argument, as found in the *Meditations* (see Chapter 2 of this Commentary).

Yet another argument against Descartes's cosmology in *De gravitatione* has its birth in folios 5 90^r and 87 131^r of the *Questiones*. In *De gravitatione*, Newton accuses Descartes of equivocating between the notions of epistemological indeterminacy and the view that extension is 'indefinite' in the

---

[187] *Ibid.*

[188] *Ibid.* "But I see what worried Descartes, namely that if he should take space to be infinite, it would perhaps agree with [*constitueret*] God because of the perfection of infinity. But in no way, for infinity is not a perfection except in so far as it is ascribed to perfect things. Infinity of intellect, power, happiness, etc., is the summit of perfection; but infinity of ignorance, impotence, misery, etc., is the summit of imperfection; and infinity of extension is as perfect as that which is extended" (based on the Hall translation).

sense that it is without limits (i.e., an extended *ắpeiron*).[189] In the latter sense of the term, Descartes can be interpreted as saying that extension is infinite. In the first sense, he can be construed as saying that extension is not defined (i.e., that it is 'indefinite'). This, in turn, admits of two interpretations: It can mean either that we are not in a position to know whether extension is finite or infinite or that it is not definable (i.e., that its limits do not exist). It is clear in both the *Questiones* and *De gravitatione* that Newton is challenging Descartes to make a genuine ontological statement. Thus, he complains in *De gravitatione* that Descartes's view of extension involves holding "that which actually is, is not to be defined [*non est definiendum*]."[190] What Newton has in mind concerning definition is the structure built into the Greek and Latin notions of *horismós* and *definio*. Both convey the notion that to define is to set bounds or limits. In defining something, we confine or limit it to a kind or species; what it is, is thus specified. But the root meaning of these traditional notions also involves reference to the fact that in reality things are said to have 'measurable' boundaries or limits. So, if it is said that the extension of the world is defined, this means that we know something about it (i.e., that it has limits in reality). If it is said "not to be defined," this can mean either that it is not known whether it is finite or infinite or that it is not definable (i.e., that it in fact lacks limits and so is infinite). It is clear that Newton accuses Descartes of playing on the ambiguities implicit in the traditional notions of the 'indefinable' and the 'indefinite' in stating his doctrine of extension. The origins of this critique are in the *Questiones*. Moreover, it cannot be doubted that Newton's examination of Descartes's views on the infinite and the 'indefinite' helped considerably in the formulation of his own doctrine of the infinity of extension.

Whereas Descartes is an important source in shaping Newton's epistemic attitudes toward the infinity of extension, Epicurean ideas played a decisive role in the formation of his cosmology. Newton claims that "all the extension that is . . . is infinite"; he refers to "all the extension which exists and not

189 *Ibid.*   190 *Ibid.*

so much only as we can fancy"; he contrasts matter moving "through an infinite space in an instant" with its moving "through a finite space," and he speaks positively about vacua and about motion "in a vacuum" as compared "with motion *in pleno.*" These phrases are culled from his essays on quantity, atoms, the vacuum, and violent motion, essays that raise certain questions regarding Newton's commitments in the *Questiones*. What does he understand by the claim that extension is infinite? Does his talk of void expanses imply an acceptance of the notion of 'noncorporeal' extension? Is Newton adumbrating a theory of space? If so, does he indicate any relationship between the existence of space and the presence of God? Does Newton think that void intervals possess a structure of indivisible parts – at least did he think this before he rejected conceptual indivisibilism?

As to the first query, it seems that Newton holds that extension is infinite in the sense that it is not intrinsically self-delimiting. The cosmological implications of this notion are expressed by Epicurus in the *Letter to Herodotus*: "Moreover, The Universe (τὸ πᾶν) is unlimited (ἄπειρόν). For that which is limited has an extreme point: and the extreme point is seen against something else (παρ᾽ ἕτερόν τι θεωρεῖται). So that, as it has no extreme point, it has no limit; and as it has no limit, it must be unlimited and not bounded."[191]

Newton knew Epicurus's argument, as it is presented in his copy of Diogenes Laertius. Charleton's *Physiologia* also contains much discussion of *"Infinite Inanity* or Ultramundane Space,"* as well as the notion of spatial infinity in relation to the presence of God.[192] Charleton does not, as such, provide a formulation of the Epicurean argument.

Given the measure of influence that Epicurus has on Newton in the *Questiones,* we can reasonably conjecture that he views the infinity of extension as follows: Extension is infinite in the sense that it lacks boundaries; anything that lacks boundaries has no limit; if it has no limit, it is logically impossible that anything other than itself exists be-

[191] *De vitis dogmatis*, Section E, lines 6–8; Section F, lines 1–3, p. 276; Usener, *Epicurea*, p. 7.

[192] *Physiologia*, Book I, Chap. VI, pp. 62–71.

yond it; it is therefore unlimited or infinite. In contrast to
infinite extension, a finitely extended thing is bounded,
because it is possible that something separate from it exists
beyond it. Thus, when Newton says that "all the extension
that is . . . is infinite," he means that it is a fully completed
being, not delimited by something other than itself. Moreover, because he links the infinity of extension with the
completeness of being, he appears to accept a non-Aristo-
telian principle: That which is unlimited is fully actual.
This is in contrast to Aristotle's views, according to which
there is a close relationship between the actual and the
limited. For Aristotle, the unlimited, or the infinite, cannot
be a completed actuality, because the "unlimited is not that
which has nothing beyond, but that which always has
something beyond."[193] Because Newton holds that infinite
extension is a completed reality, beyond which there is
nothing, his position is far removed from the Aristotelian
claim that completed actuality pertains to bounded, rather
than to unbounded, things.[194] Of course, Newton does not
deny the actuality of the finite; rather, his Epicurean position suggests that the actuality of infinite extension embodies greater reality than the boundaries of the finite.
Given Newton's actualist sensibilities regarding the existence of things, his dislike of Descartes's apparent vacilla-
tion on the nature of extension becomes evident.

This brings us to the question whether or not Newton
construes the infinity of extension as a defining characteristic of space. If so, does he see a connection between space
and the presence of God in the *Questiones*? In the first place,
when Newton speaks of infinite extension, this is probably
an ellipsis for the infinity of the universe and of the void.
There is sufficient evidence in "Of a Vacuum and Atoms"
and "Of Violent Motion" to suggest that he is committed to
interstitial vacua and the cosmic void. These commitments
are maintained in his mature thought. In "Of Violent Motion" (see Chapter 3), Newton holds that real motion can

[193] *Phys.*, III. 6. 207a1–2.

[194] *Ibid.*, 207a8–31. Aristotle's point is that the unlimited is indeterminate and
hence indefinable and unknowable. The mind cannot traverse the unlimited, as it
lacks a characteristic *péras*. Only that which has a *péras* is complete, definable, and
knowable.

take place in the extended expanses of the void. Moreover, he conceives the void to be uniform and penetrable and everywhere disposed to receive motion without resistance. Clearly, the void is not a form of nonbeing for Newton. As noted in Section 1, he appears to agree with the atomists that the real is not necessarily identical with the full or the bodily. Consequently, he is prepared to assert the anti-Aristotelian view that real intervals can exist between bodies and their parts and that a projectile can truly move through the expanses of the void. This is tantamount to the view that the characteristic of 'being extended' can be attributed to phenomena other than the physical or the corporeal. Thus, not only can the universe be said to be infinitely extended (i.e., in the sense of all that exists), it can also be said that the cosmic "vacuum is infinitely extended." And if the void interstices of the universe are infinite in number, their extension is infinite too ("Of Atoms," 3 89^r).

If Newton offers no explicit legitimization for the notion of 'noncorporeal extension,' it is probably because he accepts More's approach to the problem in *The Immortality of the Soul*. According to More, Descartes is wrong when he claims that spirits are nonextended; so, too, is Hobbes when he argues that if to exist is to be extended, then all existents are corporeal.[195] In More's view, for anything to exist it must have an extended place and thus be extended.[196] But if all that exists has dimensions, it is nevertheless not the case that everything that exists is of a bodily nature, as Hobbes contends.[197] What distinguishes the corporeal from the incorporeal is "not *trinal Dimension*, but *Impenetrability*, that constitutes a *Body*."[198] In other words, impenetrability is not part of the nature of dimensions generally understood, but only of the dimensions that inhere in corporeal things. Thus, spirits are "*Penetrable* and *Indiscerpible*," whereas bodies are discerpible and impenetrable. They are capable of "actual divisibility" or "discerpibility, [the] gross tearing

---

[195] More, *The Immortality* (1662), Book I, Chaps. VIII, IX, and X, pp. 34–43; pp. 49–74 (1659).

[196] *Ibid.*, Chap. VI, Axiome XIV, p. 27.

[197] *Ibid.*, Chap. IX, Sect. 3, p. 37; Chap. X, Sect. 7 p. 42.

[198] *Ibid.*, Chap. X, Sects. 7 and 8, p. 42; p. 71 (1659).

or cutting one part from another."[199] But like the extended nature of a spirit, the incorporeal expanse of a void is penetrable, indiscerpible, and extended.[200] More does not seem to think that the lack of resistance in a void extension is a negative property; on the contrary, it provides a positive basis for distinguishing it from the impenetrability of a body. His arguments were rejected by Descartes.[201] For Descartes, a distance is merely a dimension or a mode of extension, an attribute that itself "cannot exist without an extended substance."[202] In effect, he argues as follows: Extension is an attribute; but an attribute and its proper modes cannot exist without a substance; however, extension is the essential attribute of corporeal nature; therefore, there is no extension without corporeal substance, of which space is merely a determinate, relational mode. Thus, neither an extended void nor a continuously empty vacuum can exist, because 'being extended' cannot be ascribed to what is 'noncorporeal.' In Descartes's view, to attribute the positive characteristics of extension to a vacuum that is said to be incorporeal is to ascribe attributes to nonbeing.[203] A vacuum is merely corporeal dimensionality described under conditions in which it undergoes a change in properties, while its essentially extended nature remains the same.[204] Everything Newton says about matter in the *Questiones* commits him to a rejection of Descartes's position on the vacuum. On this score, More's doctrine of extension no doubt influenced his views on the notion of 'noncorporeal' extension.

Equally under More's influence, Newton is committed to a rejection of Hobbe's view that the corporeal alone exists, because only the corporeal can be said to have dimensions and to occupy an extended place.[205] This is, of course, an Aristotelian principle. In his essay on violent motion

[199] *Ibid.*, Chap. II, Axiome IX, p. 19 (1662); p. 11 (1659).

[200] *Ibid.*, Chap. X, Sect. 8, pp. 42–3.

[201] AT.V.267; AT.V.340; AT.V.401.

[202] AT.VIII.50, "quia omnis distantia est modus extensionis, & ideo sine substantia extensa esse non potest."

[203] AT.VII.49–50. Arts. XVI–XVIII.　　　　[204] AT.VII.49.

[205] *Elements of Philosophy. The First Section concerning Body* (London, 1656), Part 2, Chap. VIII, pp. 74–92; *Elementorum philosophiae, sectio prima de Corpore* (London, 1655), pp. 62–72. It is not clear whether Newton refers to the Latin edition or the English edition.

(53 114ʳ), Newton states that "it is true God is as far as vacuum extends, but he, being a spirit and penetrating all matter, can be no obstacle to the motion of matter; no more than if nothing were in its way." There is every reason to believe that Newton holds that vacuum extends infinitely. Moreover, there is a clear indication that he wishes to relate Divine omnipresence to the infinite extension of the void. The idea that spirits, including God, possess penetrative power (and can be penetrated) comes from More, as does the view that they are self-moving, active, and able to alter passive matter.²⁰⁶ But there is no explicit statement by More in *The Immortality of the Soul* concerning a relationship between the extension of the void and Divine omnipresence. He contents himself with saying that the *"Ubiquity* or *Omnipresence* of God is every whit as intelligible as the overspreading of *Matter* into all places."²⁰⁷ It is clear that Newton conceives God to be present in the infinite extension of the void. But he does not explicitly define the manner in which God relates to the void, such that the void can be said to be the 'place' of his omnipresence. However, nothing Newton says in the passage commits him to More's view that God is an extended spirit.²⁰⁸ It may well be that he has a Medieval doctrine in mind: God is whole and indivisibly present in each 'part' of the void, as much as in the totality of the void itself. In other words, although God is present everywhere, He is not extended in space in the sense that one part of Him is here and another there.²⁰⁹ This doctrine of Divine presence is discussed by More in connection with the traditional view that the soul is as much present in each part of the body as it is

²⁰⁶ More, *The Immortality* (1662), Book I, Chap. VII, pp. 31–3.
²⁰⁷ *Ibid.*, Book I, Chap. IV, p. 24 (1662); p. 23 (1659).
²⁰⁸ Edward Grant has argued that More and Newton make the three-dimensionality of space an attribute of God, thereby accepting that God is an extended being. See *Much Ado About Nothing: Theories of Space and Vacuum from the Middle Ages to the Scientific Revolution* (Cambridge University Press, 1981). For an opposing view, see J. E. McGuire, "Existence, Actuality and Necessity: Newton on Space and Time," *Annals of Science* 35(1978):463–508. McGuire argues that Newton has a different ontology than More's and that he recognizes levels of predication. For example, space and time are 'predicates' of pure existence; they do not signify features of reality that are 'part' of the defining nature of Divine being, as such.
²⁰⁹ McGuire, "Existence, Actuality and Necessity," p. 506.

in the whole of the body itself.[210] The notion that God's 'place' is not circumscribed in the manner of a corporeal thing is a hallmark of Newton's later view that the infinity of space is a consequence of Divine existence.[211] Newton's views also differ from those of Descartes, who holds that God's essence "has no relation to place at all."[212]

Do we see the beginnings of Newton's theory of absolute space in the *Questiones?* The answer is a qualified yes. There is more, of course, to a theory of absolute space than the conception of void and penetrable expanses that lack resistance to motion. To be sure, these features will characterize any conception of space that can be said to 'contain' the movements of corporeal existents. But more than this, the categorical nature of space must be considered. Is it a substance or an accident? Or does its mode of existing differ from the traditional categories of being? What is the connection between motions determined relative to the senses and motions with respect to insensible space itself? These are questions that Newton begins to answer in the essay on violent motion, together with the development of a proto-conception of inertia.[213] But as he stops short of stating the conception of unending motion along a right line in the void, so he stops short of positing a theory of infinite and immobile space in which various worlds can be created and moved.

This point can be brought out in another way. Absent from the *Questiones* is the notion that the existence of space (or the void, for that matter) depends on the perfection of Divine omnipresence. According to an Epicurean atomist, if visible matter is composed of quantitatively infinite and separate particles, it is necessary to postulate an infinite void in which these exist. In the atomist scheme, the full and the empty are correlatively related principles that together exist sempiternally and uncreated. They are the ground of their own existence, inasmuch as their existence at one moment refers solely to their existence at an earlier moment. But

---

[210] More, *The Immortality* (1662), Book I, Chap. X, p. 42; p. 73 (1659).
[211] McGuire, "Existence, Actuality and Necessity."   [212] AT.V.240.
[213] These questions are systematically considered in *De gravitatione* (in Hall and Hall).

according to Newton's conception of Divine nature in *De gravitatione*, God and infinite space – as well as their unending duration – exist co-eternally and prior to the creation of matter.[214] Unlike the atomists, Newton does not require that space be infinite in order to be a co-eternal principle along with matter, because matter is merely an *ens creatum*. So, in the absence of contingently existing matter, spaces and times are not empty, because there can be no times and places at which God can possibly fail to be present.[215] What is evident from the *Questiones* is the fact that Newton's commitment to the actual infinity of the void occurs before his explicit consideration of how God's presence and the extension of the void relate. If there is nothing in the *Questiones* that decides for or against Newton's acceptance of the Medieval whole-in-every-part conception of Divine omnipresence, equally nothing indicates that Newton, like More, simply identifies the three-dimensionality of the void with Divine presence so that God becomes an extended being.

This brings us to the last question. Did Newton connect his provisional acceptance of indivisibles of time, motion, and extension with a conception that void intervals possess an inherent structure of minimal parts? There is reason to suppose that Epicurus regarded space as composed of indivisible minima.[216] And it may be that Newton's metric model in "Of Atoms" is partly motivated by this picture (see Section 4). Certainly he holds that the length of a distance interval is composed of a number of determinate minima that are its ultimate units of measure. So we may conjecture that part of the purpose of the model is to provide an actualist explanation of metric properties in terms of nonarbitrary minima inherent in the distance intervals themselves. Consequently, if Newton had developed a theory of absolute space in the *Questiones* in connection with what he says about motion in 10 92^v and 21 98^r, he could well have advanced the view that the metrical properties of space are embedded in its structure. In any event, there are problems associated with this conception that Newton could hardly fail to notice.

---

[214] *Ibid.*, pp. 103–4.   [215] *Ibid.*
[216] Furley, *Two Studies*, Chap. 6, p. 101, and Chap. 8, p. 129. See also David Konstan, "Problems in Epicurean Physics," *Isis*, 70(1979):398, 402.

If the rows of indivisibles are arranged at right angles to each other, the basis of the metric is determinate, but not if the indivisibles are arranged at varying angles, or if they are different in size and shape. There is no need to pursue these questions further, because Newton says nothing about them. It is nevertheless well to notice that the logic of "Of Atoms" propels him toward such issues, especially the consequences of his view that indivisibles of quantity are in contact with each other. It is interesting to observe that in the Scholium on space and time in the *Principia*, in which Newton allegedly advances an account of space's 'intrinsic metric,' he is rightly silent on such matters.[217] If such a program did cross his mind, it was probably given up together with the rejection of conceptual indivisibilism in the *Questiones*.

[217] See J. E. McGuire, "Space, Infinity, and Indivisibility: Newton on the Creation of Matter," in *Contemporary Newtonian Research*, edited by Zev Bechler (Dordrecht: Reidel, 1982), pp. 145–190.

2

THE CARTESIAN INFLUENCE

During the first six months of 1664, Newton absorbed himself in the scientific and philosophical writings of René Descartes. He was twenty-one years of age and in his last year as a Trinity undergraduate. There is no way to state with assurance what quickened Newton's interest in Descartes. Certainly the prescribed curriculum did not demand it, nor was Newton's tutor, Benjamin Pulleyn, inclined to lead him into the paths of the new learning.[1] It is possible that Henry More of Christ College was the agent. He had had a long-standing interest in Descartes and had corresponded with him in 1649.[2] In any event, Roger North records in his *Lives* that during his undergraduate years in 1667–8 "there was a general inclination, especially of the brisk part of the University, to use him. . . ."[3] Such brisk interest in Descartes was no doubt evident during Newton's undergraduate years as well.

As noted, the edition of Descartes's works that Newton used is the 1656 Amsterdam publication by Elzevir.[4] Although not complete, it contains a wide range of Descartes's principal writings: the *Meditations*, the *Philosophical Principles* (*Principia*), the *Discourse on Method*, including the *Meteorology* and the *Dioptrics*, the *Treatise on the Passions of the Soul*, and the *Objections* and *Replies* to the *Meditations*, including the fifth set of *Objections* by Gassendi with Des-

[1] Westfall, *Never At Rest*, pp. 81ff.
[2] AT.V.267, 340, 401.
[3] Westfall, *Never At Rest*, p. 90. Westfall quotes from North's *Lives*. In Chapter 3, entitled "The Solitary Scholar," Westfall gives a fine account of Newton's undergraduate days.
[4] See footnote 24 in the Introduction.

127

cartes's reply.[5] The pages of Newton's copy are dog-eared in a number of places. They indicate an interest in such Cartesian topics as quantity, substance, place, space, the distinction between mind and body, the nature of infinity, the indefiniteness of extension, the nature of God, the vortices, the senses, the real distinction, Descartes's version of the ontological argument, and so forth.[6] That these dog-earings indicate some of Newton's interests in Descartes's writings is independently corroborated by the fact that most of the same topics are taken up in the *Questiones*. But more important is the fact that the page references that Newton gives on folios 83^r and 83^v to Descartes's philosophical views fit precisely the 1656 Amsterdam edition.

1. Newton's introduction to Descartes's epistemology and ontology

Let us begin with these references to Descartes's philosophical writings from the *Opera philosophica* of 1656. In every case Newton's citations refer to Descartes's replies to the *First, Second,* and *Fourth Objections* to the *Meditations*.[7] To buttress his analysis of Descartes's use of the distinction between formal, eminent, and objective reality, Newton cites pages 53–9; to indicate where Descartes discusses necessary being, his view of how the mind distinctly conceives the nature of infinity, as well as the distinction between mind and body, Newton cites pages 60–3. These citations fit the pagination of the 1656 Amsterdam edition, in which these and related topics are discussed by Descartes in his *Responsio ad primas objectiones*. Indeed, they exhaust the whole of Descartes's reply, a reply that discusses many of the issues that Newton mentions explicitly. But the *First Reply* informed Newton about other Cartesian doctrines that equally drew

---

[5] It is significant that Newton had access to Gassendi's objections to Descartes's version of the ontological argument. Later Newton was to adopt Gassendi's arguments to show that existence is not a defining attribute of God's nature in developing his doctrine of Divine existence in *De gravitatione*. See J. E. McGuire, "Existence, Actuality and Necessity: Newton on Space and Time."

[6] We shall indicate later the pages on which these dog-earings occur when details on these topics are discussed.

[7] AT.VII.101–121, 121–170, 218–256.

his attention: the distinction between the mind's capacity to grasp a true, immutable, and eternal nature and its power to generate an arbitrary fiction by abstraction; the causal and ontological arguments for Divine existence; the distinction between the indefinite and the infinite; the claim that we can understand that God is infinite but are yet unable to comprehend his infinity; and the doctrine that distinguishes complete from incomplete things in relation to the distinction between mind and body. Clearly, Descartes's *First Reply* contains an analysis of some of the basic principles of his *Meditationes,* a work that Newton probably read in its entirety by the middle of 1664 at the latest.

On folio 83^r Newton makes a number of references to Descartes's *Responsio ad secundas objectiones.* In this reply, Descartes defends some of the same doctrines that drew Newton's interest. The citations to pages 69, 70–3, and 80 fit the 1656 *Opera.* They refer to Descartes's analysis of the distinction between formal, eminent, and objective reality and his version of God's necessary existence. The *Second Reply* is one of the most important. It contains a defense of the real distinction between mind and body; a full discussion of God's nature and existence in relation to the epistemology of formal, eminent, and objective reality; the doctrine of the will in relation to the distinction between clear and distinct ideas; and the distinction between analysis and synthesis.

There are also a large number of references to the *Responsio ad quartas objectiones.* Newton cites pages 121–5, 127, and 129–34 (which cover the whole of the first section entitled "De Deo") of the Amsterdam edition. There Descartes discusses the ontological argument, and again the distinction between mind and body. The *Fourth Reply* is the most important of the seven that Descartes wrote. It was occasioned by a lengthy set of objections by Arnauld, the perceptive theologian and philosopher. Newton's interest in the *Fourth Reply* cannot be doubted. Besides the evidence of the pagination, Newton has dog-eared two separate pages. These contain Descartes's discussions of the real distinction, of the distinction between mind and body, and of the conception that determinates (e.g., shapes) are dependent modes of being, in contrast to a *rem completam,* a form of being presupposed by determinates.[8] Descartes argues that *res extensa* is in the second category.

Arnauld causes Descartes to defend his philosophy of mind; to elaborate his theory of God's existence, especially the claim that God is self-caused; and to defend the impact that his philosophy would have on the doctrines of the Church.[9] There can be no doubt that Descartes's replies to Arnauld's powerful objections afforded Newton deeper insights into the workings of Descartes's philosophy. Furthermore, they helped to introduce the young Trinity undergraduate to a system of thought that owed allegiance to a Platonic style of thinking, a style subtly different from the Aristotelianism of the arid curriculum he was compelled to follow in his official studies.

But Newton did not rest content with making references to Descartes's *Replies*; he analyzed two topics in Descartes's philosophy. His notes on Descartes's epistemology are not part of the *Questiones* as such; nevertheless, they were written at the same time as its earlier entries and are a significant part of his reading in the Cartesian corpus. The first is a lengthy note on folio 83[r] in which he expounds his understanding of Descartes's treatment of the representative or objective reality of an idea in the mind (as it is thought of *qua* thought of), and its causal ground in either a formal or eminent reality said to exist extramentally. As previously noted, Newton refers to Descartes's replies to the *First* and *Second Objections* in support of his analysis. In both replies, Descartes squarely relates these distinctions to the problem of how the mind apprehends Divine existence.[10]

It is not the purpose of this Commentary to provide an analysis of the difficulties in Descartes's position. Suffice it to say that Descartes accepts a tradition, which originates with the Platonists, according to which the claim that there are natures is distinguished from the claim that natured things exist. Thus one and the same nature – extension – formally exists in the world of extended things, objectively

---

[8] Descartes, *Opera philosophica*, p. 122 and p. 124. There are also dog-earings on pp. 89 and 91, which contain Propositions I–V from *More geometrico dispositae;* these form part of Descartes's *Secundus responsiones* and define his conceptions of thought, ideas, substance, and the distinction between formal and objective reality.

[9] Descartes, *Opera, Objectiones quartae*, pp. 113–20.

[10] AT.VII.102–118, 129–145.

in the content of finite minds, and eminently in the mind of God. In each case, extension has a different mode of being, and Descartes's thought reflects the ontological distinction between being a nature and having a nature, as well as the epistemic notion that a nature can have representative reality in the mind.[11] It is difficult to know whether Newton penetrated to the basis of Descartes's thought in 1664. It is clear, however, that Newton's *De gravitatione* owes a great deal to his pondering Descartes's ontology of true, immutable, and eternal natures. Indeed, Newton's theory of infinite space and time (as first developed in that treatise) is based on an ontology of natures that is indebted to Descartes.[12]

Given this analysis of the first two *Objections* and *Replies*, it is clear that Newton's note at the top of folio 83^r summarizes the core of Descartes's position on objective and formal reality. Newton borrows the terminology used by Caterus to describe the status of ideas: "An idea with respect to the object outside the mind is but a bare denomination or a mere nothing. . . ."[13] He does not, however, agree with Caterus's conclusion. For "with respect to the mind, or as the idea is objectively in it . . . it is a real entity, namely a mode of the intellect." In other words, Newton claims that ideas have a mode of being appropriate to the mind, the intentional being that something has when it is thought of

[11] This is the structure that Descartes assumes in his *Meditations*. It is clear, also, that his version of the ontological argument involves a move from the claim that 'there are natures' to the claim that 'natures exist.' This is evident in Descartes's discussion of the ontological argument in his *Primae responsiones*, in which he appeals to the ontology of a "true and immutable nature, or essence or form. . . ." AT.VII.115; *Opera*, p. 60.

[12] For example, we are told that the mind "has a very clear idea of extension abstracting the affections and properties of a body so that there remains only the unlimited and uniform stretching out of space in length, breadth, and depth" (Hall and Hall, pp. 99–100); Newton holds that "extension is not created but has existed eternally" (p. 109); he claims that extension is among those natures that "are real and intelligible in themselves" (p. 109); and, lastly, he holds "extension is eternal, infinite, [and] uncreated. . ." (p. 111). These are the sorts of characteristics that have been traditionally ascribed to natures and forms. Aristotle himself accepts that these characteristics apply to forms and species, but of course he denies that there are forms separate from matter in the Platonic sense (*Meta.*, Z. 11). Newton holds that extension is analogous to form and separate from matter in *De gravitatione*. Thus, his sensibility is Platonic, and he is clearly influenced by Descartes's ontology of natures, if not his doctrine of material extension.

[13] For Caterus, see AT.VII.92; *Opera*, p. 47. An idea ". . . extrinseca denominatio est, & nihil rei."

qua thought of. It is interesting to note that Newton does not distinguish two senses in which an idea is said to be a "mode of the intellect": (a) the sense in which an idea results from an act of mind and (b) the sense in which the idea is the representative content conceived in that act of mind. Equally it should be noted that Descartes does not preserve the distinction as clearly as he might, as witness the complaints by Arnauld in his *Fourth Objections.*[14]

To explain Descartes's theory, Newton now resorts to the analogy of the wax and seal. Descartes does not use the analogy in either the *Meditations* or his *Replies* when expounding this aspect of his epistemology. But he does use it in the *Rules for the Direction of the Mind* in order to explain the relation between things that impress the senses and the imagination and those things themselves. Because Descartes's *Rules* were never published during the seventeenth century, Newton could have no knowledge of them. Nevertheless, Descartes's use of the seal-wax analogy clarifies Newton's appeal to it; so a brief discussion will be helpful. In Rule 12, Descartes uses the seal-wax analogy twice. It is first invoked to state that the relationship between an object and the external sense is like "the way in which wax receives an impression from a seal." For Descartes, the references to the seal on the wax is more than a mere appeal to an analogy; it is in just that manner that "the sentient body is really modified by the object."[15] But the appeal that is closest to Newton's claim that "the impression is a denomination to the seal, but a mode to the wax," is also in Rule 12. Descartes tells us that "the common sense has a function like that of a seal, and impresses on the fancy or imagination, as though on wax, those very figures and ideas which come uncontaminated and without bodily admixture from the external senses."[16] However, there is no warrant in Descartes's writings for Newton's use of the seal-wax analogy as a means of explaining how ideas in the intellect relate to a causal ground external to the mind itself. In other words,

[14] *Opera*, pp. 108–13; AT.VII.206–218.
[15] AT.X.412, ". . . figuram externam corporis sentientis realiter mutari ab objecto." The *Rules* were little known during the seventeenth century and were circulated only in various manuscript versions. In 1908, Charles Adam established the first complete text in Volume X of the *Oeuvres de Descartes*.
[16] AT.X.414.

Descartes never employs the analogy to illuminate the way in which the understanding apprehends the essential nature of a perceptible object (e.g., the argument in *Meditation* II to show the mind's power to grasp distinctly and clearly that a piece of wax is an extended, flexible, and changeable nature).[17] In contrast to the understanding's power to represent the wax's essential nature, the same fact is conceived confusedly by the senses and the imagination.[18] Newton's use of the analogy to explain the status of an ideational content indicates the tension caused by two disparate strands in his epistemological orientation. As has been said, he uses a modified form of Descartes's ontology of natures (together with the claim that the understanding alone can grasp them) when advancing the doctrine that extension is infinite.[19] On the other hand, Newton is influenced by Hobbes's epistemology and its conditional use in the hands of Henry More. In the context of sensory experience, he is impressed by Hobbes's claim that cogitation, sensation, imagination, and perception are nothing but motions (see Chapter 4, Section 1). Hobbes's 'physicialism' and Descartes's epistemology of natures are strange bedfellows indeed; we shall have something to say about this later.[20] Clearly, it is the Hobbesian sensibility that motivates Newton's extension of the analogy (beyond Descartes's use of it) to explain how the intellect relates to an external causal ground. In any event, apart from Descartes's use of the analogy, it has a long history in the service of epistemological discussion, and there are many sources that would have been available to Newton.[21]

Unfortunately, Newton's extension of the analogy probably indicates that he has not fully appreciated Descartes's reasoning on the relation between objective reality and formal reality. To assimilate the way in which an idea is a mode of the intellect to the way an impression is "a mode to the wax," and to assimilate the way in which an external object acts on the mind to an impression being "a denomination to the seal," both indicate a misconstrual of Descartes's posi-

---

[17] *Opera*, pp. 12–13; AT.VII.32–34.   [18] *Ibid.*
[19] *De gravitatione*, in Hall and Hall, pp. 100–11.
[20] See Chapter 4, Section 2.
[21] See Plato, *Theatetus*, 191C8ff.; Aristotle, *De anima*, Book II. 12. 424a20ff.

tion. Descartes is concerned to explain how the understanding (not the senses and the imagination) grasps the essential nature of a physical object, such as a piece of wax. Although it is necessary that the mind experience the wax through the senses and/or the imagination, it is the understanding alone that apprehends that the wax is essentially a natured thing, a determinate bit of extension with a particular size and shape.[22] It is in connection with his claim that the understanding alone can grasp the essential nature of natured things that Descartes mobilizes the epistemology of objective and formal reality. Moreover, the causal relation between the mind's ability to grasp the nature of an object via the mediation of its inner representations and the object as it exists outside the mind rests ultimately on a theory of transcendental causation that has its basis in the Divine essense itself.[23] This is a far cry from the idea of efficient or contact causation connoted by Newton's use of the analogy.

Newton applies his analogy to another important feature of Descartes's theory in a manner that is less misleading. He speaks of a seal as being more and less "artificial." What he no doubt intends to convey is the idea that there are various ways of making an impression, an impression being no better than the seal or means that makes it; equally, "an idea by how much the more perfect so much the more perfect must its cause be. . . ." This is clearly intended as a clarification of the Cartesian principle to the effect that there must be as much reality or perfection in the cause as there is in the effect.[24] Moreover, Newton enumerates the causal types that can fall under the principle: It can be an idea, an object external to mind that has formal reality, or the nature itself as it exists eminently in the Divine mind. This is not the only place in which he encountered this principle. In his lengthy notes made about 1663 from Daniel Stahl's *Axiomata philosophica,* and entered in the same pocket book as his *Questiones,* Newton wrote the following gloss: "Everything which is in the effect pre-exists in the cause. One thing is found in another either eminently or formally. And so what-

[22] Descartes, *Opera,* p. 13; AT.VII.34.
[23] *Ibid.,* pp. 17–19, 39–40; AT.VII.40–42, 78–80 (*Meditations* III and VI).
[24] *Ibid.,* p. 88; AT.VII.165. See Common Notion IV of *More geometrico.*

ever is in the effect is in the univocal cause formally, but in the equivocal cause eminently."[25] Stahl's *Axiomata* gives a detailed account of the traditional 'causes' and their *locus* in the Aristotelian corpus.[26] Evidently, the young Newton troubled to record a traditional explanation of the principle, as well as Descartes's use of it in his transcendental account of how the mind apprehends an essential nature.

Newton's second set of notes on folio 83[r] are concerned with Descartes's conception of necessary being. The pagination that he cites refers to issues discussed in the *First* and *Fourth Replies* of the 1656 edition of Descartes's works. The main thrust of Newton's note is an explanation of the claim that a necessary being 'is the cause of itself.' Indeed, the implications of this claim greatly exercised Caterus and Arnauld in the *First* and *Fourth Objections,* respectively, which in each case drew lengthy replies from Descartes. Among other objections, Caterus raises some questions concerning Descartes's use of the term 'self-derived' or 'self-caused' in the third *Meditation.*[27] Descartes's argument is designed to show that his own existence is not self-derived, but derived from God, who contains all perfections and is the cause of himself. Caterus points out, however, that to say that something is 'self-derived' can be taken in two senses. In the positive sense, it means that a thing derives its existence from itself "as from a cause"; in the negative sense, it means simply that the thing has no cause in the sense that it is "not derived from anything else."[28] According to Caterus, it is the latter sense in which God can be said to be self-derived. For obviously anything (even God) cannot be the efficient cause of its own existence, because an efficient cause is prior to its effect, and nothing can be prior to itself.

We can now examine Newton's explanation of necessary being. At first sight, his remarks seem curious; they compare God's being the cause of himself to the way in which "a mountain is the cause of a valley, or a triangle the cause that its three angles are equal to two right ones." If there is

---

[25] ULC, Add. 3996, folio 45[v], "Quicquid est in effectu praeexistit in causa. Aliquid in aliquo reperitur vel eminenter vel formaliter. Et si Quicquid est in effectu in causa univoca est formaliter, in aequivoca causa est eminenter."

[26] Stahl, *Axiomata,* Titulus III, pp. 60–136.

[27] Descartes, *Opera,* p. 49; AT.VII.94.   [28] *Ibid.,* p. 49; AT.VII.95.

a difficulty in maintaining that God is self-caused, this is surely not illuminated by an appeal to a mountain 'causing' a valley, or a triangle 'causing' relations among its properties.

It is clear, however, that Newton could find license in Descartes's text for speaking of God as the cause of himself. But what does he understand by this? And what are we to make of his claim that a mountain causes a valley as a triangle causes intrinsic relations among its properties?

Although he uses causal language, Newton states an important qualification: that something is the cause of itself or of its intrinsic properties "is not from power or excellency, but the peculiarity of their natures." The emphasis on the Cartesian ontology of natures is important here. Newton is repeating Descartes's claim that necessary existence is a necessary and intrinsic part of God's essential nature. In other words, to say that God exists necessarily is to say that the essential nature of God is such that he cannot be otherwise than always existent. Given this, the fact that Newton speaks of God as the cause of himself can be seen as a reference to Descartes's claim that God's "existing *per se,* or having no cause other than Himself (*nullam a se diversam habeat causam*), is not from nothing, but proceeds from the real immensity of His power." So if this way of expressing the matter is permissible, we can say "that in a certain sense He stands to Himself in the same way as an efficient cause does to its effect, and hence he exists *per se* in a positive sense."[29] Accordingly, Newton's contrast between something being what it is in accordance with its nature as opposed to possessing some feature by means of "power or excellency" comes to this. To say that God's existence is entailed by his essence is not to say that God contains some power or preeminent feature by which he continually self-sustains himself. Rather, it is to claim that intrinsic to the unity of God's Divine, eternal, and essential nature is necessary existence. This is an immutable and true reality, at once grasped by the mind, not a unique causal process that ineluctably unfolds through

---

[29] *Ibid.,* pp. 57–8; AT.VII.111, *Responsio ad primas objectiones.*

time. That Newton reads Descartes in this way is confirmed by the notes that follow. They refer to Descartes's discussion in the *Responsio ad primas objectiones* of how necessary existence is included in God's essential nature, as well as how the mind is able to grasp that fact.[30] In his analysis, Descartes carefully distinguishes his version of the ontological argument from Aquinas's discussion of it by emphasizing the implications of the ontology of eternal natures on which his reasoning is based.[31] It is this feature of Descartes's thought that Newton's notes focus on, not Descartes's use of causal language in his analysis of the claim that God exists *per se*. Consequently, we need not suppose that Newton literally holds that God is the cause of himself, or that a mountain causes a valley, or that a triangle is the cause of relations among its intrinsic properties. That these are what they are is in accordance with their natures. After all, Newton could hardly fail to notice Descartes's agreement with Arnauld's opinion that it is absurd to ask for an efficient cause of why an essence is what it is. Such explanations are required only with respect to something's existence, not its essence.[32] It appears, nevertheless, that Newton has uncritically followed Descartes's use of causal language in his discussion of true and immutable natures. We shall have more to say about this presently.

The interpretation we are developing is borne out by Newton's lengthy note on folio 83ᵛ. The form of the reasoning is Newton's own, but it is clearly based on Descartes's conception of the ontological argument. Although Newton is no doubt influenced by the *First, Second,* and *Fourth Replies,* it is reasonably certain that he also draws on Descartes's formulation of the ontological argument in the fifth *Meditation*. Newton dubs Descartes's version of that argument an axiom: "It is a contradiction to say, that thing does not exist whose existence implies no contradiction, and being supposed to exist must necessarily exist." It appears that Newton's formula owes something to Descartes's view that "it is no less self-contradictory to conceive (*cogitare*) of a God . . . who lacks existence, . . . than it is to conceive of a

---

[30] *Ibid.*, pp. 60–2; AT.VII.115–119.　　[31] *Ibid.*, pp. 60–1; AT.VII.114–115.
[32] *Ibid.*, p. 133; AT.VII.243–245.

mountain which lacks a valley."[33] But the next part of his explanation summarizes a characteristic Cartesian argument that is found in the third *Meditation* and in the *Principia,* as well as in the *Replies.* Descartes's argument involves two principal premises, both of which are mentioned by Newton: first, that "a thing does not properly conform to the notion of a cause except during the time that it provides its effect, and hence is not prior to it"; second, that the infinite parts of time into which the duration of a living substance can be divided do not depend on one another. Given this, Descartes argues, it does not follow from "the fact that I have existed a short while before that I should exist now, unless at this very moment some cause produces and creates me, as it were, anew or, more properly, conserves me."[34] Moreover, we can discover no ground in our defining nature whereby continuance in existence is guaranteed beyond the moment in which we are directly aware that we exist. Thus, the ultimate ground for my continuing to exist is God, the cause that re-creates me, as it were, at each individual and indivisible moment of time.[35] As Newton summarizes the argument: "an immediate cause and effect must be in the same time and therefore the preexistence of a thing can be no cause of its past existence (also because the after time does not depend on the former time)."

Newton goes on to make the essential Cartesian point. If we ask why God exists, or continues to exist, the answer can only be that it is "from the essence of it that a thing [God] can perpetuate its existence without extrinsic help. . . ." Newton's gloss on the term 'essence' appears to raise the same difficulties that Arnauld perceived in Descartes's view of God as the cause of himself. In effect, Newton says that if an essence is sufficient to continue a thing in existence, it "must be sufficient to cause it, being the same reason of both." It is clear that Newton's use of causal language in this context owes something to Cartesian influence. But he adds an important qualification: One and the same essence is the *reason* for its being in existence as well as for its continuing

[33] *Ibid.,* p. 32; AT.VII.66.
[34] *Ibid.,* p. 23; AT.VII.48–49, *Meditation* III. See also the *Principia,* Part I, Art. XXI, *Opera,* p. 6; AT.VIII.13.   [35] *Ibid.*

in existence. The added emphasis is significant in view of Newton's insistence that an appeal to an eternal essence can alone provide the basis for claiming that God exists *per se*. It is true that Newton explains necessary existence in Cartesian terms: With respect to a necessary being, cause and effect are 'immediate.' But, unlike Descartes, he does not appear to appeal to an explanation that is 'analogous' to efficient causation. Rather, Newton's claim is that God exists, has existed, and will exist because his existence is grounded in his essential nature. And if it is asked why God exists or continues to exist, the answer is that the nature of a supreme being is such that it exists necessarily. Consequently, Newton's understanding of how God's existence and essence are related owes much to Arnauld's *Fourth Objections*, as the citations to the 1656 edition of Descartes's works amply testify.[36] There is no indication that Newton wishes to combine the conception that God possesses a true and immutable essence with the notion that he contains an 'inexhaustible power' that is his 'cause' for existing and for continuing to exist.[37] It is interesting that he has paraphrased from Stahl's *Axiomata* an argument for the traditional claim that nothing is the cause of itself: "It is no objection that God exists of himself: for he experiences no inflowing and transference from himself, even though not from another: and thus he is not from himself. Nor does it follow that he is the cause of himself just because he is not from another."[38]

It is clear that Newton had made an effort to explore the philosophy of Descartes as it is presented in the *Meditations* and the *Objections* and *Replies*. His probings caused him to single out three sets of *Replies* that are among the most im-

[36] See the foregoing analysis.

[37] Although Newton appears here to accept the conceptualist thesis that God's existence is a necessary part of his essence, in his later thought he understands this doctrine to say that at no time is it possible that God can fail to exist. See J. E. McGuire, "Existence, Actuality and Necessity: Newton on Space and Time." Newton parses the conception that God is a necessary being in causal terms (i.e., there is nothing separate from God that can alter his nature or prevent his continuance in existence).

[38] ULC, Add. 3996, folio 44ʳ, "Non tamen obstat deum esse a seipse. Non enim habet influxum & communitationem a se quamvis non ab alio: & sic non est a seipse. Non sequitur se causam suipsius esse quia non est ab alio."

portant of the seven sets of *Objections* to which Descartes replied. It is significant, too, that the 1656 edition contains the *Fifth Objections* by Gassendi, together with Descartes's reply. Gassendi's attack on Descartes's doctrine of natures, and his denial that Divine existence is entailed by Divine essence, were to have an influence on *De gravitatione*, especially on Newton's conception of how God's existence and essence relate.[39] But, equally, Descartes's doctrine of true and immutable natures is evident in *De gravitatione*. As we have seen (Chapter 1, Section 7), Newton argues that the extension of space embodies an uncreated, eternal, and infinite nature, an intelligible reality that is an object of the understanding but not of the imagination or the senses.[40] In this regard, it is well to notice that he does not speak in the manner of Gassendi. When faced with the question of spatial infinity, Gassendi characterizes infinite space as imaginary, a notion that is based on the possibility of negating the limits of sensible extension.[41] But for Newton, the infinity of extension is an object of the understanding alone. This is not to say that for Newton an experience of perceptible objects is irrelevant to the understanding's grasp of the nature of extension. Rather, it is to say that the intellect, not the senses or the imagination, is capable of understanding that the extension of space is infinite, where infinity is taken in the sense of a completed actuality. The mark of Newton's profound reading in the scientific corpus of Descartes is everywhere present in *De gravitatione*. It has not been previously noted, however, that the epistemology and ontology of *De gravitatione* owe something to philosophical issues arising from Descartes's *Meditations*. The *Questiones* clearly present evidence of the beginning of this influence. Moreover, it seems apparent that the Platonic tendencies evident in Newton's early metaphysics are in part formed by his reading of Descartes, and they probably owe more to that source than to Cambridge Platonism and Henry More. Newton's interest in 'Cartesian Platonism' is clearly indicated by the fact that he has dog-eared

[39] See McGuire, "Existence, Actuality and Necessity."
[40] See also footnote 12.
[41] *Syntagma, Opera omnia*, Vol. I, Section i, Lib. II, Cap. I, pp. 183–4.

pages 89 and 91 of the *Objections* and *Replies* in the *Opera philosophica*. These contain Propositions I, II, III, and IV of the "Reasons for God's Existence and the Distinction between Mind and Body" demonstrated *more geometrico*.[42] Here Descartes purports to demonstrate the existence of God rigorously from "the sole consideration of his nature," to offer a proof for Divine existence *a posteriori* from "the sole fact that the idea of God exists in us," to show from the fact that I cannot conserve my own existence that necessarily it is conserved by a being who has every perfection that I lack, and lastly to demonstrate again that there is a real distinction between mind and body.[43] Once again, these are all topics that Newton has recorded in his notes. And they afford further evidence of the extent of his interest in the central commitments of Descartes's philosophy.

Yet further evidence for Newton's interest in Descartes's philosophy is given by his dog-earings in Part I of the *Principia*. For example, these refer to topics such as the following: the distinctions between formal, objective, and eminent reality; the distinction between the indefinite and the infinite (the folded page points to Article XXVI); the claim that error is a function of the will, not the intellect; how freedom of the will is to be reconciled to Divine preordination (the folded page points to Article XLI); that "all the objects of our perceptions are to be considered either as things or the affections of things; or else as eternal truths" (the dog-ear points to Article XLVIII); that number and all universals are simply modes of thought (a dog-ear at the bottom of the page points to Article LVIII); lastly, a dog-ear points to Article LXXVI, which places Divine authority above clear and distinct perceptions.[44] There can be no doubt that these markings from Part I form a coherent relationship with the pagination and notes on folios 83^r and 83^v of Add. 3996.

[42] See footnote 7.

[43] Descartes, *Opera*, p. 91; AT.VII.166–170.

[44] *Ibid.*, pp. 2–3, 7–9, 11–13, 15, 22–3. Apart from the fact that Newton's markings refer to particular articles, the following are on the folded pages: VI, VII, IX, X, XI, XII, XIII, XXVII, XXVIII, XXIX, XXX, XXXI, XXXII, XXXIII, XXXIV, XXXV, XXXVI, XLI, XLIII, XLIV, XLVIII, XLV–L, LVIII, LII–LVIII, LXXII–LXXVI.

2. Newton's reading in Descartes's *Principia*

The markings from Parts II, III, and IV also closely match those aspects of Descartes's science that concern Newton in the *Questiones*. Again, if these dog-earings are folded back, they point to particular Articles in the *Principia;* once again we can be reasonably certain that they are Newton's own. From Part II of the *Principia* the markings indicate pages that contain topics such as the following: the claim that the senses do not teach us the reality of things; that the nature of body is extension alone, not weight, hardness, or color; the nature of rarefaction and condensation; the distinction between quantity and corporeal substance; the definitions of external and internal place (a dog-ear points to the former notion); how space is to be distinguished from corporeal extension; that external space is to be defined in terms of the superficies of surrounding bodies; that it is against reason to suppose that there is a space in which 'nothing' exists.[45] Apart from Descartes's distinctions between external and internal place, and between space and corporeal extension, each of these topics concerns Newton in the *Questiones*. Moreover, all of these are criticized in detail some four years later in *De gravitatione*.

Part III of Newton's copy of the *Principia* contains some important markings. These reflect an interest in Descartes's celestial optics (i.e., in questions concerning the source, transmission, and remission of light in the heavens). Among the topics in which Newton shows interest are the following: Descartes's discussion of the methodology of hypotheses; his description of his cosmogony as a fiction that is not in conflict with the Christian belief that the universe is created perfect; the claim that the laws of nature operate to create order out of the chaos in nature; how the parts of the fluid heavens become round and spherical, given that a void does not exist; how the fluid heavens generate three different forms of matter, that which forms the Sun and stars (the first element), that which forms the heavens (the second element), and that which composes the earth, planets, and

[45] *Ibid.*, pp. 26–9. The articles on these pages are VI–XVI.

comets (the third element); that the spherical particles that form the heavens tend to recede from the center of the vortex in which they move; the explanation of why the Sun and stars are round in terms of a 'centrifugal' endeavor; that the first element enters all vortices at the poles and from their equators passes into other vortices, where they touch at their poles; that the matter of the second element cannot behave in this manner; and, lastly, Descartes's cometary theory.[46]

It is striking that these topics form a coherent whole. Moreover, they comprise some of the central principles of Descartes's 'celestial optics.' Clearly, Newton had read Part III of the *Principia* closely. It is noteworthy, however, that his dog-earings do not include those pages that contain Articles LVII, LVIII, LIX, and LX.[47] These apply the concepts established by Descartes's analysis (in Part II) of how a stone in a sling tends to recede from the center of the circle it describes to the behavior of the second element in a whirling celestial vortex. The analysis involves not only the notion of *conatus,* or the tendency that matter has to recede radially from a center of constraint, but also the conception that every part of matter has a tendency to move along a right line, a notion that is essential to Descartes's formulation of the law of inertia. Interestingly enough, Newton did not make any dog-earings in Part II on the pages that contain Articles XXXVII and XXXIX.[48] These articles explain and discuss the first and second laws of nature that together provide the conceptual framework for the theory of vortical motion in Part III. We cannot conclude that Newton failed to study the articles in question, or that he failed to grasp their significance. It must be remarked, however, that there is no explicit discussion of these issues in the *Questiones.* In fact, the beginning of Newton's analysis of Descartes's three laws of nature (including his rules of impact), as well as the Cartesian theory of circular motion, was still some few months away; to be precise, it began in the *Waste Book* in

---

[46] *Ibid.,* pp. 62–7, 79, 102–109, 122–7. The articles are XLIII–LIII, LXIX, LXX, CXII–CXV, CXXXII–CXXXV.

[47] *Ibid.,* pp. 69–72; AT.VIII.108–112, *Principia*, Part III.

[48] *Ibid.,* pp. 37–9; AT.VIII.62–63.

January 1665.[49] As will become apparent, Newton's approach to mechanical problems in the *Questiones* is such that it does not presuppose a full-scale analysis on his part of Descartes's views on the problems of motion.

Newton's markings in Part IV of the *Principia* are of interest. These include folds on pages that contain the following topics: the action of the Sun and the Moon on the Earth's seas and lakes; Descartes's views on the senses; his distinction between the internal and external senses; his comments on the physiology of touch, smell, hearing, and sight.[50] Regarding Descartes's views on sensations, it appears that Newton was little influenced by them. He wrote an interesting essay on sensation, but his sources are Henry More and Hobbes, not Descartes.[51] The essay on vision owes much to Robert Boyle. And although Newton wrote headings for remarks on odors, tastes, and touch, nothing was added except a reference to Kenelm Digby on the sensation of feelings (39 107ʳ). It is nevertheless clear that Newton singled out an important topic that is representative of Descartes's approach to the understanding of the natural world: the flux and reflux of the sea. But again, as we shall shortly see, he is critical of the Frenchman's views on the cause of tidal behavior.

There are dog-earings to the *Discourse Concerning Method* in the *Opera philosophica*. In particular, Newton has marked pages 26, 27, 29, 31, 34, and 39 of Parts V and VI.[52] The first four pages contain discussions of some of the central commitments of Descartes's cosmogony. He argues that his laws of nature are compatible with the perfections of God, so that even if God had created other (successive) worlds, these laws would obtain. He adds that his laws give the mind an intelligible grasp of how the universe evolved into its present state of equilibrium and order. Moreover, they are compatible with the doctrine, received by theologians, that God preserves the world in existence in the same manner that he first brought it into existence. These doctrines are all compatible with Des-

---

[49] John Herivel, *The Background to Newton's Principia* (Oxford University Press, 1965), Part II, Chaps. II and IVa.

[50] Descartes, *Opera*, pp. 160–1, 212–15. Articles L–LVII, CLXXXVII–CXC, CXCII–CXCVI.   [51] See Chapter 4.   [52] AT.VI.564–572.

cartes's claim that there can be only one world in the sense that there is no void to demarcate one part of homogeneous extension from another. It is not surprising that Newton has marked these pages. We have seen from the analysis of the dog-earing in his edition of Diogenes Laertius that he had an interest in the details of the ancient cosmogonies. Furthermore, his markings in Part III of the *Principia* include those pages in which Descartes discusses the compatibility of his laws and cosmogony with the dictates of the Christian religion. It is clear that these dog-earings indicate a pattern of interest on the part of the young Newton. Moreover, they represent the beginning of a lifelong inquiry into the mysteries of creation and the Creator.

Page 31 of the *Discourse* contains the beginning of Descartes's general discussion of anatomy, especially his lengthy account of the structure of the heart and the theories of William Harvey. It is not entirely clear why Newton marked this discussion. But he was concerned with anatomical questions, and the views of Descartes and Harvey on the circulation of the blood and the nature of the heart were well known at midcentury. Furthermore, the methodological differences between Descartes and Harvey on these questions may well have caught Newton's interest. The last dog-eared page contains a discussion of the body construed as an automaton that, though it can be compared to a machine, is nevertheless fashioned by the creative power of God. It is altogether fitting that Newton, who had an early interest in machines, contrivances, and mechanical toys, should be intrigued by Descartes's use of those artifacts in the explanation of natural phenomena.

There are only two markings in the *Dioptrics*. These pertain to Descartes's discussion of refraction in the eye and his views on light pencils.[53] The *Meteorology* contains none, which is surprising, given Newton's extensive references in the *Questiones*. Nor is there any dog-earing in the *Passions of the Soul*, nor in the six meditations of the *Meditations*. The lack of markings in the *Meditations* is surprising, given the extent of Newton's interest in the *Objections* and *Replies*. It

[53] Descartes, *Opera*, pp. 95–9; on page 121 the fold points to Article XIV of Cap. VIII.

may be the case that Newton responded to the *Meditations* as
a whole, fully appreciating their dialectical character. In any
event, the supposition helps to explain the fact that Newton
is able to single out for comment the most important of
Descartes's responses to his critics.

We now turn to Newton's response to Cartesian science in
the *Questiones*. The main purpose of the analysis will be to
give an account of his reactions to the third and fourth
parts of the *Principia* and to the *Meteorology*. This will not
exhaust Newton's references to the scientific corpus of Descartes. Consequently, certain topics will be taken up in separate sections of this Commentary.

On folio 11 93^r Newton puts a series of interesting questions to Descartes's theory of vortical motion in the fluid
heavens. As we noted earlier, his dog-earing in Part III of
the *Principia* indicates an interest in topics such as those of
Articles XLV, XLVI, XLVIII, XLIX, L, LI, LII, and LIII.
These provide the reader with an account of Descartes's
views on the construction of the heavens based on the assumption that the latter are of a fluid nature. According to
Descartes, the heavens are fluid because the diverse motions
of their parts in all directions cause them to yield their
places easily, offering no resistance to an external force. As
Descartes put the point in Article LIV of Part II: bodies
"which are already in motion do not prevent other bodies
from occupying the place that they themselves are disposed
to leave."[54] But hard bodies, those whose hardness consists
solely in the state of rest of their parts relative to each other,
"cannot be driven from their place except by some other
force" large enough to separate their parts from one another. Given this distinction, Descartes concludes that "those
bodies are fluid which are divided into many small particles
all mutually agitated by diverse and independent motions,
and those bodies are hard, all of whose particles mutually
rest beside one another."[55] There is no doubt that Newton
had studied this doctrine. On 6 90^v he asks "whether the
conjunction of bodies be from rest?" The reference is to
Article LV of Part II, which argues that there is nothing else

[54] *Ibid.*, p. 44; AT.VIII.71.
[55] *Ibid.*, p. 45; AT.VIII.71, Article LVI.

required to explain cohesion than the notion that the parts of a body be at rest relative to one another; this is clearly an application of the general doctrine put forward in Article LIV.[56]

It is undeniable that Descartes's views are especially important to Newton, because they deny what he affirms. As we have seen, Newton advocates the Epicurean principle of atoms and the void, a commitment that made Descartes's position Newton's principal adversary, one that undoubtedly attracted his attention. It cannot be concluded from this that Newton rejected Descartes's views entirely, even though he raised queries about the conception of fluid vortices. It is now recognized that vortical components play a role in Newton's astronomical reasoning into the 1680s.[57] Moreover, various ethereal models, intended as accounts of gravitational and optical action (to be sure, explanatory media different from Descartes' fluid vortices), continue to have plausibility for Newton well beyond the period of the *Principia*.[58] Nevertheless, it cannot be denied that the conceptual basis of Descartes's matter of the fluid heavens—its identification with extension—was straightway rejected by Newton: It opposes the existence of the void and the atomic structure of matter. As noted previously, Newton's dog-earings in Parts I and II of the *Principia* indicate a keen interest in Descartes's theory of matter, space, and quantity. Among the pages indexed by Newton are those that contain Descartes's arguments to the effect that there is only a distinction of reason between the notions of internal and external place.[59] Given Newton's commitment to the principle that the parts of a quantity are distinct if and only if they are in principle separable, Descartes's views could hardly fail to escape his notice.

If matter and extension are everywhere identical, it follows not only that there cannot be an extended void but also that

[56] *Ibid.*

[57] See Curtis Wilson, "Newton and the Eötvös Experiment," in *Essays in Honor of Jacob Klein* (Annapolis: St. John's College Press, 1976), p. 194.

[58] J. E. McGuire, "Force and Active Principles and Newton's Invisible Realm," *Ambix*, 15(1968):154–208.

[59] Descartes, *Opera*, p. 28; AT.VIII.46–47, *Principia*, Part II, Articles XII, XIII.

all motions must be in a circle or ring. If a part of extension moves, another must occupy its place simultaneously, a third the place left by the second, and so on in succession, forming a closed ring of motion. This is Descartes's version of *antiperistasis* (see the discussion in Chapter 1, Section 2), and it is the basis on which he constructs his vortices, given the fluid medium of the heavens.

In Article XLVI of Part III, Descartes presents his account of the origin of the three basic forms of matter in the heavens. Invoking a principle of simplicity, he states that in the beginning God divided extension into parts of uniform size, intermediate between the sizes of those that now constitute the heavens. Two motions were equally imparted to these parts: one a rotation around their proper centers; another a disposition to rotate as a group about centers external to them. The latter disposition to rotate about an external center formed a series of rotating fluids, the vortices. It is the rotation of the parts themselves that is responsible for the primitive fluidity of the heavens.[60]

It is, of course, difficult to conceive how even God could individuate a 'part' of the homogeneous and undifferentiated nature of primitive extension. In Article XXIII of Part II, Descartes tells us that "mere division by thought alone, can change nothing; all variety of matter, all diversity of its forms, depend on motion."[61] The appeal to motion as a principle of division (as we have seen) will not do: To conceive a part as moving presupposes that it is already individuated. There is no indication that Newton was aware of this difficulty in 1664. However, in *De gravitatione* he notes the difficulty, as well as the related notion of the indiscernibility of identicals, both basic to his systematic critique of the Cartesian theory of motion.

In any event, given the rotation of the heavens, Descartes argues in Articles XLVIII, XLIX, L, LI, and LII of Part III that the uniformly angular parts of the primitive matter rotate against each other in accordance with the laws of nature.[62] The force of this rotation gradually makes some

[60] *Ibid.*, pp. 65–6; AT.VIII.100–101, *Principia*, Part III.
[61] *Ibid.*, p. 32; AT.VIII.52–53, *Principia*, Part II.
[62] *Ibid.*, pp. 66–7; AT.VIII.103–105.

of these parts round and spherical, and the scrapings that result from the rounding process fill the intervals left between the spherical parts or globules, preserving the principle that the universe is a plenum. Accordingly, there are three principal elements in the universe. The scrapings, which compose the Sun and fixed stars, are the first element. It comprises subtle particles that move with great speed. When the first element collides with the larger globules, it crumbles into innumerable parts; because it is not of any one determinate size or shape, this allows it to fill all of the intervals left by the globules. The globules, which are larger than the first element, constitute the second element, and they are the constituents of the fluid heavens. The third element, slow and large in comparison with the second element, composes the Earth, planets, and comets.

This, then, is the framework of Descartes's cosmology as it is found in Part III of the *Principia*. It views the heavens as a 'multiparticle' continuum comprising only extension and its proper modes, size, shape, position, and the disposition of its parts to be moved. It is to this view of the heavens that Newton addresses a series of questions that begin on 11 93^r. These constitute a series of interconnected queries, in the essay entitled "Of the Celestial Matter and Orbs":

Whether Descartes's first element can turn about the vortex and yet drive the matter of it continually from the Sun to produce light, and spend most of its motion in filling up the chinks between the globuli. Whether the least globuli can continue always next to the Sun and yet come always from it to cause light. Whether when the Sun is obscured, the motion of the first element must cease (and so whether by his hypothesis the Sun can be obscured). Whether upon the ceasing of the first element's motion the vortex must move slower. Whether some of the first element coming (as he confesses that he might find out a way to turn the globuli about their own axes to grate the third element into coils, like screws or cockle shells) immediately from the poles and other vortices into all the parts of our vortex would not impel the globuli so as to cause a light from the poles, and those places from whence they come.

Newton's initial query concerning the action of the first element in the production of light pertains to some interesting issues in Descartes's cosmology. In Article XLIV of Part III, Descartes explains how the first element results from the attrition of the rotating 'globuli' or parts of the second element. The minute parts of the first element fill the interstices left by the second, as the latter endeavor on all sides to move away from their centers of rotation; but owing to the smallness and speed of its parts, the first element also flows to the centers of the vortices to form the subtle matter of the Sun and the fixed stars, which are thus "very liquid bodies."[63] Newton's question comes to this: Can the first element as it agitates within the Sun press outward into the Sun's vortex to produce light and yet have its pressure continually modified by an interaction with the larger globules of the second element? In Newton's view, the first element appears to have considerable work to do, given that its force of agitation is modified and weakened as it continually fills the interstices between the globules of its vortex. But it would appear that more is at issue.

In order to evaluate Newton's worry, we need a fuller picture of the motion of the first element in relation to the second. The poles of the vortices in the fluid heavens are arranged so that they touch the equators of neighboring vortices, never at their respective poles. If they touched at their poles, they would either combine in motion or hinder each other's motion. The matter of the first element continually flows out of the vortices at their equators and into them at their poles. The matter of the second element cannot so move, because its motion is more restricted at the poles than at the equator.[64] If it were to move from the equator of one vortex to the poles of another, its constancy of motion would be violated.[65] Furthermore, the first element flows into the body of the Sun and pushes the surrounding globules of the second element equally in all directions. To explain the equality of the Sun's pressure in all directions, Descartes resorts to an analogy: Just as air blown into a mass of molten glass at once sets up an equal pressure in all directions within the glass, "so the matter of the first element, which enters the body of the sun through the poles, must press equally in all directions the surrounding parts of the second element, not only those which it en-

[63] *Ibid.*, p. 69; AT.VIII.107–108, Article LIV.
[64] *Ibid.*, p. 75–9; AT.VIII.116–121, Articles LXVI, LXVII, and LXVIII.
[65] *Ibid.*, pp. 79–81; AT.VIII.121–122, Article LXX.

counters directly, but those against which it is reflected obliquely."[66] If the resources of the analogy are taken seriously, as Descartes intends, the motion of the first element within the Sun produces a real and outward pressure in all directions from the Sun's surface on the surrounding globules present in its vortex.

It is clear from Newton's essay that he recognizes the fact that both the first and second elements contribute to the production of light. Newton's dog-earings in Part III of the *Principia* mark out neighboring pages that contain Article LV. In his discussion of this article, Descartes says:

It is a law of nature that all bodies which are put into circular motion recede from their center of motion, in as much as they can. Accordingly, I shall explain as accurately as I can that force by which globules of the second element, as well as the matter of the first element, gathered around their centers . . . endeavor to recede from those centers. For it will be shown below *that light consists in that alone. . . .*[67]

The law of nature to which Descartes refers is found in Article XXXIX of Part II. There he explains a corollary to his second law of nature, that every body moving in a circle perpetually tends to recede from the center of the circle that it describes, a phenomenon that we experience when we whirl a stone in a sling.[68] Consequently, if we consider only that part of the stone's motion in a circle that is impeded by the sling, we can say that the stone endeavors to recede radially from its center outward along a straight line, even though it is in fact impeded by the sling.[69] Similarly, each part of the first and second elements endeavors to recede from the center of motion of the vortex, but is restrained by the presence of other parts of the vortex beyond pressing against it; moreover, at the same time, all parts of the second element "are pressed by the matter of the first element gathered in the center of each vortex."[70] Thus, an endeavor to move from a center of rotation is always directed radially outward along a right line,[71] and there can be no talk of an

[66] *Ibid.*, pp. 85–7; AT.VIII.129–131, Article LXXV.
[67] *Ibid.*, p. 69; AT.VIII.108. "In ea enim sola lucem consistere, infra ostendetur. . . ."   [68] *Ibid.*, p. 39; AT.VIII.63–65.
[69] *Ibid.*, pp. 39–40; AT.VIII.63–65.
[70] *Ibid.*, p. 72; AT.VIII.112, Part III, Article LX.
[71] *Ibid.*, pp. 73–4; AT.VIII.113–114, Article LXII.

endeavor to move without the presence of a resistance to that endeavor. In the case of the parts of the second element, the endeavor to move is constrained by the parts in the vortex lying beyond them. This is simply an instance of Descartes's second law, that is, "that every part of matter, taken in itself, never tends to continue moving along oblique lines, but always along straight ones."[72] So, given that every part of the heavens is disposed to move along a right line, any opposing tendency is resisted. Consequently, Descartes views the parts of the heavens as 'contesting' one another, so that the tendency of the inner globules to move away from the center of rotation along a right line is in fact opposed by those beyond them in the same vortex. After all, Descartes's central conception in his theory of light is that light is a *conatus,* an endeavor to move, not an actual motion.

It is difficult to know how much of Descartes's theory Newton fully grasped in 1664. Although he recognized the roles that both the first and the second elements play in the production of light, it seems that Newton did not appreciate Descartes's official doctrine that light consists entirely in an endeavor to recede from a center.[73] Newton's first two queries seem to presuppose that light results from actual motions in the heavens, instead of from the instantaneous endeavor to move of all those adjacent parts of the medium arranged in a right line. Thus, Newton appears to think that the parts of the vortex push or press against one another in succession, and in so doing give rise to an outward pressure that is identified with light. If Newton's conception of Descartes's position is so interpreted, the tenor of his queries on folio 11 93^r becomes clear. If light results from the motion of the first element into the surrounding vortex of the Sun, why is it not modified and enfeebled by its encounter with the larger globules of the vortex? In the second query, Newton asks "whether the least globuli can continue always next to the Sun and yet come always from it to cause light." In other words, if the smaller globules always move nearer the Sun, how can they also move away from it to produce light? Newton's second query is probably

[72] *Ibid.,* p. 39; AT.VIII.63–65, Article XXXIX.
[73] *Ibid.,* p. 69; AT.VIII.108, Part III, Article LV.

based on Article LXXXV of Part III. There Descartes argues that below Saturn, the Sun's motion increases the motion of the globules, so that those on or near its surface move faster than those farther away. Consequently, the globules nearer the Sun are smaller than those more distant; if they were the same size, they would have a greater inclination to move and would rise above the others.[74] In any event, it is clear that Newton construes Descartes as saying that the parts of the first and second elements move, press, and push against one another, from the action of which the phenomenon of light results.

The same conception is the basis of the following queries in the essay. Newton asks "whether when the Sun is obscured, the motion of the first element must cease (and so whether by his hypothesis the Sun can be obscured)." If, as Newton understands, light is the result of a pressing or pushing of the first element, how can it be obscured? Surely the 'pressure' would be transmitted through the 'obscuring object' on the supposition that the Sun's action is construed as a real pushing against the globules of the vortex. Here Newton is directing his comments toward the *effects* of the Sun's pressure. But notice that he refers to the 'spending' of the motion of the first element. Thus, independent of his answer to the former question, Newton is also interested in the question of the cessation of motion of the first element. Thus, he asks "whether upon the ceasing of the first element's motion the vortex must move slower." This is a causal question. It may well be that Newton saw what Descartes was committed to in Article LXXV (i.e., that it is the agitation of the first element within the Sun that produces an outward pressure that in fact acts in all directions from the Sun's surface).[75] After all, Descartes admits in Article LXIV that the light emanating from the Sun not only passes in an instant from the center of the Sun along the ecliptic perpendicular to the axis of rotation but also radiates "from all points on its surface."[76] As we have seen, it is this feature of

---

[74] *Ibid.*, p. 95; AT.VIII.140, Article LXXXV.

[75] *Ibid.*, pp. 85–7; AT.VIII.129–131, Part III, Article LXXV.

[76] *Ibid.*, pp. 74–5; AT.VIII.115. ". . . id quidem secundum lineas rectas, non a solo corporis lucidi centro, sed etiam a quibus libet aliis ejus superficiei, punctis eductas."

the Sun's action that the analogy to the equality of pressure on the inner surface of a bubble in molten glass is designed to explain in Article LXXV. Here Descartes construes the spherical radiation of light from the Sun's surface in terms of a single outward pressure, which can be communicated to the adjacent vortex. It is not unreasonable to assume, therefore, that Article LXXV probably helped to prompt Newton's query, for its points to the causal basis of his question. Accordingly, if the light from the Sun is an inner pressure acting from all points on the Sun's surface, it follows that the Sun's invisibility would result from a cessation of its inner activity. Furthermore, on the assumption that the motion of the first element contributes to the behavior of globules of the vortex, if the first element were to cease, would the vortex slow down? It seems that Newton is attempting in this query to settle the individual contributions that the first and second elements make to the production of light, given that Descartes is emphatic in Article LV that they both have roles to play. If the first element ceases moving, is the motion of the second slower? If so, what is the effect of this on the production of light? Or are the behaviors of the two different elements independent of each other, as Newton implies?

It is difficult not to suppose that Newton was on the verge of perceiving a serious difficulty in Descartes's celestial cosmology. In Article LXIV Descartes makes a surprising admission: "if the body of the sun were nothing other than an empty space, we would nevertheless see its light, not, of course, as strong, but no differently than at present."[77] At this point in his exposition, Descartes considers only radiation from the Sun in the plane of the ecliptic; he has yet to consider the actual role of the Sun and stars in the production of light. But if light really consists in an endeavor of the second element to recede from its center of rotation (together with the first element), the Sun itself is not necessary for the production of light. However, when he does recognize that the Sun radiates spherically, in the course of writing the *Principia,* and offers an explanation in terms of a

[77] *Ibid.,* p. 75. ". . . si corpus Solis nihil aliud esset quam spatium vacuum, nihilominus ejus lumen, non quidem tam forte, sed quantum ad reliqua, non aliter, quam nunc cerneremus, . . ."

pressure internal to the Sun, Descartes has two separate theories of the production of light, neither of which he eliminates in favor of the other.[78] It is clear that Descartes needs the notion that light is an instantaneous inclination to move along right lines in order to preserve his commitment to the rectilinear propagation of light from the Sun. The theory that light is an internal pressure acting in all directions from the Sun's surface, if developed, would contradict the first theory, because every point (not only those of the Sun) would be a source of light. Newton's response to the Cartesian theory of light is largely based on this side of Descartes's thought. But in 1664 he had not yet seen that this approach leads to the true conception of pressure, a conception that violates Descartes's commitment to the rectilinear propagation of light from the Sun in the vortical medium.[79] Nor was Newton in a position fully to appreciate that Part III of the *Principia* contains two independent conceptions of the production of light, for he did not fully grasp the conception that light is an endeavor to recede from the center of rotation. These realizations had to await *De gravitatione* and Newton's analysis of pressure in a fluid medium and the problem of circular motion.[80]

We should not be surprised by Newton's failure to understand the Cartesian conception that light is an endeavor to motion, rather than an actual motion. Apart from the intrinsic subtlety of the conception, Descartes often speaks in the language of action, change, and motion. In discussing the behavior of the first element, he describes it as "flowing" into the poles of a vortex, "gathering" at its center, and then "passing" through its equatorial region into the poles of another vortex.[81] Descartes's language invokes the impres-

[78] For an important discussion of this and related topics in Descartes's cosmology, see Alan Shapiro, "Light, Pressure, and Rectilinear Propagation: Descartes' Celestial Optics and Newton's Hydrostatics," *Studies in History and Philosophy of Science*, 5(1974):252–3.   [79] *Ibid.*

[80] Hall and Hall, pp. 117–21. See also Shapiro, "Light, Pressure, and Rectilinear Propagation," pp. 266–84. Shapiro also has useful observations on the relationship between the *Questiones* and *De gravitatione*, pp. 266–8.

[81] Descartes, *Opera*, p. 79; AT.VIII.119–121, Part III, Article LXIX. ". . . materiam primi elementi fluere continuo versus centrum . . ."; "progredi debeat versus S . . ." Articles LXXXVI–XCIII of Part III are very physicalistic in language. In XCIII, for example, Descartes speaks of particles as being "expelled" and "extruded."

sion that the subtle matter actually moves and acts on its surroundings. Again he uses the terms *pressio* and *presso* in his discussion of light in Article LXIII of Part III and in Article XXVIII of Part IV.[82] The noun literally means 'a pressing' or 'a pressing down,' and the verb connotes the act of pressing; in both cases the sense connotes that something is being done, an action.[83] It is true that in Article LXIII Descartes explains the instantaneous tendency of the globules to move along right lines, in analogy to the way lead globes in a vessel move at once in a triangular configuration, as a globe at the apex moves through the aperture in the bottom of the vessel.[84] Nevertheless, the analogy is put forward as an explanation of light by reference to the weight of the globes, which instantaneously push each other toward the vessel's aperture. It is little wonder, then, that Newton should come away with the impression that light results from the 'action' of the parts of extension 'pushing' or 'pressing' against one another.

It is with the same sensibility that Newton asks on folio 18 96ᵛ "whether things congeal for want of agitation from the ethereal matter. Descartes." This is a reference to Discourse One of the *Meteorology*, in which Descartes speaks of the action of an ambient and subtle matter (propelled by light) on the inner structure of physical bodies (see Section 3). Descartes appeals to the agitation of this medium in order to explain a variety of effects as well as the characteristics of phenomena. Newton's query points to Descartes's claim that the less the particles of the subtle medium are able to affect those of a body, the more it can become rigid and hard.[85] Similarly, on 18 96ᵛ, when Newton asks 'Why does air moved by light cause heat or why does light itself cause heat?" the reference is to Discourse Two.[86] There Descartes speaks of the action of light agitating the subtle matter and the air; as a result of various states of compression, vapors and exhalations can become hotter. It is clear that Newton wishes to ask a prior question (i.e., whether

[82] *Ibid.*, p. 74; Part IV, p. 147; AT.VIII.114–115, 217.
[83] Shapiro, "Light, Pressure, and Rectilinear Propagation," p. 251.
[84] Descartes, *Opera*, p. 74; AT.VIII.114–115.
[85] *Ibid.*, p. 155; AT.VI.652–653, *Meteorology*, Cap. I, Article IV.
[86] *Ibid.*, p. 162; AT.VI.658–659, *Meteorology*, Cap. II, Art. VI.

Descartes's vortical theory can sufficiently explain why light as such can cause heat). For all that he says in the *Meteorology*, Descartes's account of the origin of heat is in terms of the inner agitation of the parts of bodies.[87] It may be that Newton does not think that Descartes has made clear the relationship between that account of heat and the vortical action that constitutes light.

In any event, Descartes considers himself entitled to use the language of efficient causation 'descriptively' in Parts III and IV of the *Principia*, so long as it is not confused with the fundamental principles of his natural philosophy as presented in Parts I and II. In these earlier parts Descartes argues that the physical world comprises nothing but passive extension, a nature that merely has a capacity to be moved. It is clear that this conception of matter is compatible with the conception that each of its parts has a *conatus* to move, and with Descartes's denial that true agency exists in the physical world. In Descartes's view, true causation results from intentionality.[88] Thus, only conscious agents – God and finite spirits – can be said to bring things about. In fact, all real changes in the physical world are the completed effects of God's re-creating the parts of extension, together with their various configurations, at each discrete and succeeding instant of time.[89] To grasp these subtle levels in Descartes's philosophy is no easy task. We should not be surprised that Newton failed to do so, concentrating as he did on the 'scientific' parts of Descartes's *Principia*. When he wrote *De gravitatione*, however, Newton had come to terms with Descartes's theory of *conatus* and with its relation to the doctrine of re-creationalism.[90]

Newton's last query (in the long passage quoted earlier from 11 93^r) again considers the consequence that light is the result of the parts of the first and second elements 'impelling' one another. Because Descartes accepted the phenomenon that each point on the Sun's surface radiates in all directions, an explanation had to be offered. Articles

---

[87] *Ibid.*, pp. 160–161; AT.VI.657–658, *Meteorology*, Cap. II, Art. IV.
[88] AT.VII., *Meditations* III and V.
[89] Descartes, *Opera*, pp. 22–3; AT.VII.44–45, *Meditation* III.
[90] Hall and Hall, pp. 113–14.

LXXVI, LXXVII, LXXVIII, and LXXX provide such an explanation, namely, how it is that although the Sun merely presses the subtle matter normal to its surface, the Sun's light spreads toward both the ecliptic and the poles of the solar vortex, as a result of the endeavor of first and second elements to recede from the center.[91] However, recall that although the first element (but not the second) continually flows out of a vortex at its equator, it flows into a vortex at its poles. So, in essence, Newton's query is this: If the first element is continually streaming into the Sun's vortex from the equators of adjacent vortices, and if some of the minute parts can flow into various parts of the vortex as well as to the center of the Sun itself, what prevents their pressing on the globules to cause light to emanate directly from the poles and from other regions of the vortex apart from the Sun itself? In order to assess the thrust of Newton's query, it is necessary to consider the complex motions that Descartes invokes in his consideration of the flow of the first element between the intervals among the globules. This background will also help in understanding the rather cryptic remark that Newton places in parentheses on folio 11 93[r]: "as he [Descartes] confesses that he might find out a way to turn the globuli about their own axes to grate the third element into coils, like screws or cockle shells."

Articles LXXXVI through XCIII discuss the relationship of the first and second elements in detail. In fact, Article LXXXVI reintroduces and summarizes Descartes's analysis in Articles XLVI–LII of the formation, development, and rotation of the matter of the first element and the globules around their centers.[92] He then proceeds to argue that although the parts of the first element, on entering the solar vortex, go principally toward the Sun, many are dispersed throughout the vortex.[93] As they flow around the globules of the second element, the latter are moved about their centers and in other diverse ways. The parts of the first

[91] Descartes, *Opera*, pp. 87–9; AT.VIII.131–135, Part III.
[92] *Ibid.*, p. 95; AT.VIII.142, Part III.
[93] What follows is drawn from Articles LXXXVII–XCIII, Part III, of the *Opera*, pp. 97–8; AT.VIII.142–147.

element, however, are not all of the same size; some are varied in size to allow them to accommodate the changing figures of the intervals that occur between the moving parts of the second element, and some are divided into smaller parts by the second element's action. Still others remain undivided but small enough to pass through the intervals. And because the degree of agitation of a particle is proportional to its size, the less finely divided scrapings are less rapidly agitated. Furthermore, the more angularly shaped of the particles easily adhere to one another to form assemblages that are less agitated again. Because the latter assemblages transfer their agitation to smaller ones more easily than the smaller affect them, it is the smallest parts of the first element that are most agitated. Moreover, it is the large assemblages of the parts that move in straight lines from the poles of the vortex toward its center. This is because movement along a straight line requires less agitation than that manifested by the diverse movements of the other parts.

Newton's description of particles that are shaped as cochlea refers to Articles XC, XCI, and XCII of Part III.[94] In Article XC, Descartes argues that the assemblages of the first element have to be triangular in their cross sections because they often traverse the triangular spaces formed by three globules mutually touching. They can therefore be conceived as minute cylinders that are fluted, having three twisted grooves or channels [striatae] like a snail's tail. However, the particles are not all twisted in the same direction. Some approach the Sun from the north pole and others from the south pole, as the whole vortex rotates in one direction on its axis.[95] It is to this mechanism that Descartes appeals in his explanation of the force of magnetism.[96]

To the obvious objection that the globules do not always touch to form triangular spaces between them Descartes replies that a larger quadrangular space has four angles, each of which is equal to an angle of a triangular space between three globules. However, because the figures of the spaces formed by four moving globules are always changing, "the less agitated matter of the first element placed in it

[94] Opera, pp. 97–8; AT.VIII.144–146.   [95] Ibid., p. 97; AT.VIII.144–145.
[96] Ibid., p. 195; AT.VIII.289–290, Part IV, Articles CXLIX and CL.

must flow towards one or two of these angles and the remaining space be left to the more agitated matter which can change its shape in order to adapt to all the movements of these globules."[97] This line of reasoning leads Descartes to claim in Article XCIII that, apart from the cochlea, there are innumerable particles of the first element of varying sizes owing to the diversity of the intervals through which they pass and which they fill.[98]

It must be admitted that Descartes's account is arbitrary and rather contrived. No explanation is given of the various speeds at which the parts of the first element twist as they rifle through the globular intervals. Nor is there any account of how they maintain their shapes, given that the figures formed by contiguous globules are always changing; yet at least some of the particles must preserve their shapes if Descartes is to have an explanation of magnetic polarity.

There is no evidence to show whether or not Newton had such questions in mind. What is clear from his query, though, is an awareness that Descartes's mechanism involves two different sorts of interactions between the globules and the particles of the first element. Not only does the first element impel the globules by its tendency to recede, but the globules interact with the parts of the first element itself: As they rotate on one another, they press the parts of the first element into fluted particles of varying sizes, shapes, and widths, as well as divide it into separate parts of smaller sizes.[99] Moreover, throughout these articles, Descartes speaks in the language of contact causation: terms like *extrudo* and *expello* abound. As Newton sees it, if light is the outward pressing of the first element of the globules, equally might it not arise from an interaction between the globules and the parts of the former as they enter the poles of the vortex and disperse throughout its regions? Although he does not say so, it is clear that Newton is skeptical about the presence of such luminous sources in the heavens.

But equally it is clear that Descartes is committed to these consequences. In Article LXXVI of Part III he tells us that

[97] *Ibid.*, pp. 97–8; AT.VIII.146, Part III, Article XCII.
[98] *Ibid.*, pp. 98–9; AT.VIII.145.
[99] *Ibid.*, pp. 97–8; AT.VIII.144–145, Articles XC and XCII.

the matter of the first element comprises two movements: one along the right lines from the poles of the Sun's vortex to its center and then outward to the ecliptic; another circular motion around the poles themselves that is highly agitated.[100] It is this second motion of the first element that provides the basis for Newton's query.

We can now turn to Newton's rather puzzling remark in the phrase "to grate the third element." It is tempting to suppose that Newton has confused the first element with the third. For the following reasons, this appears to be unlikely. Consistent with what Descartes tells us in Articles LXXXVI–XCIII, Newton is probably suggesting (in a compressed manner) that Descartes's account allows for the transformation of some parts of the first element into the grooved, less agitated particles of the third element. After all, the notion that particles have varying 'degrees of agitation' is in Descartes's view an important demarcator for distinguishing the elements. Moreover, it is also clear that the grooved particles differ in kind from other sorts of first-element particles, but because no observable differences are detectable, Descartes does not bother to single them out. Furthermore, when they are massed together, the grooved particles are responsible for the formation of the Sun spots, which are third-element matter.[101] In short, Newton is probably relegating grooved particles of a certain sort into what he considers their appropriate category, without mention of the Cartesian mechanism that brings them about.[102]

At folio 12 93ᵛ, Newton raises further queries concerning Descartes's cosmology in a lengthy essay entitled "Of the Sun, Stars, Planets, and Comets." He asks if the Sun moves its vortex about "by his beams," and a reference is given to page 54 of Descartes's Opera philosophica; this falls in the section that contains Part III of the Principia, and Newton's citation is to Article XXI. Descartes argues that although the matter of the Sun is highly mobile, as that of a flame, it nevertheless does not move as a whole from one place in the

---

[100] Ibid., p. 87; AT.VIII.131.

[101] Ibid., pp. 100–1; AT.VIII.150–151, Part III, Article C.

[102] We are grateful to Professor Reese P. Miller, of the University of Western Ontario, for his discussion on this point.

heavens to another. Newton's query refers to Descartes's statement that the fluid and exceedingly mobile matter of the Sun "carries with it the neighbouring parts of the heavens."[103]

This mechanism whereby the Sun's motion affects the neighboring globules in its vortex is employed in Articles LXXXII, LXXXIII, and LXXXIV.[104] Size and speed are the two factors that determine the distribution of the parts of the second element in a vortex. Those closer to the center of a vortex are smaller and move faster than those farther away, but this obtains only up to a certain distance, beyond which the speed increases with distance. The globules in the region of Saturn are the slowest in the Sun's vortex, and from Saturn's orbit outward the size of the globules remains constant. Beyond Saturn, when a globule moves faster than those around it, it immediately rises and manifests a greater *conatus*. Consequently, those globules more distant from the center move faster. But below Saturn, the Sun's motion about its center increases the motion of the globules, and those in contact with its surface move faster than those farther away. The globules nearer the Sun are therefore smaller than those more distant; if they were of the same size as those more distant, they would possess a greater *conatus* and would consequently rise above their surroundings near the Sun.

It would appear that Newton is not questioning whether or not the Sun's motion can affect the adjacent parts of its vortex. After all, he accepts Descartes's claim that light results from the pressing of the Sun on its attendant vortex. Rather, he seems to be asking if the Sun alone is capable of providing the motive force necessary to drive the vortex. This reading appears probable if we conjoin the first query with the second: "How is it that the Sun is turned about upon his axis?" Furthermore, by what mechanism is the Sun's motive force renewed? It may also be the case that Newton has an inkling of another difficulty. Will not the motions of individual vortices merge into one another over a given time? If so, will they not slow down owing to attrition? If a suitable explana-

---

[103] Descartes, *Opera,* p. 54; AT.VIII.87, Part III.
[104] *Ibid.,* pp. 91–3; AT.VIII.137–140, Part III.

tion of how individual vortical motions are maintained, sustained, and confined is not forthcoming, the stability of Descartes's cosmology is endangered. It is interesting to note that Newton concerned himself with the stability of circular motions in fluids in *De gravitatione*. Moreover, he put the same sort of question to Descartes's vortical theory in Book Two of the *Principia*.[105] In the *Principia*, however, Newton has at his disposal a sophisticated theory of how pressure acts to produce various sorts of circular motion in fluid media. In that work he is easily able to show that Descartes's fluid vortices violate the principle of the constancy of motion to which he is committed.

Newton's next two queries are concerned with Descartes's theory of the comets and the nature of their tails. The theory is expounded by Descartes in Articles CXXVI, CXXVII, and CXXXIII–CXXXVI of Part III of the *Principia*.[106] If a star is affected by the motion of a vortex, it is initially pushed toward its center by the faster motion of the parts of the surrounding second element at the circumference of the vortex. Eventually it assumes as much agitation as the surrounding globules. If equality of agitation is achieved before the star reaches the orbit of Saturn (where the globules move slowest), the star ascends and, passing into an adjacent vortex, will become a comet. Thus, according to Descartes, comets are always beyond the orbit of Saturn. They are seen by the deflected light of the Sun and by its refraction from the graduated sizes of the inner globules below the orbit of Saturn.[107]

Not only do comets move beyond the orbit of Saturn, but they move through the outer circumference of the whirling vortices, along a path that is tangential to the circle described by the globules. However, as they move through the

[105] Hall and Hall, pp. 115–21. *Principia*, edited by Koyré and Cohen, Book II, Section IX.

[106] Descartes, *Opera*, pp. 118–19, 125–9; AT.VIII.174–178, 185–190.

[107] *Ibid.*, pp. 125–6; AT.VIII.186–188, Part III, Article CXXV. In Article CXXXIV Descartes tells us that in order to explain a comet's tail, appeal must be made to a new sort of refraction not found in his *Dioptrics* ("quorum omnium rationes ut intelligantur, novum quoddam genus refractionis, de quo in Dioptrica non actum est, quia in corporibus terrestribus non notatur . . . ," p. 186). The refraction to which Descartes appeals is more like a diffusion. It is caused by the variation in size of the globules below Saturn.

fluid media, far from encouraging resistance, they are propelled by the agitation of the faster globules of the second element and are thus deflected a little by the circular motion of the globules from the straight-line paths that they would otherwise follow.[108]

Newton asks if the action of the Sun's vortex can move a comet toward the poles. In the diagram that accompanies Article CXXVI, Descartes indicates that a comet pursues a path that always moves across successive vortices at their outer circumferences away from the respective poles.[109] This is what we would expect, given the theory of cometary motion that Descartes expounds. What, then, is the sense of Newton's query? It is clear that Descartes's theory prohibits a comet from moving toward the center of the vortex. It is also clear that Descartes conceives comets to be denser than planets, because they assume agitation at a faster rate, a feature that is a function of the density of a body. Consequently, Newton may have thought that a difference in agitation could occur at the poles and the circumference of a vortex, such that the lesser agitation at the poles would allow a denser comet to move into that region of a vortical circumference. But it is difficult to square this conjecture with Descartes's claim that the faster-moving globules affect the motion of a comet as it ascends and moves in the vortex.

Newton's fourth query on folio 12 93^v asks if Descartes can explain the phenomenon of the comet's tail. This is the least satisfactory part of Descartes's theory. In Articles CXXXIV, CXXXV, and CXXXVI he explains that the reflected light of the Sun (by which comets are observed) is refracted by the parts of the second element below the orbit of Saturn. If such an account explains the presence of a comet's tail and halo, what explains the fact that Jupiter, Saturn, and the fixed stars do not exhibit tails? Surely their light ought to be bent to provide the same effects as those explained by Descartes's theory of cometary refraction in a vortical medium. Descartes fully explains in Article CXXXIX that the light of these more distant bodies (even the original light of the fixed stars) is not strong enough to allow such discrimination as it

[108] *Ibid.*, pp. 127–9; AT.VIII. 188–190, Article CXXXVI.
[109] *Ibid.*, p. 118; AT.VIII.175.

travels along the straight lines that are responsible for the appearance of the comet's body. Moreover, a comet with an apparent diameter no larger than the fixed stars has no tail, and some stars do have discernible halos, because their shapes are not sharply terminated.[110]

We can now return to Newton's query whether or not Descartes can account for the comet's tail. In the first place, Newton's reference to Descartes's account of reflection does not appear to employ the term in the sense in which it is distinguished from refraction. Rather, it is employed to denote Descartes's account of the bending of light under various circumstances. And it is clear from the tenor of Newton's remarks that he is skeptical about Descartes's refractive account of the origin, as well as the positions, of comets' tails.

Descartes recognizes that the shape, length, and disposition of the tail changes as a comet moves across the sun and that it is seen more clearly at one time than at another.[111] However, Descartes's refractive theory of the origin of the tails commits him to holding that they always manifest colored bands. This is not so, and Newton was probably aware of the fact from his observations.[112] Furthermore, if a comet's path is followed through the sky, it always appears to travel along a straight line. Recall that Descartes's vortical theory commits him to the view that a comet's path is bent (somewhat) from the right line it would otherwise traverse by the action of the vortex through which it is passing. Again, Descartes argues in Article CXXXVI that the tail of a comet can precede it, as well as follow it. The difference is to be explained by the motion of the Earth with respect to the Sun and by the changing directions in which the weaker rays from the comet are diffused as they penetrate the region of smaller globules circulating below the orbit of Saturn.

Newton's interest in comets was born early in his career and was to play a significant role in his mature system of the world. It also reveals a certain style in approaching natural phenomena—one that pursues consistency and that

[110] *Ibid.*, pp. 129–30; AT.VIII.191–192.
[111] *Ibid.*, p. 129; AT.VIII.190, Article CXXXVII.
[112] See Section 1, Chapter 7.

attempts to establish quantitative parameters where this is possible. In this regard, the contrast between Newton and Descartes is marked: The latter does not record any personal observations pertaining to comets.[113] It may well be that Descartes never observed a comet. There were, in fact, three comets observable from Europe in 1618; in 1625, another was observable; also, there was one observable in 1647 and another in 1652 that was tracked by Gassendi, Boulillaud, Cassini, Golius, and Hévélius. The comets sighted in 1647 and 1652 were too late for Descartes's purposes, but those of 1618 and 1625 were clearly available to him.[114]

Still mindful of Descartes's views on atmospheric refraction, Newton notes his claim that some Sun spots are colored like the rainbow. In Articles XCIV, XCV, and XCVI of Part III Descartes explains the origin of the spots. He views the Sun in analogy to a boiling liquid. As scum forms on the surface of a liquid that is intensely heated, so darkened spots appear on the Sun's surface that obscure its light.[115] They are initially formed by the first element rising from the interior of the Sun that, on meeting the second element of the surrounding vortex, is ground into channeled parts. They then come together on the Sun's surface and gradually acquire the opaque character of the third element, which interrupts the Sun's illumination. The spots sometimes disappear owing to the agitation of the first element beneath them. Again, to explain this, Descartes appeals to the analogy of boiling liquids on the surface of which scum appears and disappears. In Article XCVII he observes that the spots manifest colors comparable to those in the band of a rainbow. This is explained by the claim that the spots become smaller at the circumference of the Sun

[113] In Article CXXVIII Descartes mentions the observations of Horatius Grassius on the positions of the comet of 1475 as recorded in his *Libra astronomica ac philosophica*. The same comet of 1475 is mentioned in Article CXXIX (Regiomontanus is said to have observed it) and in Article CXXXIII. These references do not reveal any great second-hand observational knowledge of comets on the part of Descartes.

[114] We are grateful to Dr. Brian Marsden, of the Astrophysical Observatory of the Smithsonian Institution, who confirmed this information in a private communication.

[115] Descartes, *Opera*, pp. 99–100; AT.VIII.148–149, Article XCVI.

than in the middle region; moreover, the spots become transparent and less dense toward the circumference than toward the middle. According to Descartes, such an arrangement refracts the light that passes through it and gives rise to appearances similar to the rainbow. For a fuller discussion of the phenomenon, Descartes refers the reader to the eighth discourse of *Meteorology*.[116]

Newton's essay "Of Light" is important (folio 32 103ᵛ). It makes a number of well-founded observations of Descartes's views and subjects the Frenchman's celestial optics to searching criticism. The second observation sets the topic for inquiry: "How light is conveyed from the Sun or a fire without stops." In other words, how can we best explain how the Sun's light emanates continuously to the eye and to reflecting and refracting bodies?

Newton denies flatly that the mechanism whereby light is transmitted through the heavens results from the instantaneous 'pressing' of the globules against one another. Think what the consequences would be: It would be as easy to see at night as during the day; there would be vision with closed eyes, because the eyelids would be pressed upon; intense light would appear above our heads, owing to the downward pressing of the Sun's vortex; and "when a fire or candle is extinguished we, looking another way, should see a light."

If these observations bear the marks of youthful inquiry, others show insight into Descartes's explanatory mechanism: "The Moon and planets would shine like suns." This consequence follows directly from the fact that the Moon and the planets have individual vortices rotating about them within the larger vortex of the Sun. If light results from the outward pressing of the globules in the Sun's vortex, equally the whirling vortices of the planets and Moon ought to make them shine during the day. Again, the Sun could "not be quite eclipsed," because solid bodies in a vortex could as easily transmit light as the subtle matter of the heavens. If Descartes is right, "a light would shine from the Earth, since the subtle matter tends from the center." This is an interest-

---

[116] *Ibid.*, p. 100; AT.VIII.149.

ing observation. Newton has correctly inferred that light should arise from the Earth's center owing to its rotating vortex causing the outward endeavor of the fluid matter. And his observation takes into account the fact that according to Descartes's theory, light is an endeavor to move on the part of the second element; consequently, it ought to radiate along the plane of the Earth's equator perpendicular to its axis of rotation. Nor would the radiation be obstructed by the interposition of a body, because it, too, would endeavor to move from the center and so help to transmit the light. Newton's remaining queries follow in the same vein. There are various and changing distributions of the matter in the heavens. If Descartes were correct, there ought to be differential distributions of 'pressure' that would give rise to effects different from those we in fact observe. For example, if we consider the 'pressions' necessary for equilibrium in a vortex, and assume that light is seen in the direction of greatest 'pression,' because the "greatest pression [is] on that side of the Earth from the Sun . . . the nights should be lightest."

Furthermore, Newton naturally enough asks: Because the Sun's vortex is elliptical in shape, will this affect the action of the globules so as to send light in different directions at the same time? His reference is to Article XXX of Part III. There Descartes says that the Sun's vortex is not perfectly round, but longer than it is broad.[117] Moreover, Descartes recognizes that the planets do not move in perfect circles within the Sun's vortex. He supposes in Article CXLI of Part III that the solar vortex is distorted by the unequal pressures of the surrounding vortices on different sides, so that the globules move more slowly in the wider regions than they do in the narrower, thus causing the planets to move farther from the Sun in the former regions than in the latter.[118] Here again Newton wonders if these differences in the speeds of the globules, owing to the figures of the solar vortex, will cause differences in the intensity and direction of light.

In Article CLIII Descartes says that the path of the Moon

[117] *Ibid.*, p. 58; AT.VIII.92.   [118] *Ibid.*, p. 130; AT.VIII.193.

around the Earth is an ellipse. He is considering the lunar inequality known as variation (the only one he considers). There is no suggestion that he thinks of the Earth as occupying a focus of a geometrical ellipse.[119] On the contrary, the Earth is said to occupy the center of the orbit along which the Moon moves, as the Sun is said in Article XXX to be at the center of its vortex. If Descartes knew of Kepler's laws of planetary motion, there is no indication that he employed them in any technical sense in his vortical theories. It would be incorrect to suppose, however, that Newton is concerned to consider the compatibility of Descartes's cosmology with Kepler's laws.[120] Whereas it is true that Newton is skeptical of the Cartesian cosmology, he is in no way seeking to reject it entirely. A fully viable alternative is two decades away, with the advent of his own *Principia*.

Two further observations that Newton makes are of interest. He tells us that "there could be no refraction since the same matter cannot press two ways." He also says that "pression could not render shapes so distinct." It has been noted that the first observation shows that Newton did not conceive of Descartes's vortical medium as embodying a real pressure, that is, a force acting in all directions at any point.[121] In fact, as we have already seen, Newton uses the term 'pression' in the sense of a pressing of one thing on another. This is the same sense in which he will use it in his *Waste Book* when discussing the impact of one body on

---

[119] *Ibid.*, p. 134; AT.VIII.199–200.

[120] Newton's knowledge and use of Kepler's astronomy at various periods in his career have been topics of much debate between Herivel, Whiteside, and Russell. D. T. Whiteside, "Newton's Early Thoughts on Planetary Motion: A Fresh Look," *British Journal for the History of Science*, 2(1964):117–37; J. L. Russell, "Kepler's Laws of Planetary Motion: 1609–1666," *British Journal for the History of Science*, 2(1964):1–24; J. W. Herivel, "Newton's First Solution to the Problem of Kepler Motion," *British Journal for the History of Science*, 2(1965):350–4. It is interesting to observe, furthermore, that Newton speaks of "the aethers in the vortices of the Sun and Planets" in his hypotheses concerning light sent to Oldenburg for presentation to the Royal Society in 1675. See *The Correspondence of Isaac Newton*, edited by H. W. Turnbull (Cambridge University Press, 1959), Vol. I, p. 368. Also, in ULC, Add. 3965.14, folio 613ʳ, he can still write in 1682 that "the material of the heavens is fluid and gyrates around the center of the cosmic system according to the course of the planets" (Materium coelorum fluidam esse [et] circa centrum systematis cosmici secundum cursam Planetarum gyrare).

[121] Shapiro, "Light, Pressure, and Rectilinear Propagation," p. 268.

another.[122] Furthermore, a little later, in *De gravitatione*, he will recognize that one and the same matter can in fact press in all directions and consequently will show how Descartes's theory of light violates the postulate of rectilinear propagation.[123] But given Newton's interpretation of 'pression' in the *Questiones*, it is no surprise that we find him considering how Descartes's theory can account for refraction in a medium. The second observation indicates that Newton allows for some slight spreading of the 'pression' adjacent to the globules; this, of course, is a far cry from his later view that pressure in a medium spreads completely into the geometric shadow.[124] Although Newton is highly critical of Descartes, on the basis of this essay alone it cannot be argued that he is assuming the correctness of a 'corpuscular' theory of light. This question will be taken up when we consider Newton's essay "Of Colours" (Chapter 5).

For all that he says, it is apparent that Newton has not come to grips with the details of Descartes's theory of circular motion. Especially striking is the fact that he gives no explicit analysis of Articles LV–LXIV of Part III of the *Principia*, in which Descartes advances his theory of circular motion and shows how it relates to the laws of nature announced in Part II. It will be helpful to provide a brief account of the theory in order to place Newton's response to Descartes in proper perspective. This will help in the consideration of how that response might have differed had he come to grips with the theory in 1664, rather than later in *De gravitatione*.

As we noted earlier, Descartes applies the results of his analysis of the constrained motion of a stone in a sling to the behavior of the globules in a whirling vortex. In Article XXXIX of Part II he points out that a stone in a sling endeavors to recede along the tangential line at each instant it circulates about the center of rotation under the constraint of the sling; hence, such a tendency persists at every instant during the whole period of circulation. He immediately remarks that "we can also feel it in our hand, while we

[122] Herivel, *Background*, Part II, p. 156.
[123] Shapiro, "Light, Pressure, and Rectilinear Propagation," pp. 266–84.
[124] *Ibid.*, Section VI, pp. 284–90.

make it turn in the sling. This consideration is so important, and since we frequently use it in what follows, it is noted at this point, though it will be explained below."[125] Descartes's reference is to Articles LV, LVI, LVII, and LIX of Part III.[126] In the Latin edition of the *Principia* (the one that Newton used), the remark in Article XXXIX suffers from a certain vagueness. And in the French edition, Descartes added the following phrase to the description of the stone in the sling: "for it stretches and makes taut the cord so as to recede directly from our hand."[127] Even so, a reader could still easily conclude that Descartes is referring to the tangential tendency described previously. But, in fact, he is singling out another description that characterizes the stone's tendency to recede radially from the center of rotation, a description that Newton almost certainly failed to grasp in 1664. This goes a long way toward an explanation of the fact that Newton misses the full thrust of Descartes's use of the term *conatus* in the context of his vortical theory.

When the stone moves in a closed circle of motion, according to the second law, it is disposed to move tangentially. But owing to the external constraint of the sling, it is continuously diverted along a curved path. Descartes illustrates this by a counterfactual claim: If the stone were released from the sling, it would straightway move along a line tangent to its circle of motion. This endeavor can manifest itself only at the instant the stone is released; nevertheless, it is present at each instant of its revolution, even though the sling impedes the stone from actually moving along the tangent.[128] So much is clear from the second law of nature (i.e., that "every part of matter of itself never *tends* to move along a curved line, but always according to a straight line").[129] Thus, not only does each part of matter preserve its state, whether of moving or resting, but it always by its very nature tends to move along a right line. But this tendency can be referred to under another description. As the stone is made to turn in the sling, the string is drawn

---

[125] Descartes, *Opera*, p. 40; AT.VIII.64–65, Part II.
[126] *Ibid.*, pp. 64–72; AT.VIII.108–112.   [127] AT.IX.86.
[128] *Ibid., Opera*, pp. 69–70; AT.VIII.108–109, Article LVII.
[129] *Ibid.*, p. 39; AT.VIII.63–64, Part II, Article XXXIX.

taut. Consequently, although the whirled stone tends to move in a right line, the sling resists that tendency insofar as it retains the stone. Their interaction is the basis for describing a second effect, that the stone in the sling can be viewed as receding radially from the center of rotation, the direction of which is altered at each successive instant.[130] It is clear that these effects are considered in isolation from the stone's complete and actual motion in a circle. This point is stressed by Descartes in Article LVII of Part III, in which he argues that one and the same body can be said to tend to move in diverse directions at the same time.[131] It is clear that the only ontology to which Descartes appeals is that of the second law (i.e., that every part of matter always tends to recede along a right line). Nevertheless, in the sling example, he brings this tendency or endeavor under two different descriptions: an endeavor to move tangentially from the circle of revolution and an endeavor to recede radially from the center of rotation. It is this second description that Descartes uses to characterize the globules of the first and second elements in his claim in Part III that light alone consists in an endeavor to recede from the center of rotation of the Sun's vortex.[132] Just as the stone can be said to be restrained by the sling in its endeavor to recede from the center, so the effort of each of the globules to recede from the center of its vortex can be said to be restrained by those that are beyond it. In both cases, the endeavor to recede is balanced by a resistance.

In sum, there are three factors to consider in Descartes's account of circular motion: the resistance due to an external constraint, and an internal 'force of motion,' which can be described under two aspects. The tendency to recede radially from the center is opposed by the external constraint, so that their effects are balanced. But there is no actual motion along the string, because there is no real tendency from the center. However, the tangential tendency is not so affected by the sling, as must be the case if Descartes's second law is to hold. This way of thinking of phenomena (i.e., as manifesting diverse features simultaneously) is further illustrated by

---

[130] *Ibid.*, pp. 69–70; AT.VIII.108–109, Part III, Article LVII.   [131] *Ibid.*

[132] *Ibid.*, p. 69; AT.VIII.108, Article LV.

Descartes's analysis of the path of light through a refracting surface. He divides it into a perpendicular component and a horizontal component. After light has struck the surface, only the alteration of its perpendicular determination need be considered, because its determination parallel to the refracting surface is in no way affected.[133] Similarly, the sling is like a refracting surface. It resists the stone's radial tendency to recede, whereas the parallel component, its tangential tendency, is unimpeded.

It is clear that Descartes's reasoning is far from apparent. But, equally, it is clear that in 1664 Newton had not yet come to terms with it. We need not seek far for a reason: Newton has failed to distinguish the description of the radial endeavor from the tangential endeavor. And because he does not do so, he fails to perceive that light is not an actual motion for Descartes, but a tendency to recede from the center of the Sun's vortical motion. It is not surprising, then, that Newton interprets Descartes's theory of light as saying that light results from the concentrated effects of a series of globules actually pressing one another successively. But, in truth, Descartes is committed to the curious position that the action of light is best explained in terms of a set of nested subjunctive conditionals that jointly describe the unfulfilled tendencies of each of the linear globules, given that they can be said to impede collectively their respective endeavors to move.

In order to achieve a deeper understanding of Descartes's position, it would be necessary for Newton to study the implications of his three laws of motion and the rules of impact, to examine Descartes's definition of true motion and philosophical motion and his rejection of the 'vulgar' conception, and, lastly, to subject Descartes's analysis of circular motion to closer scrutiny, especially the role it plays in vortical motion. None of these tasks is undertaken in the *Questiones;* it is too much to expect so soon. After all, the entries of the *Questiones* represent the mere beginnings of Newton's reading in the Cartesian corpus. It is true that on folio 59 117[r] Newton provides a direct translation into

---

[133] *Ibid.*, pp. 57–60; AT.VI.591–594, *Dioptrics,* Cap. II.

English of Descartes's 'proper' definition of motion as given in Part II, Article XXV.[134] But there is no analysis of the import of Descartes's views, that is, of how his conception of motion relates to his doctrine of extension, to the second natural law, and to the definitions of internal and external place. If Newton had thought through the implications of the definition, he would have appreciated the relationship between Descartes's treatment of motion as relational mode and his denial that he is speaking of a "force or action" that transfers a body from A to B.[135] In fact, he would probably have grasped the significance of Descartes's claim that the second law applies only to tendencies to motion, not to actual motion. It is well to note also that Descartes's definition occurs as part of a list of examples of motion, as well as statements about motion, that Newton compiles. These include a paraphrase of Glanvill's wheel, a description of a snail's movement, and examples of reflexive actions that Newton construes as merely mechanical motions without volition.

Although some of the *Questiones* dates from 1665, it is only in his *Waste Book* of 1665 that Newton begins his serious investigations into mechanics, again in dialogue with Descartes. Here we find an analysis of motion employing the Cartesian conception of inertia, an attempt to shape the notion of force into a quantitative tool, and a systematic investigation of the phenomenon of impact that, though it incorporates many Cartesian concepts, transcends Descartes's achievement.[136] As in the *Waste Book,* so in *De gravitatione,* Newton appears to conceive a body's endeavor to recede as a real disposition internal to its nature. Therefore, any body moving along a circular path is made to do so by an external constraint, so that a state of equilibrium obtains between the opposing dispositions.[137] In the *Waste Book* this conception is evident in Newton's determination of the measure of centrifugal endeavor; in the model, a ball presses against a confining spherical shell, as the shell confining the

[134] *Ibid.,* pp. 32–3; AT.VIII.53–54. [135] *Ibid.*
[136] Herivel, *Background,* Part II.
[137] *Ibid.,* p. 114. "Conatus est vis impedita sive vis quatenus resistitur."

ball presses back.[138] Newton's commitment to the notion that matter possesses internal tendencies will be analyzed in the appropriate place, namely, in our consideration of Newton's essay "Of Violent Motion."

3. Newton's response to Descartes's tidal theory and the *Meteorology*

On folio 26 100ᵛ Newton begins an entry entitled "Of Water and Salt." The topics discussed draw on the *Meteorology* for the most part and will be considered shortly. However, Descartes's views on the composition of sea water, and the manner in which its salty composition may be affected by the different regions of the Earth's globe, lead Newton into a consideration of the tides and Descartes's account of them. This part of the entry begins on 47 111ʳ and continues on to 49 112ʳ, under the subtitle "Of the Flux and Reflux of the Sea."

To facilitate the analysis of Newton's criticisms, it is necessary to outline the basic principles of Descartes's tidal theory. In Article CXLIX of Part III Descartes offers a plausible account of the Earth-Moon system. He supposes that the Moon either entered the Earth's vortex before it became a planet or, given that it has the same density as the Earth, came to circulate at the same distance from the Sun, but at a faster speed, being of a smaller size. Notice that this supposition regarding the speed of the Moon is inconsistent with Descartes's view of planetary motion, namely, that a planet moves with the same speed as the surrounding vortex. In any event, on the suppostion that the Moon is as dense as the Earth, but yet moves faster, Descartes argues that it must move in an orbit around the Earth itself.[139] In essence, his argument assumes that the Earth-Moon system moves along the same path surrounding the Sun. Rather arbitrarily, Descartes then states that the Moon moves one way rather than another from the Earth's path in order to deviate "less from the straight line," that is, from the path that it would tend to follow were it not resisted by the Earth's

[138] Herivel, *Background*, p. 129.
[139] Descartes, *Opera*, pp. 133–4; AT.VIII.197–198, Part III, Article CXLIX.

atmosphere.[140] Thus, the Moon, while following the Earth's path in the motion of the Sun's vortex, generates its own small vortex around the Earth itself.

In Article XLIX of Part IV Descartes presents his basic explanation of the tides. He argues that the Earth's position in the terrestrial vortex must always be farther from the Moon, than it is from the center of the vortex itself. If this were not so, if the Earth's position in the vortex were not farther from the Moon's orbit on one side of the Earth than on the other, the Moon's presence would inhibit the surrounding fluid (which Descartes supposes to move faster than the Earth and Moon) from flowing as freely on one side between the Earth and Moon as on the other. So when the Moon is on one side of the Earth's orbit, the Earth must displace toward the opposite side, to preserve "the equality of the forces by which it is pressed on all sides."[141] From these considerations, Descartes views the Earth as moving on a small epicycle about the center of the terrestrial vortex. On the principle that the planets circulate at different distances from the Sun (the least dense being nearer), Descartes explains why the Moon always turns the same face to the Earth: That side is denser than the other.[142]

Given this model, Descartes purports to explain tidal behavior. The basic mechanism of his explanation is this. If the Moon is at B on its path around the Earth, the fluid matter of the vortex has less space to flow between both BT and TD. Consequently, in these regions the fluid matter moves more quickly and thus presses the superficies of the air and the water more toward 6,4 and 5,2. Under this pressure, the air and water in these regions are easily 'pushed' inward toward the Earth, causing a greater rise in the water at E and G than at F and H. In other words, the fluid matter ought to move more quickly and press the air and sea surrounding the facing side of the Earth and its direct opposite more than it does the other two sides. As a result of the Moon's differential effect, the height or depth

[140] Ibid., p. 134; AT.VIII.198.
[141] Ibid., pp. 158–9; AT.VIII.234, Part IV.
[142] Ibid., p. 132 and p. 134; AT.VIII.196, 198–199, Part III, Articles CXLVII and CLII.

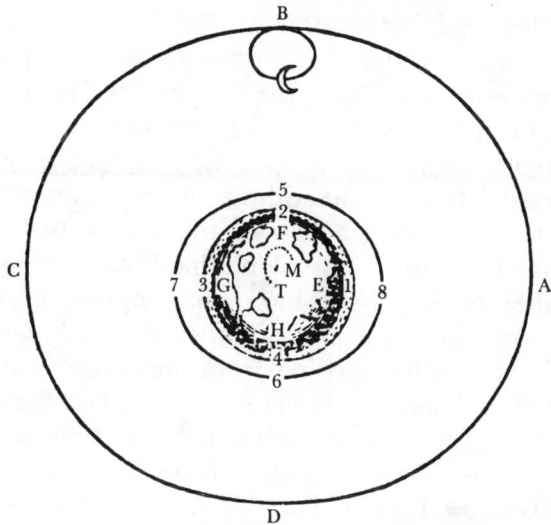

of the Earth's ambient air and sea on its facing and opposite sides ought to be less than on its other two sides. Thus, the greater 'push' on the sea at these places results in the phenomenon of the tides.[143]

Characteristically, Newton asks: Can the Cartesian theory be tested empirically? If the lunar presence pushes the air and the sea to produce the tides, should it not affect the barometer as well? May it not be the same cause that produces the same relation between the Moon pressing the water and the Earth's atmosphere pressing the barometer in a given region? On folio 47 111[r] Newton observes that "as the air is more or less pressed without by the moon, so will the water rise or fall as it does in a weatherglass by heat or cold." Is it not possible that barometric readings can be correlated with the tides? The question does not originate with Newton. The experiment was first suggested by Christopher Wren and reported in *The Spring in the Air*, by Robert Boyle, who indicated that the results were not conclusive.[144] There is no indication that Newton put Descartes's theory to the

[143] *Ibid.*, pp. 158–9; AT.VIII.232–234, Part IV, Article XLIX.

[144] See footnote 102 in the Transcription. For a discussion of this point, see Richard S. Westfall, "The Foundations of Newton's Philosophy of Nature," *British Journal for the History of Science*, 1(1962):176.

test, but he seems confident that the matter can, in principle, be settled by experimental means.

Newton observes that the "tides cannot be from the Moon's influence for then they would be least at new moons." The query is directed toward Descartes's view of what happens when the Moon is in conjunction with the Sun. There ought to be an additive effect on the tides arising from the combined action of the two bodies. In Descartes's model it is the Moon alone that is cited as the cause of the higher tides at opposition and conjunction. His discussion of the lunar inequalities does not include phenomena such as evection, arising from the eccentricity of the orbit, that were known to Ptolemy. Rather, Descartes refers to the inequality known as variation for its obvious role in any explanation of the monthly inequalities of the tides. Thus, in Article CLIII of Part III he concentrates on the shape of the terrestrial vortex as a basis for explaining these inequalities. In effect, Descartes argues that as a result of the differences in agitation and size of the globules of the vortex, the latter "is not a perfect circle but longer than broad in the form of an ellipse."[145] Consequently, the fluid matter moves quickly when the Moon is in opposition or conjunction. Article LI of Part IV appeals to Article CLIII of Part III; in Article LI Descartes gives an explicit account of "why the seas are greater when the moon is full or new, than at other times." Given the narrower space between the Moon and Earth at opposition and conjunction, the faster speed of the fluid medium, and the angle at which it strikes the seas, "the presence of the moon presses the sea more when it is full or new, causing it to rise and fall, than at any other time."[146] As is consistent with his general model, the only agent to which Descartes appeals in explaining the tides at both full and new moon is the presence of the Moon itself. Newton is certainly correct in claiming that Descartes's theory does not adequately account for the fact that the Sun combines with the Moon in conjunctive position to produce a high tide. He is wrong to suggest, however, that Descartes's theory implies that tides would be "least at new

[145] Descartes, *Opera,* pp. 134–5; AT.VIII.199–200.
[146] *Ibid.,* p. 160; AT.VIII.235–236.

moons." If anything, there ought to be no difference on Descartes's view between the tides at either full or new moon.

Newton puts further questions to Descartes's theory. If Descartes attempts to explain the behavior of the tides by the Moon's presence alone, suppose we substitute the Sun? Will its changing position with respect to the Earth allow for a satisfactory account of the difference in tidal behavior between day and night? Is it also the case that a difference in the height of the water between morning and evening would indicate "whether the Earth or its vortex press forward most in its annual motion." In other words, does the rotation of the Earth on its axis toward the Sun in the morning, together with its motion in the solar vortex, produce a higher tide in mornings than in evenings? On 49 112^r Newton asks if the Earth's diurnal motion is hindered, rather than helped, by its being carried in a vortex, "for by the same force it would move the water and air along with it, or rather faster." The next query has some interesting implications. Newton asks whether "the sea's flux and reflux be greater in spring or autumn, in winter or summer, by reason of the earth's aphelion and perihelion." Clearly he is referring to the farthest and closest positions of the Earth with respect to the Sun during the different seasons of the year. Nothing in Newton's remark indicates that he is thinking of a Keplerian orbit along which the Earth travels; rather, he is merely referring to the varying distance of the Earth owing to the shape of the Cartesian vortex. The general phenomenon of differences in tidal behavior throughout the seasons had been noted since antiquity, as well as observations of local variations. Newton's digest of Robert Moray's *Philosophical Transactions* article on the unusual tides in the western islands of Scotland, and in the Danube River, attests to his interest in the complexity of tidal phenomena.[147] It seems reasonable to assume that Newton is doubtful that Descartes's appeal to the Moon's phases can alone explain such tidal complexity. Furthermore, he seems to suggest that it is necessary to understand

---

[147] See footnote 107 in the Transcription.

the influence of the Moon and Sun jointly, as well as individually, in order to have a sufficiently broad basis on which to explain the tides. But Newton is also interested in the correlation between spring and neap tides and the phases of the Moon, as the description of the tides in Berneray indicates. Consequently, it is also probable that he is doubtful that Descartes's theory can explain these tides (especially given their tremendous variations), which do not correlate with the Moon's phases and positions throughout the year in any simple manner. Lastly, Newton makes an interesting observation concerning the main assumption of Descartes's lunar theory. If, in order to equalize pressure in relation to its distance from the Moon, the Earth is offset from the center of its vortex, should there not be a shift in the parallax of Mars as the Earth travels along its small central ellipse? It is well to notice that Newton had not yet formulated any alternative account of the tides that prompted his criticisms of Descartes. Nor is there evidence to show that he was aware of the tidal theories of Kepler and Galileo.[148] Certainly, the former's conjecture that the sea is attracted by the influence of the Moon is a conception that is consistent with Newton's mature theory of the phenomena. That no mention is made of Galileo's views is surprising, given that Newton had access to the *Dialogue*.

This brings us back to Newton's initial entry "Of Water and Salt." It largely consists of queries put to Descartes, for which again Newton probably lacks an alternative explanation; in fact, in many cases he merely recapitulates Descartes's claims in Discourse Three of the *Meteorology*, and to a lesser extent Discourses One and Two. Newton's queries are, on the whole, posed within Descartes's own framework of thought. Judging from the nature of his queries, Newton has read Descartes with care.

On folio 26 100^v Newton asks whether "salt or fresh water is more easily moved and more pellucid, refracts more, and is more easily frozen." This is a clear reference to Discourse

[148] J. Kepler, *Gesammelte Werke*, edited by Walther von Dyck and Max Caspar (Munich: C. H. Beck, 1937–), Vol. XV, p. 287. Galileo's theory of the tides and his consideration of the theories of others are to be found on pages 380–424 of the *Dialogue*, in Thomas Salusbury, *Mathematical Collections and Translations* (London, 1661). Kepler's lunar theory of the tides is derided by Galileo on page 422.

Three of the *Meteorology,* in which Descartes argues that particles of fresh water can move more easily when they "are rolled around [*circumvolutas*] those of salt than when they are alone," because in this condition they are not intertwined with one another, thus making it necessary for the subtle medium to disengage them from one another.[149] And if the particles of water in the sea are easily moved by virtue of being intertwined around the suspended salt particles, they less impede the movements of the subtle matter into their pores (given that it is agitated by the light of the Sun) and are thus "more transparent, and cause slightly greater refractions, than that of rivers."[150] Newton's query is probably prompted by the apparent arbitrariness of Descartes's account. But he goes on to deny the correctness of the way that fresh water and salt are respectively distinguished by Descartes. If fresh water did consist of "long bending parts," it would not freeze; for according to Descartes, only those bodies freeze whose structures cannot be agitated by the action of the subtle matter.[151] Yet fresh water manifestly does freeze, despite Descartes's claim that its parts are easily moved by the subtle matter, and despite the fact that they are knotted or hooked together. Newton's remark is somewhat wide of the mark, for Descartes allows that the subtle medium can lack the requisite size and power of agitation to bend the particles of water (as during the winter) so that they can lie close together to form a rigid body such as ice.[152] In other words, Newton has failed to see the positive and negative roles that the subtle matter can play in changing the inner composition of bodies like water (i.e., that it comprises larger and smaller parts that have greater or lesser agitation to affect things). Similarly, his remark that the twisting of the parts of fresh water around one another makes it soft, rather than a fluid, is not entirely fair to Descartes. Descartes is at pains to assert that the parts of water are very slippery "like little eels";[153] thus, they can move swiftly in several directions relative to one another in a manner that satisfies Descartes's

---

[149] *Ibid.,* p. 165; AT.VI.661–662, Cap. III, Article IV. [150] *Ibid.*
[151] *Ibid.,* p. 166; AT.VI.661, Article IV.
[152] *Ibid.,* p. 155; AT.VI.653, Cap. I, Article VI.
[153] *Ibid.,* p. 154; AT.VI.653, Article IV.

conception of a fluid medium.[154] It is true that water is not a perfect fluid, because some of its parts can move less quickly than those of solid bodies moving through it. But this simply means that water offers sensible resistance to the motion of bodies. Newton seems to take Descartes's claim that the parts of water are interlaced to mean that they are coiled around one another like a spring and hence are soft and yielding to the touch. Similarly, Newton's observation that Descartes's particles of water would not "admit light through them" construes them as being everywhere impacted together; moreover, to speak of them as "exceedingly heavy" views the composition of water in an atomist rather than a Cartesian manner. For Descartes, 'weight' is a derivative notion that results from the effect of the action of a vortex on the parts of objects that are contained in it.[155] It must be admitted, though, that Descartes does use atomist language; for example, he states that "it is not astonishing that salt water is heavier than fresh," because its particles "can contract [*contrahi possunt*] into less space; for it is this upon which its weight [*gravitas*] depends."[156] As is evident in his essay "Of Gravity and Levity," Newton does not analyze Descartes's theory of gravity, in which the official account of weight occurs; the omission is striking, given the importance that the topic has for him in the *Questiones*.

Next, Newton claims in point (4) (on 26 100ᵛ) that the particles of fresh water "would not refract light so well" as other bodies, because they do not, for example, "resist the pure matter as glass does." Clearly, Newton has in mind Descartes's statement that transparent bodies are the more transparent the less they impede the subtle matter in their pores; consequently, "the water of the sea must be more transparent, and cause slightly greater refractions, than that of the rivers."[157] On the other hand, Newton characterizes Descartes as holding that the "pure matter" passes "more swiftly where it finds strongest resistance, and refraction to be from hence, that the matter passes swiftest which there-

---

[154] *Ibid.*, p. 44; AT.VIII.70–71, *Principia*, Part II, Article LIV.
[155] *Ibid.*, pp. 144–5; AT.VIII.213–214, Part IV.
[156] *Ibid.*, p. 164; AT.VI.661, *Meteorology*, Cap. III, Article III.
[157] *Ibid.*, p. 165; AT.VI.662, Article IV.

fore should be in water." He certainly has in mind Descartes's claim in the *Dioptrics* that light travels faster in a denser medium.[158] This means that his reference is to Descartes's conception of salt water. According to Descartes, it is composed of thick and massive particles that "arrange themselves in less space" and can thus bend light more. Although Newton does not think that fresh water is composed of "long bending parts," he has not denied Descartes's conception of salt as consisting of "stiff and long ones."[159]

Indeed, in point (5), on folio 26 100^v, Newton thinks that Descartes is committed to a falsehood because of his view of the nature of salt particles and their relation to those of fresh water. According to Newton, Descartes's conception of salt suspended in water has the consequence that if "it [the water] has separated as much as it can of one kind of salt it could separate no more of another, which is false." Descartes's theory holds that the perpetually agitated particles of fresh water have a propensity to encapsulate those of the salt, so that they remain suspended in the watery medium. This prevents the heavier salt particles from separating out of the water and explains "why salt dissolves easily in fresh water . . . even though only a determined amount of salt will be dissolved in a determined quantity of water; namely only insofar as the flexible parts of that water [*partes aquae flexiles*] can surround those of salt by rolling around them."[160] Descartes's argument refers to his discussion of the shape of common salt particles. As the quotation indicates, he seems to hold that the 'mechanism' whereby fresh water dissolves common salt is limited in scope and that its power to do so can be exhausted. Moreover, he views the particles of common salt as being rodlike and of the same thickness at both ends, a shape conducive to suspension in water. And given that they are loosely laid across one another, the watery particles can easily encapsulate them.[161] Newton's comment is not directed at Descartes's claim that only a certain

---

[158] *Ibid.*, p. 61; AT.VI.593–594, *Dioptrics*, Cap. II, Article VII.
[159] *Ibid.*, p. 165; AT.VI.661, *Meteorology*, Cap. III, Article III.
[160] *Ibid.*, p. 165; AT.VI.661, Article IV.
[161] *Ibid.*, p. 165; AT.VI.661, Article III.

amount of a kind of salt is soluble in water; rather, he seems to imply that if Descartes's mechanism for explaining absorption is correct, having dissolved one kind of salt, water would be incapable of dissolving another. For if a given amount of water has encapsulated all of the salt that it can, it is difficult to see how the same particles of the water can at the same time suspend (presumably) differently shaped particles of another salt.

Newton is greatly interested in Descartes's conception of the particulate structure of air. In point (6) on folio 26 100ᵛ he states that the air "(because of branchy parts) would instantly quell their circular motion when they are rarefied." Newton is referring to the Second Discourse, in which Descartes describes the movements of vapor particles (which are made of the same small particles as water) away from watery bodies.[162] The thrust of his query is clear enough. If Descartes were right about the structure of the air, the rarefaction of water into vapors would not take place. It is no doubt this alleged consequence that prompted Newton to ask on 25 100ʳ if the air "consists of branchy bodies not folded together but lying upon one another. Descartes." In this, as in the previous criticism, Newton is expressing doubt concerning the ontological structure that Descartes ascribes to phenomena like water and air. On the one hand, Descartes claims that air "and most other bodies" are irregular in shape, causing them to be bound to one another relatively easily, "as are the various branches of bushes that grow together in a hedgerow."[163] On the other, he says that air's particles are small enough to be agitated by the subtle matter surrounding them, so that they can be spread out and become rarefied.[164] Descartes's descriptions of the disposition of the air are not incompatible if it is remembered that the subtle matter can act on the small particles of which it is composed. After all, Descartes says that these particles need "be only slightly intertwined" to become bound.[165] Moreover, Newton has forgotten the differential role that the medium plays in transforming the air from being "branch-like" to becoming rarefied. He has consequently

[162] *Ibid.*, pp. 158–9; AT.VI.656–657, Cap. II, Article II.
[163] *Ibid.*, p. 154; AT.VI.652, Cap. I, Article III.   [164] *Ibid.*   [165] *Ibid.*

construed Descartes as saying that the air's particles can form "a stubborn body" (26 100^{v}) that resists the circulating action of the vapors as they come off water.[166]

Newton asks "whether burning waters and hot spirits be of small spherical or oval figured parts, and have many such globuli as fire is of" (26 100^{v}). He is referring to Discourse Two, where Descartes discusses the distinguishing features of various kinds of vapors and exhalations.[167] There he depicts vapor particles as being oval in shape. They begin as water particles and have a similar interlaced structure in relation to one another. However, "when they have the form of a vapor, their agitation is so great that they rotate very rapidly in all directions [ut celerrime rotentur in omnes partes]"; this motion stretches them into an oval shape and causes them to drive one another from their respective spheres of influence.[168] In contrast, the particles of exhalations are generally of more irregular shapes; but some, such as spirits or brandies, are nearly the same shape as water, but of a finer branchy structure; "they can easily be set afire."[169]

According to Descartes, exhalations can manifest more diverse qualities than vapors, owing to the greater variety in their shapes. If their largest particles are scarcely indistinguishable from those of earth, the finer are nothing other than those of spirits or brandies, "which are always the first to rise up from the bodies which we distill." Medium-size particles form volatile salts, oils, and the fumes of these.[170] Despite Descartes's restriction of oval figures to the particles of vapors, Newton asks if "burning waters and hot spirits" are not of the same figure. He also wishes to associate them with the "globuli" of fire. This notion seems to have no clear reference in Descartes's text, except perhaps in the latter's claim that spirits are easily set afire because of the fineness of their particles. In Discourse Seven, he talks of many different sorts of fires and fiery exhalations, but does not associate fire with any particular sorts of globules as Newton

[166] Ibid., p. 162; AT.VI.659, Cap. II, Article VI.
[167] Ibid., pp. 158–9; AT.VI.656–657, Articles II and III.
[168] Ibid., p. 160; AT.VI.657, Article III.
[169] Ibid., p. 158; AT.VI.656, Article II.
[170] Ibid., p. 162; AT.VI.658, Article VIII.

does. Nevertheless, the thrust of Newton's query is clear
enough. If spirits can be easily distilled and by their agitation they "enliven men," they not only are of the oval shape
that Descartes restricts to vapors but also are composed of
"many small and solid atoms" that allow these to retain their
heat of penetration. Here Newton explicitly argues for the
atomic structure of a phenomenon, while denying that Descartes can account for it on his own terms.

Newton's next query refers to Descartes's explanation of
why water contracts and then dilates "itself before and as it
freezes." The latter illustrates his theory by reference to a
beaker that is filled with hot water and then is exposed to
freezing air. According to Descartes, the water level will first
decline and then rise until the water is frozen. His explanation is based on the differences in strength of the subtle
matter as it acts on the differently disposed particles of
water, surrounded by the colder medium of the air.[171]
From this, Descartes also concludes that "water which remains hot for a long time freezes faster than any other sort,
because those of its parts which can least cease to bend
evaporate while it is being heated."[172] This, too, is queried
by Newton, as is Descartes's claim in Discourse Three that
salt and snow help to freeze water.[173] According to Descartes, if salt and crushed ice are placed around a vessel
filled with fresh water, as they melt together, the water will
form into ice. The explanation is one of Descartes's more
elaborate. It appeals to his theory concerning how the subtle
matter acts on the particles of bodies. As the snow melts, its
particles encapsulate those of the salt. At the same time, the
subtle matter that surrounds the water particles is highly
agitated and, having more force than the matter that surrounds the snow, moves to replace it, as the snow's particles
encapsulate the salt, "for the subtle matter finds it easier to
move in the pores of the salted water than in those of the
fresh water, and it tends to pass perpetually from one body
into another, in order to penetrate those in which its movement are less resisted."[174] As a consequence, the more
highly agitated part of the subtle matter that surrounds the

[171] *Ibid.*, p. 157; AT.VI.655, Cap. I, Article VIII.   [172] *Ibid.*
[173] *Ibid.*, p. 166; AT.VI.662, Cap. III, Article V.   [174] *Ibid.*

water moves away from it, to be replaced by the less agitated, so that the water in the vessel freezes.

Newton's next questions ("Why water is clearer than vapors?" and "Whether there be more vapors when air is clearest?") are again directed toward the particulate infrastructure that Descartes ascribes to these phenomena in the Second Discourse. In the case of vapors, Descartes claims that they "can be more or less compressed or expanded, hot or cold, transparent or obscure, and moist or dry, at one time than at another,"[175] depending on their disposition relative to one another, their relationship to mountains and cloud systems, and the degree to which they are agitated by the subtle matter.[176] Newton's second query is prompted by Descartes's claim "that the atmosphere often contains as many or more vapors when they are unseen as when they are seen."[177] As an example, Descartes states that in warmer weather the Sun draws more vapors from a body of water than in cold and cloudy weather.[178] Now, Descartes holds that vapors, although they are made of the same small particles, always occupy "much more space than water."[179] Moreover, when they are in a highly agitated state in the air, they repel each other with force; as a result, they "are transparent, and cannot be distinguished from the rest of the air by sight; for since they move at the same speed and impetus as the subtle matter which surrounds them, they cannot prevent it from receiving the action of luminous bodies, but rather, they receive it with the subtle matter."[180] The sense of Newton's second query appears to be this: Under the conditions in which vapors are as transparent as the surrounding air, they are spread out to their greatest extent. Therefore, according to Descartes's conception, there are fewer vapor particles per unit of the air than when the particles are compressed together under opposing conditions, in fact under conditions in which they should be seen. What does it mean to say, then, that there are "more vapors when the air is clearest."? As to the first query, Newton

[175] *Ibid.*, p. 160; AT.VI.657, Cap. II, Article IV.
[176] *Ibid.*, pp. 160–2; AT.VI.657–659, Articles IV, V, and VI.
[177] *Ibid.*, p. 162; AT.VI.658–659, Article VI.   [178] *Ibid.*
[179] *Ibid.*, p. 159; AT.VI.657, Article III.
[180] *Ibid.*, p. 162; AT.VI.658, Article VI.

appears to ask why there should be any difference in the transparency of water and vapors if each phenomenon has the same structure of small particles. Of course, a comparison of the two states of water in this manner leaves aside Descartes's claim that the action of light is more readily communicated through the pores of bodies if the subtle matter is more highly agitated in their interstices. Moreover, on folio 18 96^v Newton asks "Why is breath and sweat seen in winter more than in summer?" This refers to Discourse Two, where Descartes explains that vapors, "like the sweat of overheated horses," become dense and visible in winter because their particles no longer "obey the fine matter so much that they can be moved by it in every way" to allow an indistinguishable state of transparency with respect to the rest of the air.[181]

Newton next embarks on a lengthy series of queries concerning Descartes's account of the nature and formation of salt and its effect on various sorts of phenomena (folios 26 100^v and 47 111^r). Descartes is adamant that salt particles are firmly fixed, so that the particles of salt water cannot be turned into vapor as those of fresh water can. A major part of Discourse Three is devoted to putting forward an account of the nature, shape, and disposition of salt particles, and of how it is that when salt forms "it floats on water, even though its particles are very fixed and heavy." In particular, he attempts to explain how salt "forms itself into small, square grains [figura quadrata], not greatly different from that of a diamond, except that the largest facet is slightly concave."[182] For this to happen, it is essential that the process take place (a) in a region of the sea that is relatively quiescent, (b) in the presence of warm, dry weather, so that the Sun's action can sufficiently evaporate the fresh water, and (c) under the condition that the water's surface is level and uniform. The last condition results from the fact that the particles of air that touch the water move among themselves in a constant manner, as do the parts of the water's surface. But the particles of air "do not move in the same manner, nor at the same speed" as the water's.[183] Moreover, the subtle matter around the

[181] Ibid.   [182] Ibid., p. 168; AT.VI.664, Cap. III, Article X.
[183] Ibid., p. 168; AT.VI.664, Article XI.

particles of the two separate phenomena moves differently. This is why their surfaces instantaneously rub and polish against one another, as if they were hard bodies, and as a result arrange their respective particles to produce this effect:

And this is also the reason that the surface of the water is more difficult to divide than is the interior, as we can see by experience from the fact that all bodies that are sufficiently small, though they are of heavy and weighty material as are small steel needles, can float and be easily supported on the surface when it is not yet divided; whereas when it is divided, the bodies sink.[184]

It is this explanation that Newton has in mind when he asks on folios 26 100ᵛ and 47 111ʳ "why the superficies of water is less divisible than it is within, insomuch that what will swim in its surface will sink in it." Here it is clear that Newton probably has no alternative explanation in mind; he is simply recording an interesting fact for which Descartes gives an explanation within the framework of the *Meteorology*'s theory of matter.

In a similar frame of mind, Newton asks, on 26 100ᵛ, "why, though salt be heavier, yet it will mix with water and gather into grains at the top of it?" The reference is to Descartes's discussion in Discourse Three. When the air is warm enough to form salt, it raises the particles of water away from the sea in vaporous form. This can be done so quickly that the particles of water come to the surface carrying the salt particles with them. There they remain, resting lengthwise on the surface, and not "heavy enough to sink anymore than the steel needles of which I have just spoken."[185] Again, Descartes's account of the phenomenon is broadly consistent with the basic mechanisms of his ontology. But it may well be that Newton is somewhat skeptical of the mixture of analogy and theory that is invoked, a mixture that rather smacks of contrivance.

When on folio 47 111ʳ Newton asks "why is salt of a square figure having a hollow and broad top and a narrow base?" he is referring to one of the most detailed accounts of a phenomenon that Descartes provides in the *Meteorology*. It comes

---

[184] *Ibid.*   [185] *Ibid.*, p. 169; AT.VI.664.

directly after the account he gives of how salt particles float
on the water's surface. Appealing to his analogy between salt
particles and steel needles, Descartes claims that the salt particles will cause hollows on the water's surface by their
weight, as do the steel needles. These hollows cause successive particles of salt to "roll and slide towards the bottom" as
they come to the surface under the influence of the fleeing
vapor particles.[186] Given the curvature of the hollows formed
on the water's surface by the initial particles of salt, and the
various ways in which subsequent particles must arrange
themselves at the surface, Descartes describes how the salt
can form a first base for producing a grain of salt. "And
when it has reached a certain size, it has sunk so much that
the newly-arriving particles of salt coming toward it" rise
above and begin to form the next square tablet.[187] The same
process is repeated again and again, each successive tablet
becoming larger than the one previous because of the
greater agitation and lack of uniformity of the water. As a
result, the grain of salt that is formed has a broad top, sloping sides, and a narrow base, whose exact characteristics depend on the degree and action of heat, the agitation of the
water, and the amount of spreading out that the salt particles
obtain.[188] If fresh water gets trapped in the spaces between
the spreading salt particles, they can expand under the action of heat to the extent that "they break their prisons at
once, with a sudden noise."[189] On the basis of this, Descartes
explains why whole grains of salt crackle and pop when they
are thrown on a fire.[190] Hence Newton's query on folio
47 111[r]; "Why grains of salt will crack in the fire, but not if
they be first rubbed apart." Obviously the latter cannot happen if the salt is reduced to powdered form, for then the
entrapped water is already released. Newton's following
query is well directed: "whether the pleasant smell of white
salt and the color of black salt proceed from some other
mixture. Descartes." Descartes makes a vague appeal to certain particles in the sea that can entangle with the forming
salt to give it the odor of violets, or the color of black salt, "or

[186] *Ibid.*　　[187] *Ibid.*, p. 170; AT.VI.664.
[188] *Ibid.*, p. 171; AT.VI.666, Article XIV.
[189] *Ibid.*, p. 171; AT.VI.667, Article XVI.　　[190] *Ibid.*

any of the variations which we can notice in salts. . . ."[191] So, too, is Newton's next query well directed: "Whether salt is melted by sudden heat because there is water in it, and not by gentle fire because that exhales the water out by degrees. Descartes." According to Descartes, salt melts easily on a fire if it has water particles entrapped, but not when it is dry and pulverized. In the pulverized state, salt particles are rigid, inflexible, and not easily bent. However, if they are surrounded by particles of water, they can be more easily affected by the agitation of the subtle matter under the action of heat.[192] This explanation prompts Newton to ask "how oil or spirits of salt (so sharp that they will dissolve gold) is extracted out of salt. Descartes, *Meteorology*. Of Salt." According to Descartes, the volatile spirit of salt is extracted from a salt base "by the violence of a very hot fire."[193] This means two things: first, that its particles are the same as those that compose the salt; second, that the rigid salt particles become flexible, and easy to bend, under the action of the fire.[194] We can well imagine Newton's wondering if Descartes's ontology is sufficient to explain the emergence of an activating principle that is strong enough to corrode gold from a base of pure salt.

Newton's remaining queries under "Of Salt" are straightforward enough. He refers to Descartes's explanation of how salt preserves foodstuffs like meat. Owing to their rigidity, parts can firmly penetrate meat and remove its internal humidity. Moreover, they can strengthen its constituent parts so that they can resist the agitation of the more pliable particles as they attempt to decompose the meat by rearranging its structure.[195] Second, he records Descartes's claim that the sea is saltier at the equator than at the poles because the Sun's action causes more evaporation there, whereupon the vapors are circulated to the poles. Newton's reference is to Discourse Three,[196] but Descartes elaborates on the phenomenon in Discourse Four, entitled "Of Winds."[197] Lastly, on folio

[191] *Ibid.*, p. 171; AT.VI.667, Article XVII.
[192] *Ibid.*, p. 172; AT.VI.667, Article XVIII.
[193] *Ibid.*, p. 172; AT.VI.668, Article XIX.   [194] *Ibid.*
[195] *Ibid.*, p. 164; AT.VI.660, Article II.
[196] *Ibid.*, p. 167; AT.VI.663, Article VIII.
[197] *Ibid.*, pp. 173–4; AT.VI.669, Cap. IV, Article II.

26 100^v, Newton asks "Why sea water is not so apt to quench fire, and why it will sparkle in the night but not if kept in a vessel?" This is again a reference to Discourse Three, in which Descartes points out that salt particles not only can augment a flame but also can cause flames by themselves. The latter can occur when salt particles are propelled from the sea into the air by the action of the waves. Being highly agitated, and isolated from the water, "they engender sparks similar to those which flints give off when we strike them together." On the other hand, sea water that has been contained for a period of time will not sparkle, because its particles are entrapped by the particles of water.[198]

On folio 45 110^r Newton asks whether oily bodies "consist of branchlike particles only, touching superficially and folded together. Descartes." The reference is to Discourse One, where Descartes states that "very rarefied and light liquid bodies such as oil and air" are composed of small branchy particles that are "simply laid on one another without being interlaced at all (or only very slightly)."[199] It is clear from his use of the term 'only' that Newton is skeptical of Descartes's account, as indeed he is of the similar structure that Descartes ascribes to air (25 100^r). In terms of his likening the relationship of the water, earth, oil, and other particles to branchlike structures, Descartes can speak of their degree of entanglement and disentanglement. Although he does not deny Descartes's explanatory mechanism, Newton contrasts the behavior of oil and water on surfaces, a difference not discussed by Descartes. According to Newton, the fact that oil spreads quickly, when dropped on bodies, might be explained by its branchy structure clinging to their superficial parts. Newton's proposed explanation appeals to an atomist picture of the structure of things, whereas Descartes's picture of the phenomenon appeals to the comparative ease of movement and agitation in their relatively disposed parts. Also on 16 95^v Newton asks why flints break on soft things easier than on hard ones, and why diamond dust is harder than the diamond itself. He is again querying Descartes's claim that things (in this case

[198] *Ibid.*, p. 167; AT.VI.673, Article IX.
[199] *Ibid.*, p. 155; AT.VI.653, Cap. I, Article IV.

hard things) consist of branchlike particles;[200] thus, Newton asks if "hard bodies stick together by branchy particles folded together." He also notes on 45 110ʳ that water does not spread so quickly over a surface, despite the fact that it is not as viscous as oil. Newton's explanation appeals to the pressure of the air to account for the crowding together of the particles of water, a process that can be disturbed if the water particles are absorbed into pores on the surfaces of bodies.

Newton again considers the air as a principle of cohesion in his essay entitled "Conjunction of Bodies" (6 90ᵛ). He begins by asking "whether the conjunction of bodies be from rest?" The reference is to the *Principia*, Part II, Article LV, in which Descartes says that nothing "joins the parts of solid bodies excepting they are at rest in relation to each other."[201] Newton retorts immediately, "No, for then sand by rest might be united sooner than by a furnace, etc." Here we shall consider only those aspects of the essay that are directly Cartesian; Newton's treatment of cohesion will be dealt with more thoroughly as part of our analysis of his approach to the phenomenon of attraction (see Chapter 6). At the beginning of his essay, he conjectures whether or not the parts of bodies cohere by the action of the ambient air, recognizing that its pressure "be but little in respect of that performed by the purer matter of the vortex, between the Sun and us, receding from the center. . . ." He clearly refers to the vortical theory of the *Principia*, as well as to Descartes's treatment of the air and winds in the *Meteorology*.[202] But by citing the evidence of Boyle's experiments on the strength of the air, he immediately rejects the hypothesis, claiming that although the air may explain the cohesion of "the parts of water," its pressure is not strong enough to give a general account of the conjunction of bodies. Also mindful of Descartes's account of the action of the ambient air, Newton asks on folio 46 110ᵛ if strong winds can drive the moisture from bodies. The references is to Discourse Two of the *Meteorology*. There Descartes attempts to explain

---

[200] *Ibid.*, Cap. I.   [201] *Ibid.*, p. 44; AT.VIII.71.
[202] *Ibid.*, p. 58; AT.VIII.92, Part III, Article XXX; pp. 173–81; AT.VI.671ff., *Meteorology*.

by the action of strong winds how it is that they carry away the moisture of bodies, and yet are more impetuous than humid winds because of their dryness.[203]

Although there are no dog-earings in the *Meteorology*, it is clear that Newton has read the first three discourses with care. It cannot be said with certainty that he was unaware of the rest of Descartes's treatise in 1664. But certainly the *Meteorology* contains a great deal that figures in Newton's later thought, especially the theory of the rainbow. It is interesting to note that Descartes's explanations of everyday phenomena have an attention to detail that is reminiscent of Newton's later observations (e.g., his explanation of how it is that a fly walks on water).[204] There can be little doubt that Descartes's attention to minutiae impressed itself on the mind of the young Newton. It is certainly an evident characteristic of his later scientific sensibility.

[203] *Ibid.*, p. 162; AT.VI.659, Cap. II, Article VII.
[204] Hall and Hall, "Partial Draft of the Preface to the *Principia*," p. 303.

3

NEWTON ON PROJECTILE
MOTION AND THE VOID

In an important essay entitled "Of Violent Motion," Newton examines three opinions concerning an object's continuance in motion (folios 21 98^r, 22 98^v, 51 113^r through 53 114^r). The strategy of the argument is to eliminate two of the opinions in order to establish the third, that is, the view that a body moves after its separation from the mover by its own "natural gravity." The first opinion that he examines is the ancient doctrine of antiperistasis, which is discussed in the writings of Plato and Aristotle.[1] Although variously formulated, the doctrine conceives a projectile's forward motion as being maintained by the propelling power of an ambient medium, such as air or water. Newton rejects the claims of traditional doctrine, and he also appears to reject a modified version advanced by Descartes.[2] The Aristotelian, Kenelm Digby, also espoused a version of the doctrine that Newton almost certainly knew of. In the first of his *Two Treatises*, Digby argues for a conception of antiperistasis in order to explain the phenomenon of violent motion.[3] On the positive side, it seems likely that Newton was influenced by Charleton's critical discussion of the doctrine; Charleton, in turn, was influenced by Gassendi's *De motu impresso a motore translato*, which Charleton indicates he is summarizing.[4] Also, Galileo's *Dialogue* was a source known to Newton. The "Second Day" of the *Dialogue* advances criticisms of Aristotle's theory of the medium; it also critically rejects

[1] *Timaeus*, 79E–80A; *Phys.*, VIII. 10. 267a21–27.
[2] Descartes, *Opera*, pp. 44–8; AT.VIII.73–77, *Principia*, Part II.
[3] Digby, *Two Treatises* (London, 1658), Chap. XII. See Harrison 516.
[4] Charleton, *Physiologia*, Book IV, Chap. II, Sect. III, p. 463.

195

the view that violent motion results from an *impetus* that is conferred on a projectile by an external agent.[5]

1. Newton's main argument against antiperistasis

Newton begins his analysis of fluid media by noting the behavior of objects carried by their currents, as well as the effects that they have on floating objects. He observes that the agitation behind an object moving through water is considerably less than the effect of the water before it, a consequence verified by floating motes. Accordingly, the pressure of the water before the projectile is greater than that behind it and should hinder rather than aid its progress. This is further confirmed by the fact that a moving thing carries along the water behind it "as in a cone." Similarly, the shape of a piece of hot lead dropped into water is rounded at front and elongated behind by the action of the water. This appearance is consistent with the object's front being 'pressed' more violently than its rear by the water. To be credible, antiperistasis theories must be able to account for these appearances coherently. Newton does not think that they can. That is, he does not think that antiperistasis can provide an explanation that fails to violate the brute fact that a medium pushes more forcibly on the front of an object (as it flows around it) than it does from behind.

Newton's first argument is straightforward. He points out that if the water (which runs quickly behind an object moving through it) were in fact to concentrate at once at its back, it would affect the still water that either follows along after the object or moves slowly away from it. But this contradicts appearance. Without arguing for parallel consequences, Newton states that the same reasoning applies to the behavior of the air around an object that is passing through it. If this conclusion is denied, the opponent must hold that an object moves forward in a medium on account of the medium's propelling power alone. But again this con-

[5] Galileo, *The Systeme of the World, in Four Diologues* in Thomas Salusbury, *Mathematical Collections and Translations* (London, 1661, 1665), 2 vols., Vol. I, Day Two, pp. 130–1, 170–3.

tradicts appearance. Water cannot act on the rear of an object in a way that is consistent with its having the power to move it forward; but neither can the air. The two statements that follow are not so much arguments as observations that Newton construes to be inconsistent with the doctrine of antiperistasis: Hot lead dropped into water has a shape inconsistent with the action of the water required by the doctrine of antiperistasis; can it be supposed that a turning globe is maintained by the action of the ambient air alone (folio 21 98^r)? Here Newton seems to refer to the contradictory properties a theory of antiperistasis would ascribe to the air in order to account for the globe's behavior. But he also assumes the correctness of the explanation he wishes to establish, namely, that the globe maintains its turning by an internal power alone. Newton's globe example is represented in the literature on the classic problems of motion. In his criticisms of the Platonic and Aristotelian theories of projectile motion, Jean Buridan argues that such theories cannot account for the motion of a grindstone or disc.[6] Their rotation continues even if the air is prevented from reaching them. Such examples were no doubt commonplace at Cambridge, and one can only conjecture about Newton's sources.

Newton's next argument involves a *reductio ad absurdum*. He supposes that a ball—and the air that surrounds it—alike manifest a disposition to move only if they are propelled forward. Given that the air immediately behind the ball can be moved only if the ball moves the air immediately in front of it, the following chain of circumstances ensues. In order that the air immediately in front of the ball may move, it must move the adjacent air on either side, and so forth, through the successive layers of air along the sides. But this means that in one and the same instant the air behind moves the ball as the ball moves the air behind, which is absurd (folio 22 98^v).

Newton next supposes that the ball and the air are constrained to rest by a force outside their framework, while they maintain the same relations as they had in their former

---

[6] A. C. Crombie, *Medieval and Early Modern Science* (New York: Doubleday Anchor, 1959), 2 vols., Vol. II, p. 66.

state of moving. Imagine that they are suddenly released
from the external constraint. Will the air not have the same
disposition to move the ball as it had when it and the ball
were in motion? Newton appears to imply that this is absurd
and contrary to appearance. It is clear that his argument
assumes that if a medium really has the power to move the
ball, it should manifest this capability at the moment that
the ball begins to move, as well as when it is in motion. But a
medium like the air does not possess a motive power according to its nature. Consequently, Newton appears to imply that an original mover must impart to the air not only
the motion but also the power to be a mover itself. Otherwise its effect on the ball will cease. But he also argues that
it is no reply to claim that the air is more compressed behind the ball than it is at the front, such that its disposition
to act on the rear of the ball is greater than its capacity to
act on its front. This is again contrary to appearance and is
open to the *reductio ad absurdum* argument. Newton com-
pletes his reasons for rejecting antiperistasis by pointing out
that a heavy object moves a greater distance with more force
than a lighter object of the same size; yet what grounds are
there for supposing that the air acts on them differently?
Again, this is a standard argument in the literature against
antiperistasis.[7]

   Newton's arguments are of traditional form, but they
have an individual character that appeals strongly to observable circumstances. He may have consulted Gassendi's *De
motu impresso*, in which accounts of projectile motion based
on the medium are discussed and criticized.[8] Although
there are similarities in the reasoning of Gassendi and Newton, this gives no warrant for supposing that Newton knew
of Gassendi's views at first hand. It is far more likely that
Newton's discussion is partly motivated by Charleton's approach to the problem. Under the influence of Gassendi
and Galileo, Charleton dismisses the doctrine of antiperistasis as formulated by the Aristotelians, as well as the second

   [7] *Ibid.*, p. 68.
   [8] (Paris, 1642), *Epistle I*, pp. 34–46. John Herivel has reproduced extracts
from "Of Violent Motion," but has not provided an extended analysis. See *The
Background to Newton's Principia*, Part II, pp. 121–217.

view that Newton examines, i.e., the view, as Charleton puts it, that "the Body Projected is carryed forward by a *Force* (as They call it) *Imprest;* which they account to be a Quality so communicated unto the body projected. . . ."[9] Although defenders of a *virtus impresso* often speak of it as a quality, notice that Charleton's language for describing this conception is similar to Newton's; i.e., he speaks of a *"Force Imprest."* His positive solution to the problem has similarities to the one defended by Newton, but we shall see that there are differences as well.

What cannot be overlooked is the fact that Newton's criticisms of antiperistasis refer in part to Descartes's discussion of the motion of bodies in fluids, an account that includes the motion of the planets and comets. We have already seen the extent to which he is critical of Descartes's account of the origin of vortical motions in the fluid heavens. In his present criticism of antiperistasis, it appears that Newton may well have in mind Descartes's discussion of fluid motion in Article LVII–LXI of Part II of the *Principia*.[10] Descartes argues in Article LVI that a body at rest in a fluid environment is in equilibrium, being pushed equally on all sides by the parts of the fluid. Consequently, unless it is pushed by some external agent, however small, it cannot move.[11] He attempts to prove this claim in the following article.

He asks us to consider a fluid in a container, into which a solid body is to be introduced. Given that the parts of a fluid are able to move in diverse ways at the same time, Descartes states that some will move in a closed chain in one direction, while others will move in the opposite direction around a common center. When a solid body is introduced into the fluid, Descartes argues that the two separate circulations of particles will become one, and together move in the same closed circle in the same direction and with the same determination. Now, if we suppose that the solid body is pushed on one side by an external force that is added to certain parts of the fluid, the body will be moved in a determinate direction, because it overcomes the influence from

---

[9] *Physiologia*, p. 463.
[10] Descartes, *Opera*, pp. 46–8; AT.VIII.73–77, *Principia*.   [11] *Ibid.*

the motion of the circulating parts on its opposite side. More-over, it so changes the direction of flow of these particles on the opposite side that they reverse themselves and circulate to the side of the body that is pushed by the action of the external agent. The parts of the fluid whose direction is reversed cannot oppose the power of those augmented; this state of affairs and the fact that the reversed flow of the particles allows the body to move more easily in the opposite direction assist rather than prevent the direction of the body's motion throught the fluid. Notice that Descartes's partial antiperistasis acts by a direct augmentation of the force behind an object, rather than by the action of a com-pressed medium circulating from the front to the rear of the object in order to push it forward.[12]

Can it be doubted that Newton has Descartes's fanciful discussion in mind? The first thing to notice is that the inertial tendency of all matter to preserve its state – a notion basic to Descartes's conception of nature – nowhere enters into his explanation of the body's motion in the contained fluid. This is so even though Descartes recognizes in Article LXI that the attribution of a small force to parts of the fluid medium violates his fourth rule, according to which a small force cannot overcome the inertial tendency of a large body.[13] Moreover, in Article LIX he argues that perfect fluidity, of the sort exemplified by the celestial matter, not only allows a body free passage but also propels it for-ward.[14] Here Descartes implies that a body's passage through a fluid is a combination of its inherent inertial ten-dency and the partial effect of antiperistasis in the sur-rounding medium. For his part, Newton opposes accounts of projectile motion that are mixed in this manner and that attribute diverse and often contradictory properties to me-dia of various sorts. Moreover, it is also clear that Newton opposes any account of fluid motion that attributes differ-ential power to one and the same medium, either in accor-dance with its own nature or as a result of the impression of an external agent. There can be little doubt that Descartes offers a suitable example of such an account, even though

[12] *Ibid.*, p. 45; AT.VIII.74.   [13] *Ibid.*, p. 48; AT.VIII.77.
[14] *Ibid.*, p. 47; AT.VIII.75–76.

he suggests that in particular cases the external cause might be the current of a river flowing in a certain direction (Article LXI).[15] Lastly, although inertia does not enter into Descartes's argument in Article LVII, nothing he says rules it out.[16] This must be the case if Descartes's abstract theory of motion is to apply to multiparticle media, an application confirmed in Article LIX (as elsewhere), in which Descartes allows that the body's inertial tendency to move through a fluid is augmented by the power it 'borrows' from the fluid itself.[17] However, the account of projectile motion that Newton wishes to establish appeals solely to the body's "natural gravity," a concept akin to Descartes's conception of an inertial tendency said to inhere in every bit of matter.[18] We shall argue later that Newton was far from taking the measure of Descartes's conception at the time of writing this essay. It is nevertheless difficult to believe that he failed to notice that certain of Descartes's arguments combine a perspective on violent motion that he wished to adopt with one that he wished to reject.

In any event, this part of Newton's essay marks the beginnings of his interest in motion through fluid media, even though his expressed purpose is a rejection of accounts of violent motion based on the medium. More specifically, it indicates an interest in Descartes's conception of fluids that, in combination with Newton's tentative exploration of the former's theory of vortical motion in the fluid heavens, marks the origins of a lifelong preoccupation with motion in fluids, especially with those theories formulated in Cartesian terms.[19]

Although Newton's discussion of antiperistasis has roots in certain of the texts that we know he read, it must also be recognized that his criticisms of the doctrine are traditional in character. Besides the observational queries that

---

[15] *Ibid.*, p. 48; AT.VIII.76.   [16] *Ibid.*, p. 46; AT.VIII.71–73.

[17] *Ibid.*, p. 47; AT.VIII.75.

[18] *Ibid.*, p. 38; AT.VIII.62, Part II, Article XXXVII, i.e., the "*Prima lex naturae.*"

[19] *De gravitatione*, edited by Hall and Hall, pp. 114–21; *Principia*, edited by Koyré and Cohen, Vol. I, Book II, Sect. VIII, pp. 510–35. See Alexandre Koyré's *Newtonian Studies* (London: Chapman and Hall, 1965), Chap. III, pp. 96–111.

he raises against antiperistasis, it is not difficult to imagine that Newton was acquainted with other commonplaces. If an object is really carried along with the air, why is it necessary that a projector touch it initially as the Aristotelians claim? Why does the brisk beating of the air alone not move an object? Such observations are found as early as the sixth century A.D. in Philoponus's commentary on Aristotle's *Physics* (Book 4, Chapter 8).[20] The stock-in-trade of those opposed to plenist accounts of the continued movement of projectiles, these observations, are represented in the traditional commentaries on natural philosophy. Given that Newton's tutor introduced him to an undergraduate curriculum in natural philosophy that was based on the Aristotelian corpus and systematized by lecture, commentary, and disputation, arguments for and against antiperistasis could hardly fail to gain his attention.[21]

Besides these traditional sources, the fact that Newton knew Galileo's *Dialogue* (in the Salusbury translation) cannot be overlooked. In the "Second Day" of the *Dialogue*, Galileo provides an extensive criticism of Aristotle's views concerning the motion of a projectile by the action of the medium. Among his many criticisms, he argues the following: that a medium impedes rather than confers motion on a projectile; that the air is unable to support, let alone maintain, the motion of heavy objects; and that an arrow shot longways would move slower than one shot sideways if the action of the air were the cause.[22]

2. Newton's arguments against the theory of impetus

The next opinion Newton considers is that violent motion continues by an external agent impressing unnatural motion on the projectile (folio 51 113^r). Corporeal and incorporeal agents are the two sorts of movers that Newton considers, and he treats them as excluding other possibilities. Given the

[20] Crombie, *Medieval and Early Modern Science*, pp. 51–2.

[21] For an account of the educational structure in Newton's time, see Costello, *Scholastic Curriculum*. See also Rouse Ball, *Notes on the History of Trinity College, Cambridge* (London, 1899).

[22] Galileo, *Dialogue*, Day Two, pp. 133–4.

nature of the opinion, if neither sort of agent qualifies as the mover, it follows that nothing at all is the cause.

In the first place, if it is said that a flow of atoms moves a projectile, the origin of their motion needs explanation. The suppressed premise is clear: A corporeal atom cannot be self-moved, nor can it originate new motion in something else, because that is inconsistent with its nature. Newton's use of this conception of what it is to be an agent of change may well have its immediate origin in his reading of Henry More's *The Immortality of the Soul*. More is emphatic that any form of gross matter (whether it is a particle or a body) is by its nature a passive principle that can never act, but is only acted on.[23]

However, if it is said that a projectile is continued in motion by an incorporeal nature, there are two candidates for such a cause, "either spirit or some quality." Concerning the first alternative, Newton's argument appears to be this. He supposes that atoms are superveniently endowed with incorporeal agents. On meeting an object, these agents, in association with the streams of atoms, transfer their active capacities to the object itself, thus causing it to move. In Newton's view, this doctrine encounters certain difficulties. What does it mean to say that an incorporeal agent is superveniently associated with a corporeal nature? How can the incorporeal be said to inhere in the corporeal? If they are said to occupy the same region or place, will the incorporeal nature not "slip through" the corporeal presence? After all, the corporeal and the incorporeal are categorically diverse. In short, what coherent ontological relationship obtains between them? Furthermore, the doctrine faces the absurdity that "every little atom must have souls in store to cast away upon every body they meet with." But surely such a state of affairs is not only improbable but also inconceivable.

Notice that Newton speaks of a corporeal and an incorporeal "efflux." This is the sort of language that is used in

---

[23] More, *The Immortality* (1662), Book I, Chap. VII, pp. 31–4; (1659), pp. 42–8. These arguments could also be based on Sextus Empiricus, *Against the Physicists*, Book I, pp. 224–8, of the Loeb Classical Library edition, translated by R. G. Bury (Cambridge, Mass.: Harvard University Press, 1936).

theories of vision in which 'visible species,' 'simulachra,' 'images,' 'species intentionales,' 'atoms,' and the like are said to emanate from gross objects. Charleton has a lengthy discussion of various sorts of emission theories in which he defends the atomic versions, especially those advanced by Epicurus.[24] In his defense of Epicurus, Charleton analyzes a text from Book X of *Diogenes Laertius,* as edited in Gassendi's version. Gassendi makes Epicurus speak of corporeal "Effluxiones" that are emitted from the surfaces of solid objects.[25] Charleton is at pains to stress that 'visible species' or 'effluxions' are substantial, not immaterial, in nature.[26] This, of course, involves him in a discussion of various sorts of theories that speak of the propagation of either corporeal or incorporeal "effluxes."[27]

It is clear from Newton's text that he is not concerned to argue for or against theories of vision that involve the emanation of species. But such theories do advance the sorts of conceptions that Newton criticizes, and they certainly involve the ontological difficulties that he queries. Therefore, Charleton's account of these mechanisms cannot be overlooked as a source for Newton's criticisms of the view that an immaterial source can explain continued motion.

Fortunately, examples of the sort of mechanism that Newton has in mind are present in traditional discussions of the cause of projectile motion. In his *Questiones in secundum librum sententiarum,* Peter Olivi argues that projectiles are not continued in motion by the direct action of either a *virtus impressa* or a medium, but by the 'mediated' action at a distance of the 'multiplication of species.'[28] In his view, 'species' or 'similitudes' are impressed by the projector on the projectile and move it after separation from the initial mover.[29] This is the sort of ontology against which New-

[24] *Physiologia,* Book III, Chap. II, Art. 4, p. 137.
[25] *Ibid.* Charleton reproduces Gassendi's version of the text.
[26] *Ibid.,* pp. 136–7, 141.   [27] *Ibid.,* pp. 138–9.
[28] See *Bibliotheca Franciscana Scholastica* (Florence: Collegium S. Bonaventura, 1922), Tom. IV, Vol. I, *Quaestiones* I–XLVIII, especially *Quaestio* XXIII, pp. 422–33, and *Quaestio* XXIV, pp. 434–8. The work is edited by Bernard Jansen. Olivi's commentary was not available in published form during the Renaissance period. For an account of Olivi's position, see Crombie, *Medieval and Early Modern Science,* p. 59.
[29] *Ibid.* Olivi, *Quaestiones* XXIII and XXIV.

ton's arguments appear to be directed. Moreover, Olivi's account appears to be an adaptation of the theory of the 'multiplication of species' found in the writings of Grosseteste and Roger Bacon.[30] These Medieval theories have their origin in early Neoplatonic conceptions of emanation. But equally important for Olivi's conception of projectile motion is the theory that each of the celestial spheres is moved by a 'soul' or 'intelligence' that impresses its alien motion on the moving spheres. This theory had the advantage that the motion of the spheres could not be corrupted by the resistance of a material medium; it thereby offered a useful model for theorists who wished to consider mechanisms for projectile motion in a vacuum.[31] It is improbable that Newton had direct access to Medieval discussions of this sort. But Bartholomew Keckermann's *Systema physicum* does allude to theories of projectile motion that appeal to 'incorporeal effluxes' and the 'multiplication of species.'[32] This source may well have been available to Newton. Although Newton is critical of using the notion of an "incorporeal efflux" to explain violent motion, he is not necessarily discarding the category of the incorporeal. Rather, he is rejecting the specific claim that a body is continued in motion by an 'efflux' of 'spirits' impressed on it, a view he takes to be incoherent.

Newton now turns to the view that a mover impresses a quality on the projectile, which then maintains its 'unnatural' motion. According to Newton, this conception violates the dictum that it is impossible for numerically one and the same quality to be transferred from subject to subject

[30] See A. C. Crombie, *Robert Grosseteste and the Origins of Experimental Science* (Oxford University Press, 1952), Chap. V, pp. 91–116, and William A. Wallace, *Causality and Scientific Explanation* (Ann Arbor: University of Michigan Press, 1972–4), 2 vols., Vol I, Part I, Chap. 3, pp. 28–49.

[31] Crombie, *Medieval and Early Modern Science*, p. 47.

[32] Bartholomew Keckermann, *Systema physicum, septum libris adornatum* (Hanover, 1612), Lib. I, Cap. VIII, pp. 46–61; Cap. IX, p. 73. Keckermann discusses 'occult' explanations of motion and violent motion at great length, in the course of which the sorts of 'mechanisms' that Newton wishes to reject are outlined. This section includes the "Theoremata de qualitatibus occultis sunt" and cites the 'magical' conception of the forms and essences of things as possessing extraordinary dispositions to affect natural objects. Thus, it may well be that Newton is also consciously rejecting a central tenet of the theory of causality accepted by magicians when he rejects the idea that bodies "cast off" little souls.

(51 113^r). If this is so, the quality and the subject are separated from the object moved, once it is moved. How, then, can they be said to be responsible for its continuance in motion? In other words, if it is impossible for a quality of the mover to impart motion to the projectile in the first place, how can it be said to continue that motion? As Newton puts the point, "how can that give a power of moving which itself has not?" Newton's conclusion that the quality of one individual cannot become the power of moving in another may owe something to Charleton. The latter criticizes the Aristotelian notion that the first principle of motion is the form of the thing moved, on the grounds that a form in itself cannot be an efficient cause of motion. Efficient causes are external principles; forms are not.[33] Similarly, Newton is saying that the quality of one individual is separate from another, and so cannot become the agent of motion in the second.

The notion that a quality cannot "*transmigrat de subjecto in subjectum*" is common enough in the Scholastic compendia of Newton's period. There can be no doubt that he became familiar with the dictum while reading Stahl's *Axiomata*. In his lengthy annotations on Chapter 16, entitled "*Circa Doctrinam Subjecti & Accidentis*," Newton writes: "Accidens non migrat de subjecto in subjectum. [i.e., unum idemque numero accidens non potest relinquere unum subjectum & recipi in alio, modo immediate in illo subjecto inhaereat, & tum subjecto illo immediate transmigrante, accidens cum illo transmigrat. Videsis quasdam objectiones in Stahlius"].[34] To be sure, in the *Questiones,* Newton speaks of a *qualitas,* not an *accidens.* But this is a minor matter. The doctrine to which he appeals is laid out in detail by Stahl[35] and paraphrased by Newton in his notes on the *Axiomata.*

If Newton's knowledge of the doctrine comes from Stahl, his awareness of 'impetus' theories of projectile motion prob-

---

[33] *Physiologia*, Book IV, Chap. II, Art. 3, pp. 436–7.

[34] ULC, Add. 3996, folio 67^r. "An accident does not move from subject into subject [i.e., numerically one and the same accident cannot relinquish one subject and be received into another so that it may inhere in an immediate way in that subject, and when by that subject moving immediately, the accident moves with it. See such objections in Stahl]." The square brackets are Newton's.

[35] Daniel Stahl, *Axiomata philosophica* (Cambridge, 1645), Regula V, pp. 336–8.

ably has many sources. The theory is discussed (though not in detail) in the works of Charleton and Digby.[36] In the *Physiologiae* of Magirus, theories of violent motion are discussed in some detail – though not the Neoplatonic explanations that employ a multiplying of the 'species.'[37] Newton's notes on Magirus are brief. Violent motion is accidental and extrinsic, and "the violent (which is produced by force) is either [in the form of] pulsion, traction, the whirling about [of a medium], [or its] collapse [to fill a vacuum]."[38] Besides these known sources, Keckermann's *Systema physicum* cannot be overlooked, nor indeed the fact that the classical problems of motion were an integral part of the curriculum that Newton followed.

If Newton's knowledge of theories of impetus and impressed forces has many sources, his specific argument against them is suggested in Galileo's "Second Day" of the *Dialogue*. Simplicius says, in support of antiperistasis, that the Peripatetics reject theories of impressed force because they involve "the passage of an accident from one subject to another."[39] Later in the discussion, Salviatus appeals to the same doctrine concerning qualities to criticize Simplicius's support of an Aristotelian antiperistasis. He states that motion cannot be impressed on the medium, "as it is impossible to make an accident pass out of one subject into another."[40] It is clear that Galileo holds that both theories violate the doctrine that qualities cannot migrate from one individual to another. In the case of antiperistasis, the Aristotelians hold that motion must be communicated to the projectile indirectly by a power impressed on the medium, whereas the 'impetus' theorists hold that a power is communicated directly to the projectile.[41] Given these considerations, it is impossible not to recognize Galileo's influence on Newton's discussion. But it cannot be ignored that Galileo repeatedly

[36] Charleton, *Physiologia*, Book IV, Chap. II, Sect. III, pp. 436–75; Digby, *Two Treatises*, Chap. 12, pp. 124–31.

[37] Magirus, *Physiologiae peripateticae*, Lib. I, Cap. V, pp. 32–40.

[38] ULC, Add. 3996, folio 17ᵛ. "Violentus (qui vi effiritur) est vel Pulsio, Tractio, Vertio, vel Vertigo." See Magirus, *Physiologiae peripateticae*, Lib. I, Cap. IV, p. 28.

[39] Galileo, *Dialogue*, Day Two, p. 130.   [40] *Ibid.*, p. 131.

[41] *Ibid.*, pp. 130–1.

uses the term *impeto* in a way that implies that it is a quality imparted to the moved from the mover. This is tantamount to saying than an *impeto* plays a causal role in the continued motion of a projectile.

It must be said that Newton's treatment of the claim that a mobile maintains its motion due to the action of an impressed force or quality lacks a little in sophistication. He discusses none of the subtle positions in the debates on violent motion during the preceding centuries, though they were no doubt taught at mid-seventeenth-century Cambridge. Thus, he makes no distinction between theories that hold that a *vis motrix* expends itself naturally and a view like Buridan's, which makes *impetus* a *res permanens*, a motive power that remains unchanged so long as it is neither diminished nor increased.[42] Thus, in Buridan's view, motion is in principle perpetual, at least with respect to celestial bodies.[43] In his attack on 'impetus' theories, Newton charges them with holding that a 'quality' can migrate from subject to subject. This implies that they reify qualities, making them real and detachable from bodies. But to conceive a quality in this manner is to treat it as if it were an entity distinct from the matter of the projectile, yet able to maintain its motion, or as if an Aristotelian 'form' acted as an efficient cause to move matter. But theorists were often at pains to clarify the nature of the power or quality left behind in the moved by the mover. They pointed out that *vires impressae* are accidental and extrinsic, not innate and permanent.[44] Unlike the particular whiteness of an object, or the heat of a fire, a *vis motrix* is not permanent; nor is it

[42] Crombie, *Medieval and Early Modern Science,* pp. 55–73; see also James A. Weisheipl, *Nature and Gravitation* (River Forest, Ill.: Albertus Magnus Lyceum, 1955), pp. 41–2. An important discussion is to be found in Annelise Maier, *Die Vorläufer Galileis*, Part II, Chap. 6, pp. 135–7, 145–50. Also see Wallace, *Causality and Scientific Explanation*, Part I, Chap. 3, Sect. 4, pp. 104–9.

[43] *Ibid.*, Crombie, p. 72, who quotes directly from Buridan's *Quaestiones in libros metaphysicae*, Question 9, Book 12. See also Maier, p. 147.

[44] *Ibid.* See Crombie, p. 60; also see Weisheipl for the views of Franciscus de Marchia and Buridan, pp. 41–2. Buridan says, "sibi [corpori] violentus et innaturalis, quia suae naturae formali disconveniens et a principio extrinseco violenter impressus, et quod natura ipsius gravis inclinat ad motum oppositum et ad corruptionem ipsius impetus." *Questiones super octo, libros physicorum Aristotelis*, edited by Johannes Dullaert (Paris, 1506), *Octavi physicorum, Questio secunda* (the work is not paginated).

transient like the process of heating or a state of moving. With questionable success, they attempted to say that an impressed quality is a 'something' intermediate between a permanent thing and a transient thing that endures for a limited time or, with Burdian, to say that the *impetus* of a body is an intrinsic principle of motion, distinct from local motion itself, so long as it inheres in the body unimpeded.[45] In any event, attempts were made to distinguish impressed 'qualities,' 'powers,' 'natures,' and the like from the sorts of particularized qualities that characterize natural objects. It is this latter sort of quality that Newton's argument seems to assume. If so, it is clear that an impetus theorist would not feel compelled by Newton's reasoning.

3. Newton's arguments for natural gravity

Thus far, Newton is satisfied that he has grounds that warrant a rejection of the three main explanations for the cause of projectile motion: the impulse of the air; action at a distance mediated by 'species'; and the 'impressed power' imparted to the projectile itself. He now turns to his positive thesis, that a projectile "must be moved after its separation from the mover by its own gravity."

Let us begin by setting out the details of Newton's argument, because it is rather convoluted and employs many stated and unstated premises. The obvious strategy is to show that we can meaningfully speak of motion *in vacuo* and thereby avoid plenist explanations of the continued motion of projectiles. Newton asks us to consider the following thought experiment (52 113ᵛ). Suppose there are three globes, each of which is partially situated in the medium of air, while the remaining half is in a vacuum. Newton reminds us that he has already shown "in the chapter *de vacuo*" that the globes "would be really separate and not touch one another." In his view, this conclusion is warranted by his argument that matter comprises indivisible parts, the separation of which can be demarcated only on the supposition that a vacuum exists. Now, Newton's imagi-

[45] *Ibid.*, Crombie, pp. 60, 70; Maier, pp. 135–7.

nary opponent will grant that the part of each globe *in pleno* occupies a place and can be said to move from one place into another. Can this be said of the half that is *in vacuo*? If it is denied that the part of the globe *in vacuo* can move with the rest of it, Newton asks on what grounds the two halves can be separated. It cannot be the vacuum, because the vacuum by its nature cannot act on the globe to separate it into distinct parts, but rather can only demarcate its parts if they are already separated. But there is nothing else that can separate the parts of the globe; so there is no reason to deny that the half of the globe *in vacuo* can move with that *in pleno*. If you deny that motion *in vacuo* is "truly motion," you beg the question in favor of the view that motion can take place only in a medium. Moreover, you must deny that globe c's disposition (as such) to globe d and to the air can be characterized coherently under different descriptions (see diagram, folio 52 113^v). Clearly, "the upper part of c has neither the same respect to the air nor to d which it had before it began to pass toward d. If this going of c to d be not motion I ask what it is." Newton concludes this part of his argument by claiming that he is not arguing for any particular theory of motion. Strive about terms if you like, it cannot be denied that coherent descriptions apply to the various dispositions of the upper part of c in respect to its surroundings.

In the next part of his argument, Newton asks why motion *in pleno* should be thought to exemplify natural motion any better than *in vacuo* (folio 53 114^r). There are, of course, obvious differences. In a medium, an object meets with resistances of various sorts that by definition are not present in a vacuum. At this point Newton interjects a commitment that is to remain characteristic of his thought. God exists in all the void places of the world and penetrates all created matter with his presence. Nevertheless, his presence "can be no obstacle to the motion of matter; no more than if nothing were in its way." Similar beliefs are expressed decades later in both the *Principia* and the *Opticks* (see Chapter 1, Section 7). Here Newton's main point is that *in vacuo* a body's motion is absolutely unimpeded, because God's spiritual presence cannot by its nature constitute any physical obstacle. Furthermore, his argument assumes that such motion exemplifies the genus of natural motion better than any other form of motion. However, because a medium impedes more than it assists motion (a conclusion drawn from the earlier argument against antiperistasis), motion "*in pleno* cannot be essential to motion." If it were essential, it would follow that natural motion most appropriately occurs where the densest matter exists, or at the place of most resistance to the passage of a body. This conclusion, as Newton claims repeatedly throughout his essay, is contrary to appearance. Here Newton, of course, assumes that he has proved the existence of a vacuum. Otherwise, a supporter of the medium could object that if the medium is an impediment to natural motion, all bodies move unnaturally, because all in fact move through corporeal media. Essentially this point is made against Avempace by Averroës in his commentary on Aristotle's *Physics*.[46] Avempace had argued (against the Aristotelians) that even in a void a body moves with a finite speed, not an infinite speed, because it moves through a distance.[47] It is clear that this principle informs Newton's claim that the motion of a mobile *in vacuo* is proper and natural. After all, the overall thrust of his argument is to show that motion across a distance can with plausibility be said to occur in the extension of a void.

The last part of his argument considers Aristotle's classic charge against the atomists: "how can there be natural motion if there is no difference according to the void and the unlimited?"[48] For "no one can say why a thing once set in motion should rest anywhere: for why should it rest here rather than here? Since a thing will either be at rest or must be moved indefinitely, unless something stronger impedes it."[49] As Newton correctly states the argument, Aristotle claims that there is no sufficient reason why a mobile should rest at any particular 'place' in the void, because it is "uniform and everywhere alike." Thus, there can be no

[46] *Ibid.*, Crombie, p. 54; Maier, p. 136.   [47] *Ibid.*, Crombie, p. 54.

[48] *Phys.*, IV. 8. 215a5–6. "ἀλλὰ μὴν φύσει γε πῶς ἔσται μηδμιᾶς οὔσης διαφορᾶς κατὰ τὸ κενὸν καὶ τὸ ἄπειρον."

[49] *Phys.*, IV. 8. 215a19–22. "ἔτι οὐδεὶς ἂν ἔχοι εἰπεῖν διὰ τί κινηθὲν στήσεταί που. τί γὰρ μᾶλλον ἐνταῦθα ἢ ἐνταῦθα; ὥστε ἢ ἠρεμήσει ἢ εἰς ἄπειρον ἀνάγκη φέρεσθαι, ἐὰν μή τι ἐμποδίσῃ κρεῖττον."

motion in a void, because there is no differentiation by which to judge change of circumstance. It is interesting to note that Newton appealed to the same argument from Aristotle three decades later in a fragment that forms part of an unimplemented revision of the *Principia* in the early 1690s.[50] His aim in the fragment is to record anticipations of the principle of inertia. He therefore construes the possibility that in a void a body would move indefinitely, as a partial expression of inertial motion.[51] Aristotle, of course, takes the suggestion to be absurd.[52]

Here Newton is concerned to answer Aristotle's main charge (i.e., that it is incoherent to speak of motion in a void). He admits that the air itself is not a possible object of perception, and he implies, *a fortiori,* that neither is the void. He states, moreover, that we judge that something is moving in relation to perceptible objects. Newton implies, however, that this is a fact about us, a fact about how we judge that simple motion has occurred or is occurring. But from this, he concludes, it does not follow that it is a fact about the way things are. Consequently, to claim that an object moves in a void is to refer to a state of affairs that can obtain independent of how we judge that an object is moving relative to another set of objects.

That this is the distinction Newton intends is borne out by the following elements in his argument. In the first place, the argument attempts to legitimize the notion that an object can be said to move across a distance in a void expanse, irrespective of whether or not we are able to judge that the object maintains a continuous identity at different places in the void. In the second place, Newton holds that the continuing cause of the motion is not some external agent (whether the action of a medium or *virtus impressa* from another object), but a condition internal to the object itself— its "natural gravity." Clearly, Newton denies the conception that movement lacks any reality apart from a moving body that can be perceived in relation to other bodies. Rather, real motion can be said to occur in a void expanse as the real effect of an internal principle of motion. In other

[50] Hall and Hall, p. 310. Newton reproduces the same passage in Greek as shown in footnote 49.   [51] *Ibid.*   [52] *Phys.,* IV. 8. 215a20.

words, Newton construes motion as a new effect that requires a causal grounding. Consequently, he rejects the view that motion is simply the different spatial relations that obtain from instant to instant between a moving object and some other object, and he affirms the view that an object can be truly said to move in relation to the void itself.

It seems that Newton begins here to adumbrate distinctions that later pertain to his doctrines of absolute and relative place. Is there not an echo of the admonition in the *Principia* that we are not to "confound real quantities with their relations and sensible measures?"[53] Is there not an implicit invocation of Aristotle's principle to the effect that what is more intelligible to us is not necessarily better known to nature?[54] In the framework of Newton's argument, the principle says that in the order of being, motion in a void is prior to its conceptualization (absolute is prior to relative), but in the order of knowing, it is not (relative is prior to absolute).[55] If this suggestion is correct, it is not implausible to suppose that Newton is replying to Aristotle's argument against motion in the void by implicitly turning one of his own commitments against him. This is a mode of arguing against opponents that occurs frequently in Newton's later thought.[56] In any event, the suggestion has an inherent plausibility, given the nature of Newton's argument, given the fact that he is under the influence of the atomist theory of motion in the void, and given that he is aware of the principles on which Aristotle rejects that theory.

Newton's conception of continued motion in a void was influenced by Charleton. The latter argues that when a pro-

---

[53] *Principia*, edited by Koyré and Cohen, Vol. I, p. 52.

[54] For an important discussion of 'on being more intelligible,' see J. D. G. Evans, *Aristotle's Concept of Dialectic* (Cambridge University Press, 1977), Chap. 3, pp. 68–73.

[55] For a discussion of this principle in Newton's thought, see J. E. McGuire, "Existence, Actuality and Necessity: Newton on Space and Time," pp. 463–508.

[56] See *De gravitatione*, Hall and Hall, pp. 91–8, in which Newton attempts to show that Descartes's theory of motion presupposses distinctions legitimized only by his (Newton's) conception of motion; also, in Section VIII of Book II, *Principia*, he argues that Descartes's theory of vortical motion necessitates his own doctrine that there are active principles in nature that replenish the sources of motion. *Principia*, edited by Koyré and Cohen, Vol. I, Prop. LII, Corols. 5 and 6, pp. 540–1.

jectile is discharged by a projector, a new motion is not impressed on the projectile, but the same motion is merely continued in the flight of the latter. In other words, the mover simply exchanges all or part of its motion with the moved, and there is no need to ask if there is "the impression of any new and distinct Force" on the projectile.[57] In Charleton's view, the principle of action is not an agent or force separate from a moving thing, but merely the fact that the thing is moving, and can impart its motion on another.[58] If it is said that there is an impressed force that remains in the thing moved, this is a misleading way of saying that the motion of the mover remains in the moved after contact.[59] Charleton goes on to argue that the motion of an object in a space that is absolutely empty must "be carried in a direct and invariate line, through the same space, and with an Uniforme and Perpetual motion"[60] unless the object meet with the same resistance. And he concludes "that *All Motion once impressed is, of itself, indelible,* and cannot be Diminished, or Determined, but by some External Cause."[61] Charleton's conception can be construed as a type of 'inertial motion,' that is, as a motion that begins as 'forced' but continues indefinitely and uniformly, once all impeding conditions are removed. This is a very different conception from the claim that there is motion in the absence of agency, that is, that motion can be said to exist and to continue independent of either extrinsic or intrinsic causes.

Charleton's discussion is little more than a paraphrase of Gassendi's view that if unhindered a projectile would move eternally and uniformly in the same direction.[62] Newton, however, does not explicitly affirm that under these conditions motion would be eternal, nor does he consider the 'mode of being' of the void in which such motion occurs. What he does accept is Charleton's view that motion is ex-

[57] Charleton, *Physiologia*, pp. 464–5.

[58] *Ibid.*, p. 465. ". . . seeing nothing else is impressed, but the very motion to be continued through a certain space; so that we are not to inquire, what motive Virtue that is, which makes the Persevering motion, but what hath made the motion, that is to persever."

[59] *Ibid.*, p. 465.   [60] *Ibid.*, p. 467.   [61] *Ibid.*

[62] There can be little doubt that Charleton's discussion, as he himself indicates is derived from Gassendi's *De motu impresso a motore translato, Epistle I*, Cap. XIII, p. 46.

changed between the mover and the moved, that bodies
have an 'inherent motility' that can be actualized by contact
with the mover. This involves holding that an impulse for
moving is exchanged between the mover and the moved, so
that in the end Newton (and Charleton) is not far removed
from the impetus position that he criticizes.[63] Although
Newton probably knew Galileo's treatment of inertia, it is
doubtful that he grasped the various and ramified turns in
the latter's argument.[64] The same is true of Descartes. In
this case, Newton's actualist persuasions clearly prevented
him from seeing that for Descartes inertial motion is not
actual motion, but rather a characterization of the disposition each part of matter has to move along a right line were
it not impeded from so doing by omnipresent external
causes in the medium.[65]

[63] *Ibid.*, p. 465.
[64] Galileo, *Dialogue*, Day Two, p. 173.
[65] Descartes, *Opera*, pp. 39–40; AT.VIII.63, Article XXXIX.

4

PHYSIOLOGY AND HOBBESIAN
EPISTEMOLOGY

Although one can make a radical distinction between a physical theory of color and a physiological theory of color, such a distinction in Newton's case can hide (and indeed has hidden) a likely source of his early physical theories of color, namely, his investigations into physiology. In what follows, we shall describe and analyze Newton's physiological experiments, examine the consequences he draws from them regarding the nature of colors, and argue that these experiments are guided in part by an epistemological program that he proposes in the *Questiones*.

1. The physicalist program

The *Questiones* contains notes and comments based on both Hobbes's *De corpore*[1] and More's *The Immortality of the Soul*. As the number and extent of Newton's references to More indicate, he read the latter's treatise on physiological topics carefully. More's book is a polemic against the 'materialism' of Hobbes, and although he repudiates Hobbes's claim that there are no incorporeal substances in the world, he accepts Hobbes's views concerning perception in a conditional form, namely, "That Sense and Perception in *Matter* in the World, supposing nothing but *Matter* in the World, is really the same with Corporeal Motion and Re-action," and "*if there be nothing but* Body *or* Matter *in the World,* Cogitation *it self is really the same thing with* Corporeal Motion."[2] Here More is claiming that intellectual cognition can be treated as if 'physiological' cognition alone exists in the world.

---

[1] *De corpore* is the commonly used Latin short title for Hobbes's *Elements of Philosophy concerning Body.*

[2] More, *The Immortality;* in *Collection* (1662), p. 5.

Newton is also impressed with the strong connection between motion and perception. He not only refers to some of More's arguments for the connection at 33 104ʳ but also remarks on, or makes use of, the connection at 35 105ʳ through 37 106ʳ, 69 122ʳ through 73 124ʳ, 75 125ʳ, 76 125ᵛ, and 86 130ᵛ. What is clear in these passages is that Newton accepts a form of the causal theory of perception, as well as a form of the sensorium account of the role of the soul in perception.[3] Motion is asserted to be an important part of the account of perception in both these theories. External objects either directly or indirectly put other bodies into motions that are then transferred to our organs of sense and subsequently are conveyed through the nerves to the brain, where the motions (or, in the case of vision, "motional pictures") are seen by the soul.

Newton believes it probable that the soul is able to perceive only by the aid of motion. In response to his own question of why it is that, given the sensorium account, we do not judge perceived objects to be in the brain, he writes: "Because the image of the brain is not painted there, nor is the brain perceived by the soul, it not being in motion, and probably the soul perceives no bodies but by the help of their motion" (86 130ᵛ). Newton also believes that the 'quantity of motion' is related to the intensity of the effect on the soul. Thus, he writes at folio 72 123ᵛ, "Though two rays be equally swift yet if one ray be less than the other that ray shall have so much less an effect on the sensorium as it has less motion that the other." External bodies are not, however, the only source of motion in the sensory nerves. Newton concludes on the basis of an afterimage experiment (to be discussed later) that "fantasy and the Sun had the same operation upon the spirits in my optic nerve, and that the same motions are caused in my brain by both." (75 125ʳ).

Thus, Newton accepts the essential role of motion (at least in the nerves and brain) in perception. As does More, he repudiates Hobbes's belief that all mental functions can be reduced to motions in the brain, and he notes More's argu-

---

[3] The sensorium account places the soul at the seat of sensation in the brain, the sensorium. The soul is then the 'observer' at the percipient center to which all sense impressions are transmitted by the nerves.

ments at 85 130^r. Thus, he agrees with More's argument that if matter were all that existed in the world, thought, perception, and memory would be *nothing but* the motion of matter. Nevertheless, although the motion of matter is closely tied to some of these activities, it cannot totally account for them. This belief strengthens Newton's view that the world does not consist of matter alone. Whereas this argument is the major thrust of More's *The Immortality of the Soul* as a repudiation of Hobbesian 'materialism,' its influence on Newton seems quite different. Newton appears to have needed little convincing of the soul's essential role in perception. What did intrigue him was the large role that the body and its motions play in perception. Thus, it may be the case that many of the things we attribute to the soul and its judgments are really products of the human body and its motions. The question for Newton is how to demarcate what belongs to each, and an inquiry into this matter is clearly both a physiological and epistemological inquiry.

We shall not claim that More is the source of Newton's question, but rather that his consideration of More's *Immortality* was the occasion of his formulation of the question. As we shall shortly see, Thomas Hobbes is also an important influence on Newton's physicalist program. On folio 28 101^v, Newton writes:

The nature of things is more securely and naturally deduced from their operations one upon another than upon our senses. And when by the former experiments we have found the nature of bodies, by the latter we may more clearly find the nature of our senses. But so long as we are ignorant of the nature of both soul and body we cannot clearly distinguish how far an act of sensation proceeds from the soul and how far from the body.

Here we have a clear formulation of what we are calling Newton's physicalist program. By "our senses" Newton appears to refer to the representations in the sensorium, as well as to the organs of sense that are parts of the body. The phrase "operations one upon another" refers not only to the operation of one material object on another but also to the operations of a material object on the material object that is the human body, and the operations of the parts of the body on each other. Only if Newton is thus understood can we readily appraise his program for understanding "how far an act of sensation proceeds from the soul and how far from the body."

2. The Hobbesian influence

It is undoubted that Newton's argument has its roots in Hobbes's *De corpore*. In the course of considering the method of analysis and synthesis, Hobbes details his theory that motion is the universal cause of all effects in the world, especially the "effects one body moved worketh upon another." He proceeds to outline his program for treating the effects of sensible qualities "such as *Light, Colour, Transparency, Opacity, Sound, Odour, Savour, Heat, Cold,* and the like; which because they cannot be known till we know the Causes of Sense itself, therefore the consideration of the Causes of *Seeing, Hearing, Smelling, Tasting,* and *Touching. . .* " is a paramount part of the investigation into the nature of things.[4] Here, in an illustration of the method of analysis and synthesis, Hobbes indicates the same order of investigation as does Newton; that is, we must establish the nature of the cause of sensation prior to understanding the nature of sensation through its causes. As Hobbes himself indicates earlier in his chapter, he is applying a version of Aristotle's distinction between what is more intelligible to us and what is more intelligible to nature. Thus, by the method of analysis, we proceed from the things that are better known to us (i.e., sensible effects) to their causes; once these are established, the order is reversed, and the effects are explained in virtue of their causes.[5] This is precisely the character of Newton's argument: Once "we have found the nature of bodies [by proceeding from their effects], by the latter we may more clearly find the nature of our senses." Newton's formulation of the Hobbesian argument is not immediately clear, because he has suppressed the analysis side of the procedure.

As will become apparent shortly, Newton also finds sup-

---

[4] Thomas Hobbes, *Elements of Philosophy concerning Body* (London, 1656), Part I, Chap. VI, "Of Method," Art. 6, p. 53; or the earlier Latin edition, *Elementa philosophiae sectio de corpore* (London, 1655), p. 45.   [5] *Ibid.*

port in Hobbes for the basic element in his physicalist ontology – motion. Hobbes can be read in the following manner: Motions preserve the pattern and configurations of the moving objects from which they come. Moreover, they excite in the sense organs motions that have similar patterns and configurations. On this view (Newton's), sensation is the conveying of determinate patterns of motion through the senses. Thus, the effects on the sensory organs are in fact motions that carry 'information' from the causal objects they resemble in structure. Ultimately, this interpretation involves a 'scientific realist' conception of how the mind relates causally to an extramental world. Similar programs are to be found in the writings of Descartes and Gassendi,[6] but there is no evidence in the *Questiones* that Newton was influenced by them on this issue.

Although there is little doubt that Newton and his English contemporaries (such as More) read Hobbes in this realist way, the latter's thought is ultimately of a different character. A key to the temper of Hobbes's position is his conception of an accident: He defines it as "*the Manner by which any Body is conceived; which is all one as if they should say, An Accident is that faculty of any Body by which it works in us a Conception of itself.*"[7] It is clear that he thinks he has brought the usual concept of an accident (i.e., that which inheres extramentally in an object) under the notion that there are general modes by which an object is conceived.

Hobbes's 'conceptualism' is nowhere more evident than in his theory of causation, in which his conception of accidents plays an important role. He defines a complete cause as "*the Aggregate of all the Accidents both of the Agents how many soever they be, and of the Patient, put together.*" He goes on to affirm that the accidents that comprise both the agent and the patient cannot be understood independent of each other.[8] Thus, the necessity of the causal connection is epistemic: It embodies the manner in which the mind conceptualizes in-

---

[6] Descartes, *Opera*, pp. 1–26; AT.X1.327ff., *Passiones sive affectus animae, prima pars*. Gassendi, *Syntagma philosophicum*, Vol. II, Sect. III, Part II, Lib. VI, Cap. 2, pp. 338–9.

[7] Hobbes, *Elements*, Part I, Chap. VIII, "Of Body and Accident," Art. 2, p. 75; *De corpore*, p. 63.

[8] *Ibid.*, Chap. IX, Art. 3, p. 88; *De corpore*, p. 74.

teractions among the natures of things, where the term 'nature' does not denote a reality existing independent of the mind's inquiry but is understood linguistically as a collection of accidents, where accidents are conceived as ways of grouping and identifying individuals. This conception of causation is a far cry from the notion that causes are extramental realities that connect the mind with the outer world. Although Hobbes is not strictly entitled to this notion, he often talks in this manner. In fact, there are sufficient passages to provide a basis for the 'realist' manner in which Newton and his contemporaries read Hobbes's philosophy of motions as causes.

But this is not the whole story. Whereas Newton is influenced by Hobbes's epistemology of 'motional' patterns in perceptual contexts, with respect to the epistemic standing of infinite extension the Cartesian influence is dominant. The nature of this influence can be brought out in the following way. Suppose that Newton *were* to hold a Hobbesian account of infinite extension. This would consist in saying that the mind can image the negation of all successive boundaries posited in a finite magnitude. On this showing, the infinity of extension is a negative idea, the idea that the magnitude of a body can be imagined to be indefinitely extended. But Newton argues that extension, rather than being merely indefinite, is positively infinite: "I mean all the extension which exists and not so much only as we can fancy" (87 131^r). Concerning extension, Newton implies that although the mind cannot imagine its infinity, it can nevertheless understand that it is infinite. By the time of *De gravitatione*, he has explicitly adopted Descartes's version of the distinction regarding what is properly an object of the imagination, as opposed to the understanding, and he applies it to his view of the mind's capacity to grasp infinite extension (see Chapter 1, Section 7).

3. Newton's physiological investigations

Having considered the roots of Newton's physicalist program in Hobbes's epistemology, we can now turn to the role that Newton's physiological investigations play in that program. Newton notes at 35 105^r that the eyes can see independently. For example, when one is pressed by the finger "their two axes respect not the same point in the object," and at such times we see two images of an object. The degree to which the eyes can function independently and yet see things on the same plane (as if at the same distance) is a surprise to Newton. We may see a part of a body in two different places at the same time, or successibely (first with one eye and then the other), and we may see two objects in the same plane successively without deforming the eyes (by parallax). Newton writes that these facts constitute an argument "either that the image in the brain is painted on a superficies, or else that the optic nerves, at their meeting before they enter the brain do sort their capillamenta by uniting there each capillamentum of the one eye to its fellow capillamentum of the other eye. So that when the eyes conspire not toward the same object but have diverse pictures painted upon the correspondent capillamenta, the stronger picture at the meeting of those capillamenta drowns the weaker" (35 105^r). Newton thus suggests that the planar character of vision, as he describes it, might be explained in one of two ways: first, that the "images" are on a planar surface of the brain and thus are viewed by the soul as planar; second, that the weaker image (Newton gives no indication at this point what would make one weaker than the other) is "drowned out" in some physical process by the stronger image at the juncture of the optic nerves before they enter the brain. In both cases the planar character is explained by some physiological factor, *not* by some judgment of the soul.

Newton next asks if fantasy is restricted, as is sight, to being planar or if we can imagine things one behind another. Newton's answer is that we are so restricted, but that we can understand, if not see, that one object is behind another. Newton does not speculate as to what physiological cause may account for this restriction on the imagination.

The fact that two eyes working together in normal vision produce a single image is another problem that Newton considers. Although this issue arises only peripherally in the foregoing, Newton considers it in detail in folios 17–20 of

Add. 3975 (Appendix), almost certainly written in 1666. Here Newton reports on a complex dissection he carried out on the tunica retina of an unspecified creature. He traced the paths of the nerves from the retina to the brain and discovered that the nerves from the right side of one retina joined and combined with the corresponding member from the right side of the other retina at the concourse of the nerves and then traveled to the right side of the brain, and a similar phenomenon obtained for the left sides of the retinas. He writes, "I conceive that every point in the retina of one eye hath its correspondent point in ye other, from wch two very slender pipes, filled wth a most lympid liquor doe wthout either interruption or any other uneavenesse or irregularity in their processe, goe along the optick nerves to ye juncture . . . where they meete . . . & there unite into one pipe as big as both of them, & so continue in one passing . . . into ye braine where they are terminated" (see text and diagram on folio 17).[9]

Newton offers a list of phenomena supposedly accounted for by his physiological findings. These include (1) why the two images in the eyes make one image in the brain, (2) why when one eye is distorted, objects appear double, (3) "why though one thing may appeare in two places by distorting the eys yet two things cannot appear in one place. If the picture of one thing fall upon A & another upon α, they may both procee[d] to p but noe farther, they cannot both be carried on ye same pipes pa into ye braine, that wch is [s]trongest or most helped by fantacy will there prevaile & blot out ye other" (folio 18), (4) why a blue seen by one eye and a yellow by the other produce a green "unlesse ye fantasy make one colour prædominant," and (5) why, if one of the branches of the nerve beyond the juncture is cut, that half of each eye on the same side as the cut will be blind, the other halves remaining sighted. Not all of these are phenomena known to Newton through experience. Surely (4) and (5) are meant as predictions, because (4) is false, and a case such as (5) must be exceedingly rare, and Newton uses

---

[9] David Brewster republished a version of these folios of Add. 3975, including a version of the diagram on folio 17, which he took from Joseph Harris, *Treatise of Optics* (London, 1775). See footnotes 14, 15, and 17–22 in the Appendix.

the subjunctive in describing it. However, (1) through (3) refer to well-known phenomena, phenomena Newton had considered in the *Questiones*. In particular, (2) and (3) refer back to the possible solutions offered by Newton to the problem of the planar character of vision and imagination. It is clear that he viewed the results of his dissection as decisive on this issue. We may also add (1) – the problem of binocular vision – to our list of phenomena to be explained physiologically rather than by the activities of the soul. It is not that the soul is confronted with two images that it then combines in an act of judgment into a single picture but that through a second concourse of the nerves in the brain a single combined picture is presented to the soul. This second joining is listed by Newton as (9) on folio 19. He writes, "9^{thly} why ye two nerves meet a second time in the brains, because ye two half images carried along IL & MK may bee united into one complete image in the sensory."

Newton reports his examination of the contents of the optic nerve on folio 20 of Add. 3975. He indicates that because light seldom strikes the parts of gross bodies, but rather passes between the parts and so goes through such bodies, the motion of light is transferred to the "diversity of æthers" that lie among the parts. This motion is then transmitted through the optic nerves to the sensorium. But through which medium within the nerves is this done? Not through water, "for water is too grosse for such subtile impressions," and not through animal spirits, for "though I lyed a peice of ye optick nerve at one end & warmed it in ye middle so [to] see if any aery substance by that meanes would disclose it selfe in bubbles at the other end, I could not spy the least bubble; a little moisture only & ye marrow it selfe squeezed out" (folio 20). Newton concludes that the motion is transferred through ether that fills the pores of the optic nerves but (one supposes) is far too subtle to be seen by any technique of dissection.

These findings regarding the operation of the eye and optic nerves remained little changed into the *Opticks* in Query 15. The fullest discussion concerning vision in which Newton was involved in the intervening years was with William Briggs (1642–1704) during 1682–5. Briggs had apparently been friendly with Newton at Cambridge, having been a Fellow of Corpus Christi College in 1668. In Newton's earliest extant letter to Briggs, which is dated 20 June 1682, he addresses Briggs as a friend ("to be free with you as a friend should be"), indicating an earlier relationship between the two.[10] Furthermore, there is some reason to think that Newton and Briggs may have carried out dissections on the eye at some earlier date, perhaps in the 1660s.[11] In a letter to Briggs dated 25 April 1685, which was prefixed to the Latin edition of Briggs's *Theory of Vision* (London, 1685), Newton writes, "Your skill and artistry in its [the eye's] dissection once, I remember, afforded me no small enjoyment. You neatly displayed the motor muscles in their natural positions and so disposed all the other tissues before us that we could not so much understand as perceive the functions and services of each, with the result that there is no refinement that I have not long expected of your knife."[12]

Briggs's "Theory of Vision" had been read (Briggs was not present) at a meeting of the Royal Society on 15 March 1682 and printed in the *Philosophical Collections*, where Newton read it.[13] In his letter of 20 June 1682 he informs Briggs

I have perused your very ingenious Theory of Vision in wch . . . there seems to be some things more solid and satisfactory, others more disputable. . . . The more satisfactory I take to be your asserting yt we see wth both eyes at once, your speculation about ye use of ye *musculus obliquus* inferior, your assigning every fibre in ye optick nerve of one eye to have it's correspondent in yt of ye other, both wch make all things appear to both eyes in one and ye same place and your solving hereby ye duplicity of ye object in distorted eyes and confuting ye childish opinion about ye splitting ye optick cone. The more disputable seems your notion about every pair of fellow fibres being unisons to one another, discords to ye rest, and this consonance making ye object seen wth two eyes appear but one for ye same reason that unison sounds seem but one sound.[14]

Briggs's account of the pairing of the capillamenta of the eyes is regarded by Newton as correct in its broad outline

[10] *The Correspondence of Isaac Newton*, edited by H. W. Turnbull (Cambridge University Press, 1960), Vol. II Item 261, pp. 377–8.

[11] Possibly the very dissection described on folios 19 and 20 of Add. 3975 (see Appendix).

[12] *The Correspondence*, Vol. II, Item 280, pp. 417–19 (Turnbull's translation, p. 418).

[13] *Philosophical Collections*, 6(1682):167.   [14] See footnote 10.

and incorrect in its details. We have seen that Newton uses this pairing *and* their consequent joining as his solution to the problems considered by Briggs. Although Briggs has noted the pairing, he does not use their juncture to explain the phenomena, but rather invokes a notion of tuned unison strings to do so. He has observed, as did Newton, that the nerves from the corresponding parts of the eyes are of the same length and, Briggs supposes, of the same tension, whereas those of noncorresponding pairs are of different length and tension. This 'tuning,' Briggs believes, will make corresponding nerves vibrate in unison, thus forming not two different sets of vibrations, but a single set and so a single image in the sensorium.

In a letter to Briggs dated 12 September 1682 Newton details his objections to Briggs's account, then raises the question of how one might resolve the problem, and writes the following curious passage:

If when we look but with one eye it be asked why objects appear thus and thus situated one to another. The answer would be because they are really so situated among themselves and make their coloured pictures in ye Retina so situated one to another as they are and those pictures transmit motional pictures into ye sensorium in ye same situation and by the situation of those motional pictures one to another the soul judges of ye situation of things without. In like manner when we look with two eyes distorted so as to see ye same object double if it be asked why those objects appeared in this or that situation and distance one from another, the answer should be because through ye two eyes are transmitted into ye sensorium two motional pictures by whose situation and distance then from one another the soule judges she sees two things so situate and distant. And if this be true then the reason why when the distortion ceases and ye return to their natural posture the doubled object grows a single one is that the two motional pictures in ye sensorium come together and become coincident.

But you will say, how is this coincidence made? I answer, what if I know not? Perhaps in ye sensorium, After some such way as ye Cartesians would have believed or by some other way. Perhaps by ye mixing of ye marrow of ye nerves in their juncture before they enter the brain, the fibres on ye right side of each eye going to ye right side of ye head those on ye left side to ye left.[15]

---

[15] *The Correspondence*, Vol. II, Item 264, pp. 381–5. Newton's reference to the Cartesians is meant to be ironic. Descartes describes all the nerves of the left eye as going to the left side of the brain, and those of the right eye as going to the right side of the brain, in his *Dioptrics* (*Opera*, pp. 81–2, 90–1; AT.VI.129, 138–143). The Cartesians do not offer a 'way' in which this is done in the sensorium.

Although Newton seems reluctant to 'speculate' about the problems posed and is tentative in giving his answer, we know that twenty years earlier he was convinced of the final solution he mentions here. Nevertheless, he soon warms to his task, and even offers a new argument for this solution. He writes in the same letter, "if you say yt in ye Camealion and Fishes ye nerves only touch one another without mixture and sometimes do not so much as touch; 'Tis true, but make altogether against you. Fishes looke one way with one eye ye other way with ye other: The Chamaelion looks up with one eye, down with tother, to ye right hand with this, to ye left with yt, twisting his eyes severally this way or that way as he pleases. And in these Animals which do not look ye same way with both eyes what wonder if ye nerves do not joyne?" Newton's solution to these problems, as found in Add. 3975 (with the omission of the "drowning out" hypothesis), and his new argument offered to Briggs together constitute the content of Query 15 of his *Opticks,* published in 1704, forty years after they were first considered in the *Questiones.*

Notice also that Newton repeatedly uses the phrase "motional picture" in the letter to Briggs. From the character of his arguments it is clear that he is still invoking the physicalist program that he derives from Hobbes. To be sure, he pretends that he has not considered these problems previously, and he even mentions the 'solution' of the Cartesians. However, there is no doubt that Newton is committed to a program not unlike that which he adumbrated in the *Questiones.*

Before turning to Newton's afterimage experiment, there are two short entries that should be considered. The first concerns another phenomenon that Newton noticed in the course of pressing his eye with his finger. He asks, "why does the forcible turning of the eye one way with the finger make the object seem to move the other way, but not the voluntary turning of it?" (35 105$^{\rm r}$).

Newton offers no answer to his question, and indeed an answer has only recently begun to be understood, involving one of the most complex feedback mechanisms in the human body.[16] The question itself is interesting, however, in

[16] See Emilio Bizzi, "The Coordination of Eye-Head Movements," *Scientific American,* 231(1974):100–6.

its relation to Newton's physicalist program. In the phenomenon, a physical modification of the body (brought about by pressing the eye) is specifically contrasted with a voluntary action (i.e., one involving the operation of the soul). The difference between these cases is likely to be critical, for what the soul observes in one case – the opposite motion of the object – it does not observe in the other.

The second involves a perceptual fact that Newton probably became aware of in his study of drawing and painting; it is suggested in Bate's *Mysteryes of Nature and Art.*[17] Newton writes (36 105^v):

> That dark colors seem farther off than light ones may be from hence: that the beams lose little of their force in reflecting from a white body because they are powerfully resisted thereby, but a dark body, by reason of the looseness of its parts, gives some admission to the light and reflects it but weakly. And so the reflections from whiteness will be sooner at the eye. Or else, because the white sends beams with more force to the eye and gives it a fiercer knock.

Newton here presupposes two possible ways in which, all other things being equal, closer objects may differ from those farther away. One is that the light from those that are closer reaches us in less time than that from objects farther away; the second is that light from closer objects gives the eye a "fiercer knock" than that from objects farther away. The first is too trivial to require comment. The second is far more interesting. It is difficult to see why a single particle of light (and although Newton here writes of "beams" of light, he thinks of light as a stream of particles throughout the *Questiones*) would give a 'lesser knock' over a longer distance than a shorter, unless, of course, the particle were to slow down as it moved farther. What is more likely is that Newton thought of a beam as a stream of particles that would contain fewer particles as the light fanned out in all directions. To say the *beam* gives a 'fiercer knock' would mean that more particles per unit of time would strike the eye, rather than that each particle would strike it harder individually.

---

[17] John Bate, *The Mysteryes of Nature and Art* (London 1634), p. 123. Newton made extended notes from Bate on such topics as drawing, painting, and the mixing of colors. These notes are to be found in a notebook Newton used in the period 1659–61 that is now in the Morgan Library in New York.

One can only characterize Newton's first proposed answer to his original question as a momentary lapse, for although it is true that the light from a closer object takes less time to reach us, there will be no characteristic in what reaches the eye to inform us of that fact. The second proposal is in fact the one adopted by Newton, and it occurs in various forms in other places in the *Questiones*. It is interesting that he never makes explicit what is involved in his 'beams of greater force,' but if we are right in assuming that it consists of more particles per unit time impacting on the eye, then it is related to a notion of pressure on the eye, and this is perhaps too close to the Cartesian theory he has so effectively attacked at folio 32 103$^\mathrm{v}$.[18]

We turn now to Newton's afterimage experiments, described by Newton on folios 43 109$^\mathrm{r}$, 75 125$^\mathrm{r}$, and 76 125$^\mathrm{v}$ of the *Questiones*. These afterimage experiments were also described by Newton in a letter he wrote to John Locke, 30 June 1691.[19] Newton looked at the Sun in order to produce an afterimage in his visual field. He then looked at various objects to see what effect the afterimage had on the perception of colors. In (3), on folio 43 109$^\mathrm{r}$, he writes, "after the motion of the spirits in my eye were almost decayed, so that I could see all things with their natural colors . . . ," which suggests that a necessary condition for seeing a given color is the presence of a specific degree of motion of the spirits. When the spirits are moving 'rapidly,' after just looking at the Sun, light-colored objects appear red and dark-colored ones blue. But it is not only the interaction of our organs of sense with external objects that can produce these motions of the spirits. Newton goes on to tell us that when the effects of the afterimage had ended, or at least were no longer manifest (i.e., "almost decayed,"), he was able to restore them by closing his eyes and 'heightening' his "fantasy" of seeing the Sun. The result was as if he had recently looked at the Sun, from which he concludes "that my fantasy and the Sun had the same operation upon the spirits in my optic nerve, and that the same motions are caused in my brain by both" (75 125$^\mathrm{r}$). That the eyes play no role in this

---

[18] See Chapter 2, Section 2.
[19] *The Correspondence*, Vol. III, Item 365, pp. 152–4.

mental 'calling up' of the Sun's image is shown by Newton in the claim that he was able to call up that image with his eyes open (in the dark) or with them closed.

Newton repeated his experiments, this time comparing the effects of the Sun on a closed eye and an open eye. Closing the eye that had looked on the Sun (what Newton calls the "distempered" eye), and opening the eye that had not looked on the Sun, he discovered that he could see the afterimage almost as well as by looking with the distempered eye. Here, too, Newton is establishing that the afterimage does not reside in the eye itself. The eye is a transmitter of sorts that sends motions along the optic nerve to the brain, where the soul perceives. These perceptions in the case of afterimages are 'superimposed' on the other images producing the usual afterimage effect. This is not to claim, as we have already seen, that the eye is totally passive; indeed, Newton makes it sound as if the color changes are produced at (in?) the eye. He writes, "When I opened my right eye [the distempered one], all objects would appear colored as when I had first seen the Sun. But I could not perceive any such motion in the spirits of my left, for all objects appeared in their right colors, except when I fixed my eye on them for then the Sun appeared" (75 125^r). We thus have Newton's answer to the question of where the effects take place. The superimposition of the Sun's image on other images, such as white sheets of paper and clouds, takes place in (on?) the brain, where the soul views the one image superimposed on the other. In the case of color changes, where objects of one color appear to be another, the seat of the effect is the eye itself. According to Newton, the motions of the spirits in the eye produced by the Sun apparently become 'mixed' with the motions produced by viewing a colored object, which produces a resultant motion in the optic nerve and brain characteristic of some other color. This resultant motion in the brain is then 'seen' by the soul.

Just how physicalist an account of perception Newton is offering can be gathered from the fact that he speaks of perceiving the motions. He writes, as we have seen earlier, "I could not perceive any such motion in the spirits of my

left [eye] for all objects appeared in their right colors"
(75 125^r). Newton also writes, "I had been in a dark room
for two or three hours and my eyes were made tender
thereby, so the motion made in them would be more easily
conserved and consequently more uniform" (76 125^v).

The physiological theory that appears to emerge is as
follows:

1.   External objects interacting with the organs of sense
produce a motion in the spirits that are present in the
nerves associated with the sense organs.

2.   These motions are normally characteristic of the external object. In the case of vision, for instance, there are
degrees of motion corresponding to various colors.

3.   These motions are transmitted to the brain by the
nerves in question, where they produce the same(?) motion
in the nerves that make up that region of the brain associated with that sense (e.g., in the case of vision, the optical
chiasma at the back of the brain).

It is clear that Newton thinks these motions can affect
each other, because his account of the afterimage experiment seems to indicate that the motions caused by looking
at the Sun change the colors we see when we look at various
objects. That it is not just a matter of the motion caused by
the Sun 'overpowering' and thus 'replacing' the motions
caused by other objects emerges when Newton notes that
light-colored objects look red and dark-colored ones look
blue. The objects thus continue to contribute some element
to the motion produced by the Sun.

Newton is convinced by More's arguments (and his own)
and does not identify these motions in the brain with the
perceptions, imaginings, thoughts, and so forth, to which
they are related. Instead, these motions are in some way
present to the soul in the sensorium. Although the manner
of their presence is not explicitly discussed by Newton, he
raises a number of queries and observations on folio 86 130^v
that indicate a naive view on this issue. He asks why objects
do not appear in an inverse position, and he answers that
"the mind or soul cannot judge the image in the brain to be
inverted, unless she perceived external things with which
she might compare their image." Here Newton speaks of

the motion of the spirits in the brain as being itself an image. It is also clear that these 'images' are all that the soul sees, because there is no direct perception of 'external things.' Thus, it is evident that Newton is a modified dualist. He does not fully accept Hobbes's identification of thoughts with motions in the brain. Also, his talk of "motional pictures" as mental representations should not be equated with Hobbes's position. Although Newton is much influenced by Hobbes, like More he still wishes to find a role for the soul in a dualist schema.

The next question that arises from the previous response is why objects then appear to be outside our bodies (brain), that is, appear to be external things. Newton's response is that an image of our body (or, more properly, some parts of it) appears among the other images and changes its spatial relationship to them at the direction of our will. In this case, what responds to our will is not the motion that is the image of our bodies but our bodies themselves, which, of course, on Newton's account would result in a change in the images. The third and final query in this series again arises from the two responses, and this response fully reveals the naiveté of Newton's position. Newton asks why these objects are not judged to be in the brain. His reply is "because the image of the brain is not painted there, nor is the brain perceived by the soul, it not being in motion, and probably the soul perceives no bodies but by the help of their motion." Given Newton's account, that the image of the brain is not there follows from the fact that we do not see it. This is due to the fact that the brain does not move as a whole, and in relation to the soul that 'resides' within it. It is therefore curious that Newton should use 'probably' and write as if the image were different from the motion. But let us look more closely. The soul sees motions in specific regions of the brain as images, not as objects. It may well be that only moving things can interact with our organs of sense to produce the motion of the spirits within the nerves that ultimately produce local motions in the brain (or images, as Newton calls them). Thus, it may be that a motion of the brain as a whole is a precondition for perception of the brain, but a further precondition is an organ of sense that can produce a motion in a relevant region of the brain (i.e., an image of the brain). In fact, the brain must have the general motion spoken of, because a trepanned individual can, with a mirror arrangement, see his own brain. Newton's confusion arises from his consideration of the possibility of the soul's directly perceiving the brain's motion without the intervention of an organ of sense. If this were indeed possible, we would realize Newton's conjecture that "were the brain perceived together with those images in it we should think we saw a body like the brain encompassing and comprehending, ourselves, the stars, and all other visible objects."

The last of the physiological experiments we shall consider are those that involve pressing the eye in a manner that produces the sensation of seeing colors. Here again we shall compare an experiment as it is described in the *Questiones,* as it is described in Add. 3975 (Appendix), and finally as it appears forty years later in the Queries to the *Opticks.*

At folio 72 123^v, entry 9, Newton describes and interprets one of his eye-pressing experiments. By pressing the side of his eye, while looking away from the pressing finger (thus pressing a part of the eye nearer the back), he was able to produce an image of a red circle with a blue center. He notes that the part of the image that falls on the more forward part of his eye is "duller" than the part that falls nearer the back. This is a phenomenon that he notes again in Add. 3975, folio 16, where he concludes "that vision is made in the retina appeares because colours are made by pressing the bakside of the eye: but when ye eye turns towards ye pressure soe y^t it is pressed before ye colours cease."

Newton is apparently explaining why the red circle has a blue center when he writes, "for the capillamenta are more pressed at *n* and *o* and round about the finger than at *a* toward the midst of the finger." (see the diagram on folio 72 123^v). The pressure of the finger near the back of the eye thus produces the sensation of color, and the degree of pressure determines the color we see – blue where the pressure is less, and red where the pressure is greater. This, as we shall see in the next chapter, is related to Newton's view in the *Questiones* that blue rays are slow and red rays are swift, which means, all other things being equal, that the particles of red light strike the eyes with greater force than the particles of blue light.

The description of Newton's second eye-pressing experiment begins at the end of folio 72 123^v and runs onto folio 73 124^r. Here Newton describes pressing his eye much nearer the back by pressing a brass plate between the eye and the bone of the orbit, and thus making "a very vivid impression." The impression produced is recorded as differing in several ways from that produced by the pressure of his finger. He reports that the image produced by the pressure of the brass plate is elliptical, as opposed to the round image produced by his finger. This he attributes to the shape of the plate, "because the edge of the plate with which I pressed my eye was long and not round like my finger." Newton also reports that the image produced by the pressure of the plate is far more complex than that produced by his finger. The image is purple at its periphery and, moving inward, then passes through blue, green, yellow, and a flaming red. These are the colors in the order of the spectrum. Continuing inward, the spectrum is repeated, but in the opposite order; that is, after the red there follows yellow, green, blue, and purple, which shades off into a lighter blue; finally, at the center of the image, there is green.

Newton does not attempt to account for the impression of all of these colors, though it is significant that such a wide variety of colors, most particularly those belonging to the spectrum, is produced by the pressure of the brass plate. Newton describes the difference in results when doing this eye-pressing experiment in the light and in the dark. The eyes are closed in both cases, but Newton correctly supposes that some light gets through the eyelids when the experiment is done in the light. In that case, if the pressure were not very strong, the colors apart from the central green and purple would be blue. Newton describes the central colors of the image when the experiment is done in the light as being lighter when the pressure is against a more forward part of the eye rather than more to the back. If the pressure of the plate is strong, then the total impression produced in the light is as vivid (or nearly as vivid) as in the dark. With respect to the visual field, in the region outside the color and toward the pupil the light is stronger than toward the back of the eye. This, Newton explains, is "because the pressure helps the motion from without but is not strong enough to turn it to color" (73 124^r). Here again he assumes that the colors are due to pressure and that the result is the same whether the pressure is from light "entering" the eye or from pressure applied at the back. Indeed, the pressure from the back may augment the "motion from without." Even more interesting is Newton's claim that the pressure sufficiently far from the brass plate's point of contact will be too little to produce colors or to turn "it [the light coming through the lids] to color." This seems to imply that less pressure is required for mere light (white light) than is needed for colors, and as we have seen, various degrees of pressure produce various colors.

In Add. 3975 (see the Appendix) Newton discusses yet another eye-pressing experiment that dates from about 1666. Here he presses his eye with a "bodkin," probably a large needle with a rounded blunt point made of ivory or bone. Once again the instrument is placed between the eye and the bone, so that the eye can be pressed on its surface as far back as possible (see diagram on folio 15). Despite the similarity of this experiment and the one with the brass plate described earlier, several interesting new elements arise in Newton's description. First, it is not pressure alone that produces the colors, for the image fades after a time and can be renewed by moving either the eye or the bodkin. Newton writes that the circles are most plain when he continues to rub his eye with the point of the bodkin. Second, he notes that when the experiment is done in a light room, and then a dark room, the blues become red and, in general, the bright circles become dark and the dark ones bright. Newton also notes a similar phenomenon in the case of afterimages. He writes, "white objects looked red & soe did all objects in ye light but if I went into a dark roome ye Phantasma was blew" (folio 16 of the Appendix). Third, Newton relates the circles of the image with the degree of curvature and strain at the back of the eye around the point of pressure produced by the bodkin. Explaining the results of the experiment done in the dark room, he writes, "where ye curvature of ye Retina at ma & fn [see diagram, folio 15] began & was but little ye blew colour tv has caused; at ab & ef where ye Retina was most concave, ye bright circle ts was caused: at bc, & de where ye Retina was not much incurved nor strained ye dark blew circle sr was caused & at cd where ye Retina was stretched & made convex ye light spot r was caused." The results of the experiment done in the light room are explained: "ye spirits were perhaps strained out of ye Retina at ab, ef, & cd or otherways made incapable of being acted upon by light & soe made a lesse appearance of light y^n ye rest of ye Retina."

The complementarity phenomenon (i.e., the behavior of colors in light and dark rooms, as described earlier) poses a serious problem for Newton's earlier simple identification of pressure as the cause of the colors in the images. It is clear that a single factor will not do; so Newton introduces the curvature and accompanying strain of the retina. In the dark room, the 'strained' areas give rise to bright circles or light spots, and 'unstrained' areas to dark blue. No indication is given as to why this should be so. In the light room, the 'strained' areas are dark, and we are told that perhaps the spirits that undergo the vibrations at the back of the eye are 'strained' out of those stretched areas. It is difficult to see why this did not happen in the dark room, unless the mechanism by which we see light when the eye is pressed is different from when it is not, but this is what Newton has denied all along. The motions are caused in different ways, but they are motions of the same material and indistinguishable from each other. Newton gives no indication here that he sees the difficulty posed by his explanation of the complementarity phenomenon.

That the general character of Newton's account of the eye-pressing experiment remained with him can be gathered from Query 16 of the *Opticks*, where Newton writes:

When a Man in the dark presses either corner of his Eye with his Finger, and turns his Eye away from his Finger, he will see a Circle of Colours like those in the Feather of a Peacock's Tail. If the Eye and Finger remain quiet the Colours vanish in a second Minute of Time, but if the

Finger be moved with a quavering Motion they appear again. Do not these Colours arise from such Motions excited in the bottom of the Eye by the Pressure and Motion of the Finger, as, at other times are excited there by Light for causing Vision? And do not the Motions once excited continue about a Second of Time before they cease? And when a Man by a stroke upon his Eye sees a flash of Light, are not the like Motions excited in the *Retina* by the stroke? And when a Coal of Fire moved nimbly in the circumference of a Circle, make the whole circumference appear like a Circle of Fire; is it not because the Motion excited in the bottom of the Eye by the Rays of Light are of a lasting nature, and continue till the Coal of Fire in going round returns to its former place? And considering the lastingness of the Motions excited in the bottom of the Eye by Light, are they not of a vibrating nature?[20]

It is interesting to note not only that the eye-pressing experiments of the *Questiones* and Add. 3975 enter into Query 16 but also that Newton makes use of his examples of the circle of fire and the struck eye that "sparkeleth" from folio 35 105ʳ of the *Questiones*.

All of the foregoing shows how much of vision is determined by the physiology of the eye and the optic nerves. For Newton, the eye is not the simple *camera obscura* that leaves everything to the observer's (the soul's) interpretation. Colors are mixed, images are drowned out, double images become one, and so forth, all before the final "motional pictures" are presented to the soul.

Vision is not the only faculty that is considered in Newton's physicalist program. Although he does not report any other experiments on other organs of sense, he does note the relevant comments of others. At folio 59 117ʳ, Newton notes Glanvill's reflection from *The Vanity of Dogmatizing* that "An Artist will play a lesson not minding a stroke, sing neither minding nor missing a note, and a man may walk without thinking of it."[21] This example fits neatly into Newton's physicalist program. It is an indication of how much may be moved out of the operation of the mind and into the physical workings of the body. Newton does make reference to the other senses, such as his reference to the sense of touch at folio 39 107ʳ, where he writes, "A man has been deprived of his feelings," something he attributes to Kenelm

[20] Isaac Newton, *Opticks* (New York: Dover Publications, 1952), based on the London, 1730, edition, Query 16, p. 347.
[21] See footnote 117 in the Transcription.

Digby. This was probably taken by Newton to be an indication that the complex of nerves covering the body have some union before they convey their motions to the soul, for how else could they be deadened everywhere without 'dispatching' the person? Newton probably also viewed it as a confirmation of his position that the senses are capable of intensification and diminution, a view that is reflected in his interest in the gypsy scholar story noted at folio 43 109^r. Ways of heightening the "fancy" are noted at the same place, and notes concerning the heightening of vision and hearing are found in Add. 3975 at folio 22. Glanvill tells of a student at Oxford University who was forced by his financial condition to join a "company of Vagabond Gypsies, whom occasionally he met with, and to follow their Trade for a maintenance." He was soon taught their "*Mystery,*" at which he grew very proficient. Having met two former friends, he "told them, that the people he went with were not such *Impostours,* as they were taken for, but that they had a *traditional* kind of *learning* among them, and could do wonders by the power of *Imagination,* and that himself had learnt much of their Art." To convince them of the truth of this, he left them to talk together and then returned to tell them what they had said in his absence. He then explained that "what he did was by the power of *Imagination,* his Phancy *binding* theirs; and that himself had dictated to them the discourse, they had held together, while he was from them: That there were warrantable wayes of heightening the *Imagination* to that pitch, as to bind others."[22]

Newton's descriptions of heightenings and lessenings are generally associated with illness, diet, and habits and do not concern the secret arts alluded to in the gypsy scholar story. This suggests that Newton attributed these heightenings not so much to the state of the soul or mind but to the state of the body and the organs of sense. It is true that according to Newton the soul can cause motions in the nerves that perfectly mimic those caused by external objects, as discussed in the afterimage experiment. The soul can also make one motion greater than another by augmenting it through the

---

[22] Joseph Glanvill, *The Vanity of Dogmatizing* (London, 1661), pp. 196–8.

"fancie," as in Newton's discussion of the different images from each eye in Add. 3975 (folio 18). But even the soul has its work made easier by the state of the organs of sense, as Newton indicates at folio 76 125^v, where he writes, "when the impression of the Sun was not too strong upon my eye, I could easily imagine several shapes as if I saw them in the Sun's place, whence perhaps may be gathered that the tenderest sight argues the clearest fantasy of things visible, and hence something of the nature of madness and dreams may be gathered."

At 41 108^r, Newton considers the operation of memory and locates it in the soul, not the body, though here, too, he associates changes in the power of memory within the body. Newton's collection of 'facts' concerning memory at 41 108^r shows him considering evidence that would bear directly on the major accounts of perception (and thought) of his day. The principal accounts he has in mind are those of Hobbes, Descartes, and More. The emphasis is on the relationship between the external occurences and the internal states.

Physical trauma (a blow with a stone, a fall from a horse, a wound) or illness (a fever, the great pestilence at Athens) can cause a loss of memory. Memory can also be affected by other factors. Things dreamt can remind us of other dreams, as can events in our waking life. In this connection, Newton lists what appear to be items that might have come from a book on the art of memory, namely, a concatenation of things to be recalled by association, such as words uttered and things seen together, or things that go naturally together such as length and breadth, meditations and actions (presumably meditations concerning the actions). Newton seems to express the view that things not coming directly through the senses, such as meditations, thoughts, dreams, and recollections, are distinctly remembered (i.e., cases where the mind or soul is chiefly involved). The recollection of something, such as an external object, is aided by thought about that thing. Newton is thus presenting these 'facts' as an argument that memory is located in the mind or soul. These arguments, which are Morean in character, are interesting insofar as they show us how Newton argues that some things lie outside the physicalist program. Memory cannot be moved outside the soul to anything near the degree that the senses can, and this is clearly the contrast being drawn by Newton. Things not thought about are forgotten; "A man cannot remember what he never thought upon, as a blow or prick or noise in his sleep, the things and sounds which he hears and sees but minds not."

Newton denies the possibility of accounting for memory by "characters in the brain," that is, that there exist any signs in the brain that designate, and thus allow us to recall, past occurrences, and so forth. Such theories are question-begging, or lead to infinite regress, because, as Newton puts it, "the soul . . . must remember those characters." In repudiating physicalist views of memory, Newton does not spare Hobbes, the source of his own physicalist program. On folio 85 130^r he writes:

Hobbes, Part 4th, Chapter 1st. Motion is never the weaker for the object being taken away, for then dreams would not be so clear as sense. But to waking men things past appear obscurer than things present, because the organs being moved by other present objects at the same time, those phantasms are less predominant.[23]

Newton raises two further objections to the Hobbesian account. First, if memory is a motion in the brain, our memories must remain constantly before us until that motion ceases; and once stopped, we will be unable to remember, because we have "no principle within us to begin such a motion again." His second objection is that we would be unable to distinguish between "sense and fantasy," because they could be constituted of equal motions. This second objection appears to allow that a quantitative difference in the motions that are sense and memory will serve to individuate them, a view in conformity with his position that their quantity of motion can be the distinguishing feature among colors and sounds. His objection simply asserts that if these motions were in fact equal in a given case, then sense and memory could not be distinguished.

[23] See footnote 144 in the Transcription.

5

THE ORIGIN OF NEWTON'S
OPTICAL THOUGHT AND ITS
CONNECTION WITH
PHYSIOLOGY

One of the sciences closely associated with Newton's name is optics, and the *Questiones* provide ample evidence of his early interests. Of significance is the fact that they allow us to construct a plausible account of the paths that led Newton into optical theorizing. We shall advance the following thesis: Newton's interest in the central questions of optics emerges from his physicalist program in physiology and perception. To be specific, his optical speculations began in the physiological context of color perception, and he moved on to consider the phenomenon of colors as seen through the refracted light of a prism. A detailed defense of this thesis will be given in Section 2. That section will also examine Newton's first 'mechanical' account of the production of the spectral colors, as well as his initial treatment of the conception of "unequal refractions." In this regard, we shall indicate that the prismatic spectrum that Newton records on 69 122^r is the one produced by the 'boundary colors,' a fact that Newton never recognized even in his subsequent use of the prism experiment. In the third section we shall consider the information on colors that Newton gleans from Boyle. Of significance is the close relationship between the optical entries in the *Questiones* and a manuscript entitled "Of Colours" that dates from 1666 (see the Appendix). In this manuscript, Newton sets forth the first account of his mature theory of colors, the origins of which we shall attempt to explain.

1. Some optical observations

At 14 94^v, 23 99^r, 25 100^r, 32 103^v, and 34 104^v Newton records observations on the nature of light. Some of these entries appear to have no obvious origin in any of the texts that Newton is known to have read in early 1664 (Charleton, Descartes, and Boyle), whereas others can be traced to one or other of these writers. Under the heading "Of Perspicuity and Opacity" Newton asks why different substances and media transmit light more readily than others, why being in a wet or dry state makes a difference in a substance's transparency, and why oil that is "less diaphanous than water, yet . . . makes a paper more diaphanous than it." It is probable that these queries were suggested by Newton's response to Descartes's views on transparency in the *Meteorology*, for he again asks why water is clearer than vapors.[1] On 23 99^r Newton continues in the same vein, now asking about the way in which different surfaces and media reflect and refract light. Some of these queries have their origin in Boyle's *Experiments Touching Colours* and *The Spring of the Air*. Why does the temperature of water make a difference to the refraction of light? Why does the silvering of a surface allow it to reflect light better? Is there reflection from a "clear glass" in a vacuum? Because there is refraction in a vacuum, is it caused by the same subtle matter that refracts light in the air? Does glass placed in Boyle's receiver refract light as it does in the air?[2] Again he asks if light has smaller parts than the air, and whether or not the air consists of branchy particles that merely touch one another (25 100^r). These questions are clearly suggested by the *Meteorology*.[3]

On 32 103^v he asks the sort of question that is characteristic of his later optical investigations. If light is absorbed less by white paper than by black paper, why is it reflected more readily from the former? The fact that a black surface readily absorbs light is easily verified: "For hold a paper between you and light with a black spot on it and it is blacker when towards you than when to the light." Has light

---

[1] Descartes, *Opera*, p. 162; AT.VI.658–659, Cap. II, Article III.
[2] See footnotes 56 and 57 in the Transcription.
[3] Descartes, *Opera*, p. 154; AT.VI.652, Cap. I, Article III.

the pressure to move an object as the wind can move a sail? Does light move with a finite speed? On this score, Newton suggests a simple if naive experiment: Observe how long after the flame of a candle is extinguished its image disappears in a fixed glass that is observed at a distance through a telescope. Newton's interest in the speed of light may well have been aroused by Descartes's claim that it propagates instantaneously.[4]

The entry entitled "Of Species Visible" on 34 104ᵛ probably has its origin in the Cartesian corpus. Newton's "globulus of light" and the reference to the subtle matter are reminiscent of Descartes's second element of the *Principia*, and his description of the globule as a ball seems to suggest the *Dioptrics*.[5] The purpose of the entry is clear enough: to offer an explanation for the propagation of light. Newton's mechanism is a straightforward antiperistasis. The subtle matter that is displaced by the forward thrust of the globule flows behind the globule to propel it forward. If Newton's account was suggested by Descartes, he again interprets the latter's approach to light within the framework of mechanical causation in a fluid medium. It is also clear that Newton took antiperistasis seriously when he wrote this entry. He also appears to consider antiperistasis as a plausible mechanism on 59 116ʳ. There he reproduces Glanvill's example of the wheel. If the wheel is "divided into 24 parts by the 24 letters, *a* cannot move before *b* nor *b* before *c* and so forth to *z*."

Our analysis of Newton's handwriting indicates that these entries are among the earlier in the *Questiones*. As we have seen, Newton's acceptance of antiperistasis was short-lived. "Of Violent Motion" is a systematic diatribe against it in favor of motion in a void. This means that when Newton wrote "Of Violent Motion" he had given up antiperistasis as an explanation of projectile motion, and also as a means of

---

[4] *Ibid.*, pp. 50–1; AT.VI.585–586, *Dioptrics*, Cap. I, Article III. Remember that for Descartes, light is a tendency to motion. Thus, he does not hold that light is an instantaneous motion, but rather an instantaneous communication of movement. See A. I. Sabra, *Theories of Light from Descartes to Newton* (London: Oldbourne, 1967), Chapters I and II, for a discussion of Descartes's theory of optics.

[5] Descartes, *Opera*, p. 69, AT.VIII.107, *Principia*, Part III, Article LIV. *Opera*, pp. 53–4; AT.VI.588, *Dioptrics*, Cap. I, Article VIII.

explaining the propagation of light. But it does not mean that he rejected antiperistasis as a plausible mechanism by which to explain 'localized' motions that involve the action of a subtle matter. On 71 123^r he appeals to antiperistasis notions in explaining the behavior of colored rays in the pores of bodies. Entry 6 speaks of the "elastic power of the subtle matter whereby the motions of the rays are conserved. . . ." And in entry 7, Newton speaks of the subtle matter not being "on either side of it [a globule] to press it toward its hinder parts, so much as it is pressed before." It seems reasonable to suggest that in these entries Newton is invoking an antiperistasis model in order to explain 'localized' motions within bodies. Not only is there evidence that some of Newton's optical ideas are among the earlier in the *Questiones,* but there is the possibility that he had some early knowledge of Descartes's theory of color (i.e., the theory as found in the *Dioptrics,* in contrast to the celestial optics of the *Principia*). As we shall see, this possibility involves some complications.

The entry "Of Colors" on 36 105^v is important. It is again among the earlier in the *Questiones,* given the evidence of the hand. We have already discussed Newton's two possible explanations for the appearance of dark objects as farther away than light ones (see Chapter 4, Section 3). Notice here that he associates a 'light' body with a 'white' body, which seems to indicate that he had already begun to dissociate whiteness from color. The true significance of this is apparent at 69 122^r. If we are right in suggesting that Newton conceives whiteness as the measure of the amount of light (i.e., the number of particles per unit time that strike the eye), he has already rejected it as a property of reflected light. The next entry, concerning some plausible causes of color, may have been suggested by Charleton's survey of various ancient and modern opinions on the subject in the *Physiologia.*[6] In particular, Newton's statement that colors may be caused by "parts of the body mixed with and carried away by light" (36 105^v) is probably indebted to Charleton's discussion of 'visible species.'[7] That is to say, Newton's talk

---

[6] Charleton, *Physiologia,* Book III, Chap IV, Sects. I–III, pp. 182–97.
[7] *Ibid.,* Chap. II, Sect. I, pp. 136–45.

of the 'parts' of a body seems to refer to the ancient notion that 'simulachra' or corporeal 'effuxions' are given off by objects, and then propagated through a medium. We have previously noted Newton's interest in such mechanisms in his essay on violent motion (see Chapter 3, Section 2). The other two suggested causes also have their origins in traditional discussions, the first being a statement of the claim that colors arise from modifications of light and dark.[8] What is strikingly evident is the fact that Descartes's theory of color is not cited, even though Charleton gives a brief account of it.[9] This is the more surprising given Newton's interest in Descartes's scientific writings and the fact that Descartes's views on light and color are so evident in the *Meteorology* and the *Dioptrics*. It seems that Newton had not yet come to terms with Descartes's theory when he made this entry, even though "Of Species Visible" appears to indicate some knowledge of the *Dioptrics*. It may be that it is a comparatively later entry, if compared to the earliest on optical matters.

The following observations are important (36 105ᵛ). Among other things, Newton distinguishes the effects of splendor and dullness and those of color, blackness, and whiteness. Although he begins by relating them to the possible causes of colors he cites, Newton quickly decides that the first two effects are the result of more and less perfect reflection, his example chosen to illustrate the distinctions he wishes to make. Newton asks: "Why are coals black and ashes white?" If colors are produced by "stronger and weaker reflections," coal ought to reflect light more strongly than ash, and so ought to be white and the ash black. But this is against the observation of the phenomena. Similarly, Newton denies that colors result from a mixture of "pure black and white." Were this so, any combination of light and shade, such as an ink drawing or printing on a light surface, ought to produce the sensation of colors. In the midst of examples designed to raise difficulties for the first two accounts of the causes of colors, Newton provides examples to indicate the difference that refracted light makes to the colors with which objects are perceived. Again, it is surprising

---

[8] *Ibid.*, Chap. IV, Sect. III, pp. 191–3.   [9] *Ibid.*, p. 197.

that there is no mention of Descartes's theory of refraction. At this stage, it is clear that Newton had not made much progress toward a theory of color. Nevertheless, the objections he raises against the received view are well founded.

2. The boundary-color phenomenon and the causes of color

Folios 69 122^r and 70 122^v contain a report of a prism experiment performed by Newton and his attempt to explain the phenomenon he observed. The reference to a two-prism experiment to determine if the casting of two spectra on top of each other in a reverse order of colors will produce a white is probably drawn from Boyle (see folio 74 124^v and footnote 134 of the Transcription). Judging from the handwriting and the way in which it is squeezed onto the top of the page, the entry was probably entered above the existing diagram at a later date. In the experiment described on folio 69 122^r, Newton used a single prism as pictured (it is likely that he owned only one prism at the time).

The experiment consists of the observation of the horizontal boundary between two differently colored objects as seen through a prism. What Newton observed is the phenomenon now called boundary colors, which was to become the center of Goethe's subjective color theory of 1810.[10] Boundary colors arise whenever lighter and darker objects abut each other in the field of vision and their interface is viewed through a prism. There results an area of warm colors (red, orange, and yellow) over that part of the lighter object near the boundary, and the cool colors (blue, indigo, and violet) over the adjacent dark area. The colors are displaced by refraction, and green is absent. This is characteristically different from what is now called the "Newtonian spectrum." In this case, white light is passed through a prism, and the resultant colors are cast onto a surface; characteristically, green is among the colors present. From Newton's experimental arrange-

[10] See Arthur G. Zajonc, "Goethe's Theory of Color and Scientific Intuition," *American Journal of Physics*, 44(1976):327–33. Zajonc gives a good account of the relations between the 'Goethean spectrum' and the 'Newtonian spectrum.' See also P. J. Bouma, *Physical Aspects of Color* (New York: St. Martin's Press, 1971), Chapter VII, pp. 126–34, for a fine account of the nature of the boundary colors.

ment on folio 69 122^r it is clear that the prism is used to collect light as it comes *from* a surface.

By systematically reproducing this experiment, we have been able to verify all but one of Newton's results as they are reported beneath the diagram and in the table of folio 69 122^r. The case of blue above and black below should indicate red, not "Greene or Red," against it. Given the cancellation in the two preceding cases, it is possible that Newton intended to cancel the green. It is to be noted that *eodc* (and *cdqp*) are extremely narrow bands and are grossly exaggerated in Newton's diagram. The same experiment is described on folio 4 of Add. 3975 (Appendix), though the results given there are more general and summarize the results of folio 69 122^r by dividing them into two cases: (1) lighter above the boundary and (2) lighter below the boundary. The first gives rise to a red boundary (*eodc*), the second to a blue.

Newton writes at 70 122^v that slowly moving rays are refracted more than swift ones, and he attempts to explain the results of his experiment in terms of this 'fact.' In Newton's account it is the prism that separates the slowly moved rays from the swift ones, so that "two kinds of colors arise, viz: from the slow ones blue, sky color, and purples; from the swift ones red, yellow." Although this account probably originated with Newton, the association of color with speed may have been suggested by his reading of Descartes. It is true that the latter believed the production of colors to be associated with the differing rates of rotation of the corpuscles of the subtle medium; nevertheless, when he discusses refraction, Descartes associates the greater or lesser speeds of corpuscles with the condition of density in a given medium in order to account for refraction.[11] So it is not im-

[11] For the theory of colors in connection with the rotation of the corpuscles, see *Opera, Meteorology*, pp. 216–18; AT.VI.702–703, Cap. VIII, Article VI. Descartes holds that the rotation of the globules along the red ray is accelerated, while that along the blue ray is retarded. The other colors are accounted for by supposing that these rotary speeds range between those of red (the greatest) and blue (the slowest). For Descartes's account of refraction, see *Opera, Dioptrics*, pp. 62–3; AT.VI.594, Cap. II, Article IX. For a discussion of Descartes's claim that light is communicated faster through a dense medium like water or glass than through air, see Sabra, *Theories of Light*, Chap. 4, pp. 105–16, and Chap. 12, pp. 299–302.

probable to suggest that Newton is attempting to simplify
the Cartesian approach to the production of colors by using
the parameter of speed to account for both refraction and
colors. In any event, he would have every motivation for
focusing on speeds, given the central role that motion plays
in his physicalist program for the physiology of vision and
perception (see Chapter 4, Section 3). On 70 122^v Newton
asserts that green arises from those rays that are neither
very swift nor very slow, and from a mingling of the slow
and swift there arise white, gray, and black. As has been
pointed out, there is in fact no true green in the spectrum
of colors produced by the boundary-color phenomenon. A
green can be seen when the prism is drawn back and the
bands of yellow and blue widen sufficiently to overlap.
There is no evidence, however, that Newton produced a
green in this way. But he did get a green with an interface
of black above and blacker below, a result that we have
confirmed by reproducing the experiment. We also noted
that at the boundary *eodc*, both green and a dark nonbril-
liant red can be observed. Newton does not appear to be
aware of the great difference between the phenomena of
colors as produced by looking through a prism and the
phenomena of colors that are cast by a prism through which
white light has been sent. He seems to think the only difference is the order of the colors, because he writes on folio 4
of Add. 3975, "Prismaticall colours appear in ye eye in a
contrary order to y^t in wch they fall on ye paper." But in
Newton's view, this does not constitute a real difference,
because nothing more is involved than looking at the colors
from first one side and then the other of the paper.

Nonetheless, Newton's general notion of how the phenomenon arises is correct. He thinks that the degree of
refraction makes the rays appear to come from a location
different from their actual source. Newton writes, "If *abdc*
be shadow and *cdsr* white, then the slowly moved rays coming from *cdqp* will be refracted as if they come from *eodc*."
The slow-moving rays (the blues and purples) are refracted
more, so that they appear to come from above their actual
source. On the other hand, the swift-moving rays (the reds
and yellows) are refracted less, so that they appear to come

from below their actual source. Consequently, if a white surface (or a light surface) is viewed through a prism, the blues and purples are seen above, whereas the reds and yellows are seen below. And it is apparently Newton's view that the same holds for a black surface (or a dark surface). When the results of Newton's experiment as reported on folio 69 122^r are checked, the case already mentioned of black above and blacker below yields a green, as well as the expected dark red, the green predominating the red. This fails to fit his hypothesis—because there is nothing to indicate that greens are to be expected.

When he wrote the entries on folio 4 of Add. 3975, Newton had found a simpler formulation of his hypothesis (i.e., one that treats all cases of a dark color with a lighter, or a darker with a dark, or a light with a lighter, as cases of black with a white). It is clear from the analysis thus far that Newton's commitments in no way constitute the view that colors result from a modification of light and dark, a point to which we shall return. In fact, Newton is surprised that the generalization of folio 4 is possible. In particular, he finds it strange that even an extreme case like black and blacker can be dealt with as white and black. He writes on folio 4 of Add. 3975, "And this will happen though ye colours differ not in species but only in degrees, as if acdb bee black, & cdsr darkness or blacker y^n abdc ye edge dc will bee red and much more conspicuous y^n ye black, wch is strange." It is interesting to note that Newton has conveniently omitted the green that, as we have indicated, was so troublesome in this case.

At the bottom of folio 69 122^r Newton writes, "The more uniformly the globuli move the optic nerves, the more bodies seem to be colored red, yellow, blue, green, etc. But the more variously they move them, the more bodies appear white, black, or gray." This passage, in connection with his claim on folio 70 122^v that the sources of specific colors are to be identified with specific speeds of the rays, and white, blacks, and grays with rays of mingled speed, strongly supports the view that at the time these folios were written, Newton had *a* theory of white light (and gray and black) as a commingling of colors, or, put properly, that the cause of our sensations of white light is the commingling of the causes of sensations of colors. Furthermore, he already believed that the prism separates these 'causes' by differential refraction. There is, of course, no notion that the property of the ray that causes the color is intrinsic to it. Indeed, because that property is speed, it is plainly *not* intrinsic.

It must also be noted that the principle that blue rays are slow and red rays are swift appears to have emerged from Newton's physiological investigations of vision, particularly from his eye-pressing experiments in which greater pressures produce redder images and lighter pressures produce bluer images.[12] This, combined with the observation that blue light is more refracted than red, led Newton to his additional principle that slow rays are more bent by prisms than swift rays. Thus, Newton's physiological investigations are a major source for his theory of colors in the *Questiones*.

In entry 3, Newton suggests an experiment that should not be misinterpreted. If a blue-and-red thread is placed against a dark background and viewed through a prism, the thread appears to be divided or broken, the blue segment of the thread appearing to be displaced more than the other, "by reason of unequal refractions." This experiment is merely a further instance of the principle that the more slowly moved rays are refracted more. Thus, it provides no evidence for the view that Newton already conceived white light as a heterogeneous aggregate within which colors are present intrinsically[13] (that is the mature theory in which the rays that produce various colors are immutable); here,

[12] It is, of course, true that Descartes holds that red rays are the swiftest and blue the slowest. See footnote 11. But there is no need to suppose that Newton is indebted to Descartes's *Dioptrics* for this point. Given that he is already commited to a physiological program based on motions and speeds as causes, the experiments he performs are sufficient (given his knowledge of refraction) to confirm the principle in his mind. Also, if he had really been involved in Descartes's framework for explaining colors at that time, it is doubtful that Newton's program would have the conceptual basis that it in fact has in the *Questiones*.

[13] Ever since A. Rupert Hall's pioneering paper on the *Questiones*, "Sir Isaac Newton's Notebook, 1661–65," *Cambridge Historical Journal*, 9(1948):239–50, scholars have tended to see this experiment as an intimation of Newton's mature theory of refraction. Recently, Richard S. Westfall has come close to affirming this claim in *Never At Rest*, Chap. 5, pp. 159–61. It is actually affirmed by John Hendry in "Newton's Theory of Colour," *Centaurus*, 23(1980):230–51. Sabra, *Theories of Light*, emphatically denies the claim (Chap. 9, pp. 246–7).

Newton simply associates rays with speeds, a parameter than can be altered if they interact with bodies.

We have seen at 32 103^v and 36 105^v that Newton is concerned with colors as they appear to arise from the surfaces of bodies and as they are affected by the inner structures of objects. This is clearly a concern of the present entries, and at entry 4 of 70 122^v Newton states a hypothesis concerning the manner in which colors "are made in bodies." His hypothesis is no doubt prompted by Boyle's distinction between colors as they come from the surfaces of objects, in contrast to colors as seen through an object.[14] A body manifests red or yellow because the swift rays that produce these colors are not hindered as much by the body as are the slower-moving rays. The opposite is true when the body is blue, green, or purple. It is clear that Newton supposes bodies to have the capacity to alter or 'modify' the speed with which rays are reflected. This should not be seen as evidence that he adheres to a modification theory of colors. On the contrary, the colors that are produced as light affects the eye are the result of rays moving with various speeds. But these speeds are affected by the textures of bodies, so that there is no unique speed always associated with any given ray.[15] But as we have just emphasized, this is not tantamount to Newton's later claim that color is immutable, original, and connate with the ray.[16] Newton's belief that the colors of bodies are "a little mixed" (folio 4, Add. 3975) leads him to propose the view that "in some bodies all these colors may arise by diminishing the motion of all the rays in a greater or less geometrical proportion. . . ." For example, if we take the speeds of the incoming rays to be in the proportion $2 : 3 : 4 : 5 : 6$, and so forth, and reduce this by the geometrical proportions $\frac{1}{2}, \frac{1}{4}, \frac{1}{8}, \frac{1}{16}, \frac{1}{32}, \ldots$, there results a ratio of the speeds of the rays $1 : \frac{3}{4} : \frac{1}{2} : \frac{5}{16} : \frac{3}{16}$, and so forth. This reduces the

---

[14] See entry 47 of folio 94 134^v and entry 51 of folio 95 135^v and footnotes 176 and 179 in the Transcription.   [15] Sabra, *Theories of Light*, p. 247.

[16] Add. 3975 presents Newton's mature theory of colors (1665–6). However, the first time that he uses such descriptors for his theory occurs in 1669–70 in his *Lectiones opticae*, see Lectiones 6, folio 55^v, in *Isaac Newton's Cambridge Lectures on Optics, 1670–72*, edited by D.T. Whiteside (facsimile of ULC ms. Add. 4002) (Cambridge: The University Library, 1973).

differences among the speeds of the rays constituting the colors of the body such that its predominant color is produced by speeds close to unity.

Entries 5, 6, and 7 on folio 71 123ʳ attempt to give a physical account of the causes of colors in bodies, and in entry 8 Newton attempts to provide a quantitative model of the interaction of incoming rays with the particles that exist in the pores of bodies. In each of these entries Newton conceives bodies to have interstitial pores within which there reside particles and a subtle matter. As we have already indicated, Newton conceives the action of his subtle matter in terms of an antiperistasis, restricted to a body's interstitial pores. It is not improbable to suggest that Newton's 'medium' is modeled on Descartes's subtle matter as outlined in the *Meteorology*.[17] But, unlike Descartes, Newton characterizes his subtle matter as having an "elastic power." It is well to recognize that although Newton provides hints in other entries, this is his first attempt to give a detailed physical explanation of what causes the colors of natural bodies.

But it should also be recognized that Newton's model for explaining the causes of colors relates to the nature of his physiological investigations of vision. Notice that in the present entries Newton refers to the effects produced on the subtle matter by the narrowing of the pores of bodies, and he also refers to the constricting presence of the loose particles within their pores. These factors, in combination, have a characteristic physical effect (specific to each body) on the incoming globules (corpuscles) of light. It is this that also constitutes the colors of the body when viewed by a normal observer.

What does Newton require of a normal observer? As has been noted, the motion of the incoming ray is transferred to the retina, where, in turn, the motion is transmitted along the nerve tubes to the sensorium (see Chapter 4, Section 3). It is here that the "motional pictures" are 'viewed' by the soul. It is essential in this process that the speed that characterizes the color as it comes off the body be conserved, because any modification will alter the color seen. Newton

[17] Descartes, *Opera*, pp. 154–6; AT.VI.652–653, *Meteorology*, Cap. I, Articles III, IV, V, and VI.

writes on folio 20 of Add. 3975, "granting mee but that there are pipes fill'd wth a pure transparent liquor passing from ye ey to ye sensorium & ye vibrating motion of y^e æther will of necessity run along thither. For nothing interrupts that motion but reflecting surfaces, & therefore also y^t motion cannot stray through ye reflecting surfaces of ye pipe but must run along (like a sound in a trunk) intire to ye sensorium." In the normal observer, according to Newton, the tubes of the nerves are open and unrestricted, allowing for the conservation of motion.

We have already established that Newton merges the physiology of vision with a theory of color in the *Questiones*. The foregoing passage is evidence that the interrelationship between physiology and optics continues to direct Newton's thought. It should be noted that in the case of natural bodies, colors are carried by the motion of globules, whereas in the tubes of the nerves they are carried by the vibratory motions of the ether. Nevertheless, both sorts of motions are viewed by Newton under the same model. In each case, uniformity of the passageways is required for uniformity of the motion. Further evidence that Newton applies the same model to the physiology of vision and to natural bodies is implied by the following passage: "From ye whitenes of the brain & nerves the thicknesse of its vessells may be determined & their cavitys guessed at. And its pretty to consider how these agree wth the utmost distinctnesse in vision" (folio 20, Add. 3975). Here there is every indication that a narrowing or constriction of the nerve tubes will lead to a diminution of vision. And still further evidence of the use of the model is to be found in the *Questiones,* where on folio 35 105^r Newton writes, "Dimness may come from the deficiency of these spirits, and the optic nerve obstructed." Although there is no indication here of a change in color as a result of a diminution of motion, we have seen earlier that Newton does not believe that every change in speed produces (or constitutes) a change in color, but only a sufficiently great one. Lesser changes may only alter the amount of light or, more properly, the intensity of the effect on the sensorium. The model employed, however, is clearly the same.

It should not be supposed that the interrelationship that Newton perceives between physiology and optics is merely contingent. On the contrary, he believes that the study of light can be established by a study of the organs of vision and that the nature of light helps to explain the workings of sight. On folio 81 128^r we are told that the design of creatures cannot happen by "fortuitous jumblings of atoms," and the design of the eye in sighted creatures is stressed. This teleological sensibility is present not only in the *Questiones* but also in Newton's mature thought. In a manuscript written in the 1690s he says that the God who "framed ye eyes of all creatures understood the nature of light & vision. he that framed their ears understood ye nature of sounds & hearing, he y^t framed their noses understood ye nature of odoars & smelling, he that framed the wings of flying creatures & ye fins of fishes understood the force of air & water & what members were requisite to enable creatures to fly & swim. . . ."[18]

Let us turn now to the details of Newton's account of the causes of colors in natural objects. In entry 5, folio 71 123^r, he conjectures that if the elastic power of a body is unable to return the whole of the incoming ray, "then that body may be lighter or darker colored according as the elastic virtues of that body's parts is more or less." In other words, if the sensation of color is a matter of which rays predominate in the sensorium, their speeds are in some measure a function of those 'elastic virtues' of the parts of the body that reflect the various rays. Similarly, in entry 6, Newton suggests that if the elastic power is impeded in the same way, depending on the nature of the obstruction, the speed and hence the color of the rays will be affected. He indicates that an obstruction can occur by the constricting of a pore or by the presence of "loose particles."

Entry 7 is of greater interest and must be read in conjunction with entry 8. Here Newton attempts to explain both reflections and transmissions in terms of his model of bodies as porous objects into which rays of various speeds can penetrate. At this point he makes explicit reference to

[18] ULC ms. Add. 3970, folio 4 79^r.

the capacity of the subtle matter (which resides in the pores of bodies) to act by local antiperistasis on an incoming globule of light. Newton tells us in reference to his diagram that *ce* is "too much straitened." It is clear that the pore is 'constricted' at that place, in order that the capacity of the subtle matter be characteristically affected to act on a globule that is passing through. As the globule passes through the constricted pore, it separates the subtle matter before and after, leaving "no matter on either side of it to press it toward its hinder parts, so much as it is pressed before." Given these conditions, there are two possibilities. If a globule has "force" enough to get through the constricted pores, its motion will be reduced by the subtle matter that is compacted in its path; if it has not, it will be reflected without any considerable loss of its incoming motion. Newton attempts to illuminate the second possibility by drawing an analogy with the behavior of rays as they pass from glass into air. Owing to the oblique 'grazing' angle at which they enter the air, they lose little motion along their reflected paths.

Newton next considers a related situation. He tells us that "bodies full of such straitened passages in their pores must be of dark colors. . . ." But here he considers constriction to be a function of the size of the "loose particles" that lie within the pores. And he has in mind pieces of glass (which are otherwise the same), in one case being stained or colored blue, and in the other tinted red. According to Newton, the difference in their colors is to be explained by the sizes of the particles that lie in their pores. The interstitial particles that help to produce a blue tint in a piece of glass are larger than those that are involved in imparting a red or yellow tint. Thus, the blue tint results from the larger particles in the pores slowing down the passage of the incoming rays, so that the glass appears blue if it is looked through. But if the pore is a little larger, the "slowly moved particles designated *o* in the diagram above pass through with loss of most or all their motion." In this case the more swiftly moving rays will pass through easily, and the glass will have a red or yellow tint. Newton likens this situation to "a glass whose pores are full of smaller particles of the tincture than are those of the blue glass."

Although the features of Newton's model are straightforward enough, it fails to give him control over the phenomenon that he wishes to explain. Constriction can arise when the size of a pore is reduced, either by narrowing the pore itself or by increasing the size of the particles residing in it. At the beginning of the entry Newton refers to the way in which antiperistasis is affected by a narrowing of a pore, whereas in his explanation of the blue and red tinctures of glass he appeals solely to the way in which the size of the interstitial particles of the pores affect the transmission of the rays. It seems clear that he wishes to consider both 'mechanisms' in his explanatory arsenal. But whichever is considered, it is not clear why the incoming particles of certain rays should lose "most or all their motion" in passing through *ce* at the same time as the swifter rays pass more freely through the enlarged pore. The passage of the pore is enlarged because the interstitial particles are smaller in size. Presumably the rays that cause the blue tint are slowed because of the action of the subtle medium, although Newton does not invoke the medium in this part of his argument. They certainly cannot be hindered by the smaller interstitial particles. But if it is the action of the subtle medium that slows the rays that cause blue, what effect does it have on the faster-moving rays that cause red? All that Newton states explicitly is the role of the smaller particles in allowing the faster rays to pass "more easily." Again it is not clear why the presence of smaller particles should slow those rays that are responsible for blue. It is clear that Newton's explanatory model suffers from vagueness. What is a ray of light? If rays are composed of particles or globules, what is the relationship between their size and speed? And what is the relationship between their size and speed and the size of the interstitial particles that they meet in the pores of bodies? Newton begins to consider these questions in entry 8.

Whereas entry 7 (folios 71 123^r and 72 123^v) considers the roles of differently sized interstitial particles, entry 8 begins by stating that if "two rays be equally swift yet if one ray be less than the other that ray shall have so much less an effect on the sensorium as it has less motion than the other." It is evident that a ray is 'lesser' than another just in case its incoming particles are smaller in size. Equally evident from Newton's argument is his concern to give a quantitative account of the relationships between the sizes and speeds of the rays before and after transmission, given that their incoming speeds are equal. Again, Newton is concerned to calculate the effect on speed and size, on the supposition that there is a reduction in the size of a body's pores. In the case he considers, he takes the particles in the pores to be the same size, so that each pore is reduced or constricted by the same amount. Given that the incoming rays are of the same speed, but composed of differently sized particles, what will be the individual effects on their speeds after transmission, that is, after each has collided with or squeezed by the interstitial particles of the same size? Newton tells us that the particles in the pores bear "a proportion of the greater rays as 9 : 12; and the lesser globulus is in the proportion to the greater as 2 : 9." Let the particles in the pores be designated a, and let the particles of the larger and smaller incoming rays be b and c, respectively. We are told that $\frac{a}{b} = \frac{9}{12}$ or $\frac{27}{36}$ and that $\frac{c}{d} = \frac{2}{9}$ or $\frac{8}{36}$; consequently, $\frac{c}{a} = \frac{8}{27}$. Newton tells us that b, "by impinging on such a particle, will lose a $\frac{6}{7}$ part of its motion, and the lesser globulus will lose $\frac{2}{7}$ of its motion. . . ." If both the incoming rays c and b have the same speed, then in each case this can be designated $\frac{7}{7}$, or unity. However, c, on colliding with an a, loses $\frac{6}{7}$. It follows that the remaining motion of c is $\frac{5}{7}$ and that of b is $\frac{1}{7}$. Newton then notes that $\frac{5}{7} : \frac{1}{7} :: 9 : 1\frac{4}{5}$. In other words, there is an inverse relationship between the sizes of b and c and their remaining speeds after colliding with an a; in fact, the b's are nearly four times the size of the c's, but their speed is roughly five times less. In this case, a great deal of speed is lost, so that the ray is now slower-moving, "and such a body may produce blues and purples." However, if the interstitial particles a are of the same size as the incoming c's, the latter will lose all of their motion, but the b's will lose only $\frac{2}{11}$ of theirs; in this case, very little is lost, and "such a body may be red or yellow." It is not clear why, if they are the same size as the interstitial particles, the incoming c's will lose all of their motion. Newton may have some sort of balance principle in mind, whereby the motion of the c's is entirely counteracted by the inner particles. If so, there is no more reason to think that they will be stopped than to think that they will be reflected back with a speed equal to the incoming speed. To say that the b's will lose only $\frac{2}{11}$ parts of their motion is arbitrary, given Newton's model. He probably chose the fraction to illustrate the small resultant loss of motion in the b's after colliding.

Despite its apparent simplicity, Newton's model suffers from a certain vagueness. He speaks of a "greater" ray and a "lesser" ray, of a "greater" and a "lesser globulus," and "motion" is used as a vague umbrella term. One conclusion is certain: Newton is not appealing to determinate rules of collision, such as those found in Descartes's *Principia*.[19] In the absence of evidence in the *Questiones* to show that he had begun to analyze Descartes's theory of motion in detail, this is what we would expect. Nor is Newton attempting to reduce the colors of bodies to a theory of elastic collisions of his own devising. At that time, he was not in the possession of such a theory, and its creation was over a year away.[20] In any case, the concept that he is working with in entry 8 is merely the simple notion of an inverse relationship between the size and motion of a ray after it has squeezed by an internal obstacle. It is clear, too, that Newton conceives the distinction between a "greater" and a "lesser" ray in reference to the sizes of the globules that compose them. In view of this, it is clear enough that the term "motion" is used to refer to the combined effect that size and 'speed' have in the production of a perceived color. In other words, the resul-

---

[19] See the analysis in Chapter 2, Section 2. The character of Newton's handling of colliding bodies in entry 8 provides further evidence that he had not yet formulated rules of impact, whether based on Descartes or otherwise.

[20] This claim is made by Zev Bechler, "Newton's Search for a Mechanistic Model of Colour Dispersion: A Suggested Interpretation," *Archive for the History of Exact Sciences*, 11(1973):1–37. Bechler quotes part of entry 8 and then says "I wish to draw attention to the significance of the fact that being, at that very date, already in possession of the basic law of elastic collisions . . . Newton at once set out to construct a model of the invisible realm of light corpuscles. . . ." Bechler also suggests that Newton is drawing on the mechanics of the *Waste Book* (later in time), but there is no evidence in the *Questiones* that Newton possesses any laws of elastic collision. Given the nature of the argument in entry 8, none is needed; physical intuition is sufficient, something Newton has in abundance.

tant 'motion' of a ray includes in its description the impact
that it has on the sensorium.

This raises an interesting question: Is a specific color in any
way associated with the size of the globules that comprise an
individual ray? In order to answer this, let us survey Newton's argument thus far. He begins on folio 69 122^r by considering various colors as they are reflected from the illuminated surface *abrs*, and he views them through his prism. As
we have seen, he records the various combinations of colors
that arise from the displacement action of the prism. In
entries 1–6 Newton speaks of the different speeds of the rays
as they interact with bodies, but he does not explicitly associate globules or particles with them. Swift-moving rays produce red or yellow; slow-moving rays produce blue or
purple; those of intermediate speed are responsible for
green; slow and swift rays intermingled produce white or
black or gray. As regards bodies themselves, Newton speculates that they interact with light to produce their natural
colors either by the "elastic virtues" of their parts or by the
presence of either subtle matter or "loose particles" in their
pores. But it is also evident that Newton all along associates
particles with the rays of light. This is nowhere more evident
than in those entries in which the actions of rays on the eye to
produce colors are described in terms of "pressure," "force,"
or the giving of a "fiercer knock." At 36 105^v Newton distinguishes the lightness and darkness of colors by the number
of particles that strike the eye per unit of time. In entries 7
and 8 he considers two basic combinations of parameters in
his attempt to explain the natural colors of bodies as a result
of the interaction of rays on either the subtle matter or the
"loose particles" in their pores: (1) that the globules of the
rays are of the same size, but get reflected or transmitted
with various speeds; (2) that they hit the "loose particles" with
the same speed, but are transmitted with speeds inversely
related to their sizes. It is clear that Newton is already considering two important 'mechanisms' that could be responsible for the colors of bodies – reflection and transmission.
And if light colors are to be distinguished from dark by the
number of particles that strike the eye in a unit of time, it is
not surprising that Newton should want to consider the role

that the size of the globule plays in the production of specific colors.

The model in entry 8 is geometrical in character. It considers the changes in motion of differently sized incoming globules after they squeeze past an internal obstacle of certain size lodged in the pores of a body. On the hypothesis that the rays are initially "equally swift," the emphasis is on the way in which *individual motions* are altered, given that they are composed of 'greater' and 'lesser globules.' Thus, the size of the globule (as well as the size of the body's internal pores) is a factor only because it affects the resultant motions of the rays differently. It is clear, therefore, that 'size' is just one among those parameters that Newton attempts to introduce in his search for a more fine-grained account of what causes the colors of bodies. At this stage there is no point to the question whether he associates color with either the speed or the size of the rays.[21] It is not an 'either–or' situation. What is certain is that there is no unique speed always associated with a given ray. Nor is there anything in entry 8 to indicate that globules of a particular size are invariably associated with a given ray.

We can now return to our claim that Newton's theory of color in the *Questiones* is not a variant of the modification theory in any ordinary sense. It is tempting to see entries 1–8 as stating a dualist theory, according to which blue and red are in some sense principal colors, of which other colors are simply various mixtures or dilutions.[22] To be sure, Newton notes that the "slowly moved rays are refracted more than the swift ones," so that he can speak of two extreme colors, blue and red, respectively. But he also speaks of intermediate colors, which he associates with different speeds (such as green, which arises from rays that are neither very slow nor very swift), and of white or black or gray, which he associates with the intermingling of slow and swift speeds. The only sense in which Newton thinks of colors as being modified is in his belief that contingent cir-

---

[21] See Bechler, "Newton's Search," for a discussion of this problem in Newton's optical thought.

[22] This is the manner in which Westfall describes Newton's position in the *Questiones*. See *Never At Rest*, p. 171.

cumstances (pertaining to the structure of bodies) can alter the speed and size of the rays in reflection and transmission. This picture is very different from Hooke's dualist conception in the *Micrographia*, which can be classified as a modification theory. According to Hooke, "Blue is an impression on the Retina of an oblique and confus'd pulse of light, whose weakest part procedes, and whose strongest follows." Red is the impression of "an oblique and confus'd pulse of light, whose strongest part procedes, and whose weakest follows."[23] The various pulses that are focused between the extremes of blue and red produce by combination the effects of all the other colors, so that Hooke can say they result from the modification ('admixing' and diluting) of blue and red. Hooke's description of the pulses of light as possessing degrees of 'strength' and 'weakness' is based on his conception of the behavior of pulses as they are refracted through an optical medium. It is precisely this sort of descriptive language that does not apply to Newton's conception of swift and slow rays. Or, if it does apply, it refers to the number, speed, and size of the globules that impinge on the retina, a conception intimately connected in Newton's mind with his eye-pressing experiments, in which greater pressures produce redder images and lighter pressures produce bluer images. We have already noted that Newton probably derived the principle that blue rays are slow and red swift from these experiments. In this regard, entry 9 (73 124^r) is significant, because it argues that the degree of pressure determines the color we see: blue where the pressure is less, and red where it is greater (see Chapter 4, Section 3). Thus, when he distinguishes the reds, yellows, and greens produced when the globules move the optic nerves uniformly from those produced when they "more variously . . . move them" (69 122^r), Newton employs this principle, mindful of its foundation in his physiological experiments and guided by his conception of the causes of colors. Given this line of interpretation, there is no need to

---

[23] Robert Hooke, *Micrographia* (London, 1665), p. 64. For a discussion of Hooke, see Alan E. Shapiro, "Kinematic Optics: A Study of the Wave Theory of Light in the Seventeenth Century," *Archive for the History of Exact Sciences*, 11 (1973):134–266.

suppose that Newton is consciously contradicting some established theory of color such as Hooke's, according to
which light comprises pulses, and colors are confused impressions on the retina.[24] As a matter of fact, there is no
evidence that Newton had read Hooke when he wrote the
entries on 69 122^r to 73 124^r. These entries were written
into the *Questiones* about mid-1664 and were certainly written before the publication of Hooke's treatise in 1665. They
largely reflect Newton's own speculations on his experience
and experiments, together with the information he derived
from reading Boyle's *Experiments Touching Colours*. Although
there is reason to suppose that Newton's association of color
with the speed of the rays may owe something to Descartes's
treatment of refractive colors in the *Dioptrics*, at this stage he
had not begun to consider the phenomena of refraction and
dispersion as they are produced by light in passing through
a prism, or when it is transmitted through media of different densities. If seeking for the causes of colors is physical
optics, Newton begins his optical theorizing in that tradition. This is what we would expect on the basis of his physiological program, and given his adherence to the method
of analysis and synthesis. Newton's firm conviction that the
techniques of geometry are essential to a science of optics
was some years in the future.[25]

3. Newton's notes from Boyle and a third essay "Of Colours"

It is evident from the first entry "Of Colours" (36 105^v)
that Newton is interested in the conditions under which
light is reflected and refracted, and how the texture of a
body affects its behavior. We have seen that he clearly
distinguishes the properties of splendor, dullness, light,
and dark and denies that any color can arise from a mixture of "pure black and white." Already he had a richer
body of data to explain than was normally considered in
regard to the colors of natural objects. The 41 entries that

[24] Westfall thinks that Newton is consciously opposing Hooke's position in the
Questiones; Never At Rest, Chap. 5, p. 158.
[25] *Cambridge Lectures on Optics*, Lectiones 3, p. 23.

Newton has taken from Boyle comprise data concerning the ways in which the colors of objects are changed under a wide variety of circumstances. They record the effects of heat, the characteristics of various sublimates, acids, and precipitates, the ways objects change in various lights and positions, the effects of dyes, of solutions, and salts, and the changes wrought on colors by various combinations of these 'instruments.' Although Newton's aim is to increase his basis of information, the entries are more than a haphazard miscellany. Each bit of information pertains in some way to either the difference in the appearance of a body's color when it is looked on, in contrast to when it is looked through, or the ways in which a body's color can be changed. Thus, "Of Colours" and the notes from Boyle support the claim that Newton's interest in optics begins with the question of how natural objects are colored, and the physiology of how they are perceived.

This is how matters stood in 1664. Newton had satisfied himself that the data did not allow the colors of natural objects to be explained on the hypothesis that there is a graduation in the dilution of color, ranging from red (the color with the least admixture of darkness) to blue (that with the most, and the last stage before the extinction of light by darkness). Moreover, he had made progress in integrating into the same model the physiology of perception and the causes of colors. But it is obvious that he could not have remained satisfied with his progress. It did not provide consistent control over the data, and his attempt to quantify did not rest on any settled view regarding the parameters he used. There were also other facets of the "celebrated Phaenomena of Colours" to which Newton had not yet directed his attention.[26] In the first place, he had not yet considered the colored images that confuse the focusing of light in telescopic observations—what is now called chromatic aberration. How are they to be explained, and can they be eliminated? Second, he was still to consider Descartes's derivation of the sine law of refraction in the

[26] This phrase is from 1671–2, "New Theory of Light," *Isaac Newton's Papers and Letters on Natural Philosophy,* edited by I. Bernard Cohen (Cambridge, Mass.: Harvard University Press, 1958), p. 47.

Dioptrics.[27] This involves the projection of a prismatic spectrum on a screen in order to examine the colors.[28] Descartes had argued that spherical lenses do not refract parallel rays to a perfect focus, but that hyperbolic and elliptical lenses would do so, given his sine law of refraction.[29] However, these were among the questions that occupied Newton's mind as he turned to a more systematic examination of Descartes's views on refraction during the year 1665.

His source was again the 1656 compendium of Descartes's works, the edition from which Newton had gleaned his first knowledge of the Frenchman. As yet, the mature theory of color, according to which the individual rays of light are immutable in their properties and always manifest the same specific degree of refrangibility, had not been formulated. But the carefully noted 'fact' that "slowly moved rays are refracted more than swift ones" had deeply impressed Newton's mind.

Newton's first detailed treatment of refraction and his notes on the grinding of lenses are found in an early mathematical notebook.[30] These investigations are not dated; but they follow, and are in the same hand as, his annotations on Wallis's *Arithmetica infinitorum,* which can be dated in the winter between 1664 and 1665.[31] There is, of course, a close connection between any attempt to treat the optics of lenses mathematically and the problem of minimizing the effects of the differing refractive powers of incident light. Both investigations were inspired by Descartes's *Dioptrics,* and Newton computed a table of "proportions of y^c sines of refraction of the extremely heterogeneous rays" between different media, connecting the different refrangibilities with the 'speed' of the rays.[32] The table is based on Descartes's model in the *Dioptrics,* according to which globules of light in their "tendency to motion" at the interface of media of different densities preserve that tendency in the direction parallel to the interface, whereas in the perpen-

[27] Descartes, *Opera,* pp. 56–63; AT.VI.590–595.
[28] See Westfall, *Never At Rest,* p. 164.
[29] Descartes, *Opera,* pp. 107–29; AT.VI.622–634, *Dioptrics,* Cap. VIII.
[30] ULC ms. Add. 4000, folios 26^r–33^v.
[31] See Whiteside, *Mathematical Papers,* Vol. I, pp. 91ff.
[32] Add. 4000, folio 33^v.

dicular direction they lose or gain a determinate speed.[33] On the basis of these assumptions, Descartes shows how Snell's law of refraction can be derived. Concerning the grinding of lenses, Newton puts his knowledge of the conics to good use and describes in detail several devices for producing hyperbolic and elliptical contours.[34] It is at this point in his investigations that he probably began to think about Descartes's claim concerning nonspherical lenses and the sine law of refraction for colors (i.e., that with nonspherical lenses, chromatic aberration could be eliminated). His mind no doubt went back to his earlier theory of the causes of colors in natural objects and the roles that the fast- and slow-moving rays played in his explanatory model.

With this background in mind, we can now turn to Newton's third essay entitled "Of Colours" (Appendix), which is found in another early notebook (Add. 3975). The essay is remarkably free of corrections, and it provides the first systematic account of the mature theory of color. It can be dated from 1666. Apart from the optical entries in the early mathematical notebook, there is no further substantial evidence on which to construct a picture of Newton's researches into the nature of color from the period of the *Questiones* to 1666. Although the essay "Of Colours" of Add. 3975 lacks the sophistication of the *Lectiones opticae* (itself the product of research conducted in 1668 and 1669), it records a series of investigations whose results go beyond anything to be found in the *Questiones*. Given the paucity of evidence, we can only speculate on the path that Newton followed to his final theory of color. This is the conception that the direct light of the Sun is not homogeneous in character, but a heterogeneous composite of rays whose colors are inherent and immutable. Thus, colors are now conceived by Newton to be ingenerably present in the rays, and they are made manifest by any process that separates them from the heterogeneous composition of white light.[35] To be sure, he does not work with these ideas with the clarity

[33] Descartes, *Opera*, pp. 62–3; AT.VI.590–595, *Dioptrics*, Cap. II, Articles IX and X.   [34] Add. 4000, folios 26ʳ–33ᵛ.
[35] See *Papers and Letters*, pp. 53–5.

evident in the *Lectiones opticae*,[36] but they *are* present in "Of Colours," and they inform the conception of its investigations. The question is this: How did these ideas come to have their first, if tentative, expression in this essay "Of Colours"?

In what follows, we shall concentrate on some significant similarities between the *Questiones* and "Of Colours" in order to construct an answer to this question. The first thing to notice is that the initial five entries repeat information already recorded in the notes from Boyle in the *Questiones*. Newton has one of the same purposes in mind as in the *Questiones:* to distinguish the difference between a body's color when looked through and its color when looked on. "And perhaps there are many coloured bodys wch if made so thin as to bee transparent would appear of one colour when looked upon & of another when looked through." In each case Newton is concerned to give examples that illustrate the heterogeneous structure of light. Moreover, he states that reflection, as well as refraction, can analyze white light into its component parts, given that bodies are disposed to reflect or transmit one color more than another.

On folio 4, Newton repeats the same prism experiment as the one he records on 69 122^r of the *Questiones*. As we have pointed out, his results are more general and divide the table of 69 122^r into two categories (viz., a lighter color above the partition boundary, and a lighter below). In the first case we have a red boundary (*cd*), and in the second, a blue. It is important to notice that the experiment is placed after experiments 7 and 8 of folios 2 and 3. These record Newton's first trials with colors produced by projecting a prismatic spectrum on a wall some twenty-one feet from a prism positioned to collect the Sun's light as it passes into a darkened room through a small hole. This is the first version of an experiment designed to shake the received view that white light is homogeneous, and we shall have something to say about it shortly.[37] But if Newton now treats the shape of the spectrum produced in experiment 7 as an indi-

---

[36] *Cambridge Lectures on Optics,* Lectiones 6 and 7, folios 48^v–63^v.

[37] Newton's experiment entered the optical literature when he sent his 1671–2 "New Theory of Light" to the Royal Society. See *Papers and Letters,* pp. 47–50.

cation that white light is a heterogeneous aggregation of colors, does he now view his prism experiment in the *Questiones* as being of a significantly different character? The answer is no. Newton treats the spectrum produced in the eye by light that passes through the prism simply as a reversal of that produced by prismatic projection on a wall. At least, that is how he conceives the matter in "Of Colours," an interpretation that he probably then placed on the experiment as recorded in the *Questiones*. For him, the only difference between the two uses of the prism is that "Prismaticall colours appear in y^c eye in a contrary order to y^t in wch they fell on y^c paper." Given this conception, it is not surprising that Newton considers the earlier experiment to be compatible with his new interpretation of colors as arising from immutable properties inherent in the rays. It is important to notice that he shows no awareness of the fact that there is no true green present in the spectrum that results from viewing colors *through* a prism. Clearly, Newton did not recognize what today is called the boundary spectrum. In any event, the reappearance of the initial prism experiment in "Of Colours" strengthens the case for saying that in the *Questiones* Newton already conceives white light as a heterogeneous mixture of commingling colors. As we have seen, there is no conception in this early manuscript that the colors are intrinsic to the rays.

But how did Newton move from the account of the causes of color in the *Questiones* to the first expression of the mature theory in "Of Colours"? In answer to this, we can only speculate, but the following reconstruction appears plausible. It seems certain that the computing of refractive indices and the development of techniques for grinding nonspherical lenses played no decisive role in the emergence of Newton's mature theory. After all, according to Descartes's sine law of refraction, rays incident at different angles are refracted at different angles; thus, any spectrum that is cast by prismatic projection ought to be accounted for. It is only if one has reason to suppose that a spectrum can be produced that is not readily accounted for by the received laws of refraction that an experiment is contrived to show it. The same observation holds for Newton's denial that 'chromatic aberration' can be eliminated by the use of nonspherical lenses.[38] That claim also presupposes the conception that rays of unequal refrangibility are innate to the Sun's light. Accordingly, the conception that light rays possess an immutable degree of refrangibility and the experimental projection of an elongated spectrum are artifacts of Newton's theory, not necessarily factors that initially suggested the theory. To a large extent, what is involved here is the fact that in "Of Colours," Newton employs the same sort of evidence, and many of the same experiments, he used in the *Questiones*. Apart from the initial prism experiment of the *Questiones*, which has already been discussed, the colored-thread experiment reappears in "Of Colours," together with a closely related variant of it (folio 2 of Add. 3975). This suggests that the shift from the *Questiones* to "Of Colours" is as much a shift in the ontological basis of Newton's explanation of color as it is a response to an experimentally produced domain of new evidence.

As already indicated, one of the most important experiments for the new interpretation of colors that is not found in the *Questiones* is on folios 2 and 3 of Add. 3975. It is the first version of the famous prism experiment of 1672, in which Newton tried the "celebrated Phaenomena of Colours." According to the received laws of refraction, a beam of light passing through a round hole into a darkened room and then refracted through a prism ought to produce a colored circle "were all ye rays alike refracted, but their forme was oblong terminated at theire sides r and s wth straight lines" (see diagram on folio 2). If the received theory predicted that a circle of light colored at its edges ought to be perceived, what Newton observed on the wall some twenty feet from the prism was a spectrum five times as long as its breadth. It is clear that this is not a chance observation, despite the impression that he later contrives to give in the 1672 paper on light and colors.[39] Evidence of design is further confirmed by the fact that on folio 2 Newton immediately answers the most serious objection that could

[38] For a thorough discussion of this issue, see Zev Bechler, " 'A Less Agreeable matter': The Disagreeable Case of Newton and Achromatic Refraction," *British Journal for the History of Science*, 8(1975):101–26.

[39] *Papers and Letters*, pp. 47–8.

be raised against his theory. According to Newton, the shape of the spectrum can be accounted for only by supposing that the Sun's light is heterogeneous and comprised of immutable rays of unequal refrangibility. Because in Newton's experiment the Sun fills a visual angle of 31 minutes, the rays incident on the prism could not be parallel. If, by the sine law, different angles of incidence result in different angles of refraction, could Newton's spectrum not be an unusual product of the sine law itself? In experiment 8 (folio 2) Newton placed a board with a hole in it some twelve feet from the shutter to narrow the incident beam. As a result, the maximum angle of incidence was "lesse y^n 7 minutes, whereas in ye former experiment some rays were inclined 31 min." The dimensions of the spectrum produced were reduced by an equal amount, and its elongation further accentuated, rather than being diminished as might be expected. Newton also records "y^t ye Red and blew rays wch were parallel before refraction may be esteemed to be generally inclined one to another after refraction (some more some lesse y^n) 34^{min}." Thus, in his first record of the prismatic projection, Newton notes that the rays are equally refracted on both sides of the prism; this is the condition necessary for minimum deviation of the rays, and it is precisely the condition under which an elongation of the spectral image was not to be expected. But an image of this sort can be produced when the prism is turned to the requisite position. So far was the effect from being an unexpected occurrence that the production of equal refractions at both faces of the prism was a feature designed into the experiment. Although it is not stated explicitly, the proposition that rays of light differ both in intrinsic color and in degree of refrangibility is clearly a presupposition of the experiment's design. At this stage in the development of his theory, Newton has not yet seen the connection between the claim that the Sun's light consists in rays of unequal degrees of refrangibility and the claim that specific colors are immutably present in its individual rays.[40] Examples of such experiments are 10, 44, 45, and 46, of Add. 3975, which use

[40] For a fine discussion of this and other complexities in Newton's changing views, see Alan E. Shapiro, "The Evolving Structure of Newton's Theory of White Light and Color," *Isis,* 71(1980):211–35.

the prismatic arrangement in experiments 7 and 8. In experiment 10, Newton painted "a good blew & red colour on a piece of paper neither of wch was much more luminous y^{n} ye other. . . ." When viewed in "Prismaticall blew" and "Prismaticall red," the patches appear to have the color of the incident light, the blue being fainter in red light, and the red fainter in blue. Newton notes "y^{t} ye purer ye red/blue is ye lesse tis visible wth blew/red rays." Further evidence for the unalterability of the colors of the rays is produced in experiments 44 and 45, in which Newton shows that the "blew rays" are refracted more by a second prism than are the "red rays." His conclusion is manifest: Blue and red rays are each identically refracted by each prism. In experiment 44 we have the earliest version of the *experimentum crucis*. It was not recorded in its most sophisticated form until the drafting of the *Lectiones opticae* in 1669.[41] But neither here nor in the *Lectiones* does the experiment carry the theoretical burden that it is made to bear in the "New Theory" of 1672.[42] In experiment 46 (folio 12), Newton analyzes the joint spectra of three prisms arranged so that the green or yellow of *A* overlaps with the red of *B*, and the red of *C* overlaps with the green or yellow of *B*. The red of *A* and the blue of *C* will appear at either end of the band of colors, but "where ye reds, yellows, greenes, blews, and purples of ye several prismes are blended together there appeares a white." Newton also argues for the recomposition of colors to make white light in experiment 47, which follows, and in the earlier experiment 22. It is clear that he now believes that neither refraction nor reflection can alter the properties inherent in a ray of light. Colors arise from the heterogeneous aggregation of white light, and, once separated, they can be recombined experimentally to form the composition of the original light.

We do not know when Newton persuaded himself that the properties of the rays are inherent and immutable, but it is clear that prismatic analysis played an important role in

---

[41] For a thorough discussion of this experiment, see Shapiro, *Ibid.*, pp. 212–21. Shapiro discusses the difference between the presentation of the experiment in the *Optical Lectures* and in the "New Theory" of 1671–2.
[42] *Papers and Letters*, pp. 50–1.

this realization. It was necessary to show whether or not this conception had a clear relation to the colors of natural objects, a first explanation of which he had already attempted in the *Questiones*. In the early manuscript, Newton had satisfied himself that refraction and reflection are involved in the production of natural colors; a body, for example, is disposed to reflect some rays more than others, its appearance being the colors it best reflects. Thus, under conditions of normal lighting, bodies are disposed to exhibit those colors that are characteristically associated with them. But also in the *Questiones,* as in "Of Colours," Newton is concerned to explain how it is that specific rays can fall on the retina without modification of the colors they 'carry.' And in entry 8 (folio 72 123^v) of *Questiones* he attempts to develop an account of the ontology of light that will explain how characteristic colors are exhibited by bodies and transferred in an unaltered state to the optic nerves. Their speed is the ontological parameter of the explanation, together with the ways in which it can be altered by the parts of bodies, or by "loose particles" or subtle matter present in their pores. As Newton left the model in the *Questiones,* it embodied too many contingent factors to marry well with the notion that specific colors are in some manner preserved as they travel from bodies to the retina. Thus, even in the *Questiones,* Newton is disposed to search for some immutable property in light that can be associated with color. At this point, two alternatives spring to mind: Either some property other than speed determines the color of a ray, so that its definition remains, though the speed varies, or speed itself is conceived as fixed according to some explanatory mechanism such that specific colors can be said to be the products of determinate speeds. We can well imagine that an experiment like 10 on folio 3 of Add. 3975 settled the matter in Newton's mind. When a red patch and a blue patch were painted on a white sheet of paper, in the projected light of "prismaticall blew," both appeared blue, though the red was fainter; with the casting of "prismaticall red," both appeared red, but the "painted blew afforded much ye fainter red." Here, in one experiment, are combined the notion that rays are immutable and the notion that bodies are characteristically disposed to exhibit some rays more than others. Experiments such as these on the immutable properties of colors and consideration of what might be the bearers of such properties must also have further impressed the corpuscular conception of light on Newton's mind, a conception that in any event had guided his thought from the beginning.

4. Newton on the ring phenomenon and the influence of Hooke

It has been claimed that Newton came upon the phenomenon of rings on reading Hooke's *Micrographia*.[43] Although Newton did read the *Micrographia* sometime in 1666, we believe that he came upon the phenomenon in the course of an earlier series of optical experiments that were motivated by issues raised in the *Questiones*. These issues involve several questions: what happens when one color falls on another; the relationship between reflection and refraction; and the question "if two prisms, the one casting blue upon the other's red, do not make white." Issues of this sort led him to a series of experiments carried out in 1666, usually believed to have been performed at Woolsthorpe during the plague years.[44]

The experiments of 1666 (Add. 3975 folios 2–14) begin with an examination of differential refraction suggested by the broken-thread experiment of the *Questiones* (folio 2). This is followed by the first description by Newton of the true 'phenomenon of colors.' He next proceeds to an experiment suggested by the boundary-color phenomenon of *Questiones* (folio 3) and then repeats the boundary-color experiment itself (folio 4). Next, a number of experiments are attempted that use a 'four-square' vessel filled with water to act as a refractive agent. It is not surprising that Newton should consider this arrangement, as it involves refractions, double refractions, and reflections, an interest already exhibited in the *Questiones*. He notes that the results of these experiments differ considerably from those done with a

[43] Westfall, *Never At Rest*, p. 173.
[44] *Ibid.*, pp. 159–60. Westfall argues that the experiment could have been performed at Cambridge.

single prism (see folio 5). He then suggests that "some of ye Phænomena may bee tryed by tying two Prismes thus together" (i.e., base to base). After two experiments involving a single prism (folio 6), he embarks on a series designed around the two prisms bound together base to base (experiments 24–32). It is evident that in experiments 24–26 Newton once again examines the phenomena involved in his investigation with the four-square vessel, as he himself had suggested. In experiment 27 (folio 8) and those that follow he considers a phenomenon that he has observed in the course of doing the immediately preceding experiments, namely, the presence of a white circle surrounded by "several circles of colours" projected from the bases of the two prisms. As he inclined the prisms more to the rays of light, the red color, together with the colored circles, vanished; yet the white circle remained. In experiment 29, folio 8, he determines the following:

By variously pressing ye Prismes together at one end more y^n at another I could make ye said spot R run from one place to another; & ye harder I pressed ye prismes together, ye greater ye spot would appear to bee. [Soe y^t I conceive ye Prismes (their sides being a little convex & not perfectly plaine) pressed away ye interjacent aire at R & becoming continguous in y^t spot, transmitted ye Rays in y^t place as if they had beene one continuous peice of glasse; soe y^t ye spot R may bee called a hole made in ye plate of aire (ef)].

In a series of experiments following 29, Newton determines a relationship between the thickness of the air between the bases of the prisms and the width of the colored bands, the bands being broadest where the air is thinnest. In order to more clearly observe this phenomenon, Newton tied a convex lens to a plate of glass in order to make measurements of the width of the bands in relation to the thickness of the air (see experiment 35, folio 10). From what has been said, it is apparent that Newton's discovery of the rings need not owe anything to Hooke's *Micrographia*. That he should encounter this phenomenon is a result of a connected series of experiments that have their initial motivation in the *Questiones*.

But after he had written folios 1–20, there is no doubt that Newton read Hooke's *Micrographia*, the notes from which appear on folios 1^r–4^r of Add. 3958.[45] There he found material that was closely connected with his own work on the rings. At that point he apparently made an addition that refers to Hooke's theory (experiment 35, folio 10): "Soe y^t (el) y^e thicknesse of y^e aire for one circle was $1/64000^{inch}$, or 0.000015625 [wch is y^e space of a pulse of y^e vibrating medium] by measuring it since more exactly I find $1/83000$ = to y^e said thicknesse." Thus, it is clear that Newton discovered the phenomenon of colored rings independent of Hooke. This claim is, of course, consistent with noting that Newton's later analysis owes something to Hooke, especially the recognition that the phenomenon involves periodicity.[46]

[45] ULC ms.

[46] See Westfall, *Never At Rest*, p. 173, for a brief account of Newton's debt to Hooke on the question of the periodicity of the phenomenon.

6

GRAVITATION, ATTRACTION, AND COHESION

1. A mechanical theory of gravitation of the *Questiones*

A theory of the material cause of gravity (i.e., a mechanical theory of gravitation) is offered by Newton in the *Questiones*. Gravity is to be explained by the action of a stream of particles traveling upward and then downward. Newton notes that "the matter causing gravity must pass through all the pores of a body" (19 97^r). It is not clear whether he maintains this notion or rejects it in what follows. Because, as we shall see, the major feature in Newton's theory is the swiftness of the stream of particles (the streams of particles moving swiftly downward and slowly upward), the downward stream will pass through the pores, but not the upward. The upward stream does not "sink into every pore" but rather "yield[s] from the superficies of a body with ease so as to run in an easier channel as though it never strove against them" (19 97^r). The particles of the upward stream are grosser; and with respect to them it is as if the bodies have no pores, thus leaving no way in which the particles can 'fasten' on the body. Consequently, they stream between the bodies, and so have little effect on their motion.

Newton also notes that the streams, once having come down to the Earth, must ascend again (i.e., there is a recycling of the particles that cause gravity). If they did not, the Earth would swell with the accretion of particles, unless it had large cavities into which the particles could fall. But even so, the Earth would grow, for the total of all the particles that have "borne down the Earth and all other bodies to the center . . . cannot if added together be of a bulk so little

275

as is the Earth" (19 97^r). So even if all the Earth were a cavity, it would have been filled with the descending particles before Newton's time.

The major principle directing Newton's account (as, indeed, any effluxial account) is that "it must ascend in another form than it descends or else it would have a like force to bear bodies up as it has to press them down, and so there would be no gravity" (19 97^r). The difference Newton proposes is that the upward stream of particles moves slowly and is of a "grosser consistence" than the downward stream.

As we have seen, the grosser consistency of the upward stream will keep it from entering the pores and from making a purchase on the body. The slowness of the ascending stream will keep it from "strike[ing] bodies with so great a force as to impel them upward" (19 97^r). Newton is aware that the grossness of the ascending stream produces a more "weighty force" than would a less gross (thinner) stream, but he believes this to be outweighed in effect by the slowness of the stream. He writes, "if it should ascend thinner it can have only this advantage: that it would not hit bodies with so weighty a force, but then it would hit more parts of the body and would have more parts with which to hit with a smarter force, and so cause ascension with more force than the others could do descension" (19 97^r).

As the descending stream encounters the ascending stream, it presses it "closer" and makes it "denser," and "therefore it will rise the slower." Likewise, the descending stream will grow "thicker" as it comes closer to the Earth, but will lose no speed until "it finds as much opposition as it has help from the flood following behind it." The "following flood" refers to that portion of the stream above the part being considered, and whose 'weight' (or pressure) presses on that part. The descending streams apparently never become very 'thick,' because they pass through the Earth and meet at its center, where they press together in a "narrow room." Either they are turned back and return in the direction from which they came or they press through the streams moving in the contrary direction, becoming quite "compacted" (67 121^r). As a consequence, they are held down by the newly descending streams, until "they arise to the place from whence they came. There they will attain their former liberty" (67 121^{r}).

Newton's streams are streams of particles, and it is worthwhile to consider how the various properties and activities of the streams translate into the properties and activities of the particles that compose them. Both the direction and speed of the streams translate directly into the direction and speed of the particles. The notion of "grosser consistence" is rather more problematic. There are two possibilities to be considered: that grosser consistency means (1) that there are more particles per unit volume of the stream or (2) that a stream is composed of larger particles. Newton refers to the streams as if they had various degrees of density. For example, he writes of them as being more or less thin, as having more parts when thinner, as growing thicker, as being closely pressed together, and as being compacted. On the one hand, the ascending stream must not have particles small enough to enter the pores of bodies; on the other hand, the particles of the descending stream are said to crowd into those pores bearing the objects down to the Earth. Thus, the ascending particles are larger in size than the descending, which would seem to indicate the correctness of interpretation (2). In the same way, we are told that a thinner stream will hit a body in more parts, and have more parts to hit with, which implies that a thinner stream is composed of smaller particles than a thicker, denser, or compacted stream. On Newton's account it is the particles that are compacted and pressed together to form larger particles that are now farther apart, because he implies that there are now fewer of them to hit any given body. This interpretation is also consistent with Newton's claim that a denser stream will rise more slowly, whereas interpretation (1) is not. If the particles in a stream were simply moved closer together, why should they move slower? In fact, because whatever force causes the stream to rise is now acting on a smaller cross section, one would expect the stream to increase its speed. Indeed, fluids flowing through pipes flow more rapidly at constrictions than elsewhere, because the same volume has to flow through in the same time at every cross section. If, however, we are referring to the larger 'weightier' particles, as in interpretation (2), any increase in their size and weight will slow them down. This is not to say that Newton employs anything like the laws of his mature physics. Only a rudimentary familiarity with the streaming of fluids is necessary, as well as some physical intuition, something that Newton never lacked even at this early date.[1]

Interpretation (2) is not consistent with everything Newton has to say about his streams. When the streams meet at the center of the Earth and are "compressed into a narrow room, closely pressed together, and thus very much opposing one another either turn back the same way that they came, or crowd through one another's streams with much difficulty and pressure, and so be compacted" (19 97^r and 67 121^l), it would seem that the 'compressing' and 'pressing' together does not lead to larger particles that are more widely spaced, but simply to more particles per unit volume in the streams. How else could we account for their difficulty in 'crowding through' one another? Furthermore, it appears that Newton believes that it is as a result of this compressing that the rays are so compacted. It is unlikely that he is making the trivial claim that the rays are compacted by virtue of their being pressed together. It is more likely that the foregoing passage is meant to be an account of how the finer particles of the downward streams are turned into the grosser particles of the upward streams, that as a result of the crowding through each other with such difficulty and pressure, the finer particles are compacted together to form the larger particles of the upward stream. As the newly arising stream travels through the descending stream, the particles remain compacted. It is only when they reach the top of their path and are free from the pressure of the descending streams that the particles become free of each other and again regain the 'degree of freedom' that is characteristic of a newly descending stream. Thus, both notions of what it is for something to increase in density are used by Newton in this effluxial theory.

[1] Newton also had available to him D. Benedictus Castellus, *Discourse of the Mensuration of Running Waters* in *Mathematical Collections*, edited by Salusbury (London, 1661) Vol. I, Part II.

Thus far we have seen that the thinner descending stream has a greater purchase on bodies by virtue of its penetrating the pores of those bodies. At the same time, the grosser particles of the ascending stream (being unable to enter the pores of bodies) simply slip around them. There is, however, a further feature in Newton's account that requires careful consideration, because it raises profound problems for the effluxial theory. There is a clear appeal to collisions in Newton's account, as well as to a simple pressure of the particles of the stream in the pores of the body. It is possible, of course, that Newton's pressure account was meant (at a deeper level of explanation) to be reduced to the collision account, but that is made unlikely by the vivid picture Newton paints of the grosser particles of the ascending stream "yield[ing] from the superficies of a body with ease so as to run in an easier channel as though it never strove against them." It thus seems likely that both accounts are meant to have independent explanatory power.

In the case of collisions, the openness of the pores of bodies to the thinner descending stream allows for more collisions between those collections of particles that constitute the stream and those collections of particles that constitute the body. Thus, the descending particles strike more of the particles that constitute the body because they strike both the upper particles of the surface as well as many interior particles, but the ascending particles strike only the surface particles on the bottoms of the bodies. As Newton writes of thinner and faster streams, "it would hit more parts of the body and would have more parts with which to hit with a smarter force."

Furthermore, even if the particles of bodies were identical atoms, how could they be arranged so that *every* atom in the body would offer the same cross section to the descending stream, no matter how the body was turned? Newton does not claim that all the particles of a body are struck by the descending thinner stream, but only that "it would hit *more* parts of the body" than the denser ascending stream.[2] The effluxial theory as offered by Newton in no way provides

----

[2] Emphasis added.

the quantitative details needed to show that it conforms with Newton's additional claim on folio 68 121$^\text{v}$ that "the gravity of bodies is as their solidity."[3]

Newton saw other consequences of his effluxial theory that concerned him and ones he may have thought of as a *reductio ad absurdum* (or at least a disconfirmation) of his theory of gravity in the *Questiones*. These consequences are found on folio 68 121$^\text{v}$, where Newton writes: "try whether the weight of a body may be altered by heat or cold, dilation or condensation, beating, powdering, transferring to several places or several heights, or placing a hot or heavy body over it or under it, or by magnetism. Whether lead or its dust spread abroad is heaviest. Whether a plate flat ways or edge ways is heaviest."

Newton suggests all, or most, of these experiments as tests of his effluxial theory; or, because some of the results of the proposed 'experiments' were probably known to Newton when he proposed them, they are recognized by him as disconfirmations of (or, at best, problems for) that theory. They all consist of changes that do not affect the "solidity" of bodies (i.e., the total number of particles remains the same, while changing their configurations, distances, etc., in the bodies that they constitute). These are changes that are likely, under the effluxial theory of the *Questiones*, to alter the effects of the streams on the bodies.

The possible effects of heat and cold on gravity seem to involve the motions of the particles that make up the body or to involve the particles of the stream. Although Newton does not explicitly identify heat and cold with the motions of the particles of a body in the *Questiones*, such a notion is present in the atomists he was reading, for example, in the writings of Boyle and Charleton.

In the case of "dilation" and "condensation," the problems are clearer than those of heat and cold. If a body is dilated so that its density decreases and its interstitial pores are increased in size, within certain limits, then the number of particles likely to be struck by the descending stream

[3] For another view concerning Newton's streams, see Curtis Wilson, "Newton and the Eötvös Experiment," in *Essays in Honor of Jacob Klein* (Annapolis: St. John's College Press, 1976).

increases. On the other hand, if a body is condensed so that its density increases and its pores become smaller, it is as if the particles of the stream become grosser (i.e., unable to enter the pores), and so the body will fall slower. Clearly, both cases are contrary to the fact that "all bodies descend equal spaces in equal times" (68 121ᵛ). The same problems arise for the next items on Newton's list, because "beating" and "powdering" can be understood to be cases of "condensation" and "dilation," as can "lead or its dust spread abroad."

Newton's question, "whether a plate flat ways or edge ways is heaviest" clearly shows that he recognizes that on the effluxial view a plate in different positions would not be affected in the same way by the streams, whereas the principle of solidity requires that it fall at the same rate no matter how situated.

Newton's reference to magnetism as a possible way of altering the weight of a body probably arises from his consideration at the time of the *Questiones* of an effluxial theory of magnetism, which will be discussed later. It will suffice at the moment to note Newton's query on folio 29 102ʳ, where he writes, "Whether magnetic rays will blow a candle, move a red hot copper or iron needle, or pass through a red hot plate of copper or iron." It is not surprising, then, to suppose that the magnetic efflux might add its effect to that of the ascending or descending streams, and thus affect the fall of bodies.

Newton considers one other difficulty with the effluxial theory (i.e., whether or not 'gravity' can be reflected or refracted). Streams of particles (e.g., light) can be both reflected and refracted. Why should the same not be true of the gravitational stream, and for that matter the magnetic stream? If it were possible to reflect and refract the gravitational stream, then two possible perpetual-motion devices could be constructed, as depicted on folio 68 121ᵛ. The first device would consist of a heavy wheel on a horizontal axle with one half of it shielded by a material that could reflect the gravitational stream. The other half, being impacted by the stream, would overbalance the wheel. Would it not then turn perpetually? The second device would consist of a wheel mounted on a vertical axle. The construction of the wheel may be interpreted in at least two ways. It may be a representation of a pinwheel or turbine blade configured like a windmill sail. The gravitational stream, on being re-flected by the angled blades, would cause the wheel to turn. Another possible interpretation construes the wheel as hav-ing prismlike devices mounted on it that would refract the stream, thus changing its direction and causing the wheel to turn. Despite the fact that such prismlike devices would ob-viously be of an unknown character, the second possibility is more likely the correct one; Newton precedes his descrip-tion of the two devices with the question whether or not the streams can be reflected or *refracted,* and he offers the de-vices as a consequence of an affirmative answer.

We are not claiming here that Newton is arguing for the falsehood of the effluxial theory from the impossible conse-quence of perpetual motion. There is no reason to believe that he thought perpetual motion was impossible at that time, but he did have ample reason to think it at least unlikely.[4] His reading of John Bate and John Wilkins had introduced him to various ways of feigning perpetual mo-tion, some of which the young Newton had made note of in the Morgan Notebook. It is unlikely that he missed Wilkins's admonition in *Mathematicall Magick,* where Wilkins wrote of an account of perpetual motion, "it sounds rather like a chymicall dream, then a Philosophical truth."[5] Any account of gravitation that had a perpetual motion as a consequence might also have seemed more a "chymicall dream, then a Philosophical truth."

Despite the difficulties that he perceives in effluxial the-ory, in the course of considering the concepts involved, Newton attempts to clarify some notions that appear in his later thought. Newton's various notions of density, includ-

---

[4] We thank H. S. Thayer for suggesting that the role of the gravity machines might be as a counterexample to the efflux theory.

[5] Wilkins, *Mathematicall Magick* (London, 1648), p. 22. On the other hand, among the methods of human transportation, Wilkins lists being carried through the air by birds and witches (pp. 199–203). For a discussion of seventeenth-century views on perpetual motion, see Henry Dircks, *Perpetuum Mobile; or, Search for Self-Motive Power During the 17th, 18th, and 19th Centuries* (London, 1861), Chap. III, pp. 60–84, and Chap. IV, pp. 85–105.

ing his claim on folio 68 121^v that "the gravity of bodies is as their solidity, because all bodies descend equal spaces in equal times," all refer back to the notion that bodies are conglomerates of atoms all of which are the same. The mass of an object and its density are sometimes taken to be the same thing by Newton. Consider his device for finding which of two bodies is the more dense, described and illustrated on folio 13 94^r of the *Questiones*. The two bodies to be considered are projected from each other by a spring that is free to move toward either object after they are projected. Newton writes that the two objects "receive alike swiftness from the spring if there be the same quantity of body in both." If the "quantity of body" is not the same in both, the spring (and an attached pointer) will move toward "the body which has less body in it." What is being measured is mass, the resistance to being projected, which shows itself in the degree of recoil of the spring. Thus, this device can tell which of two bodies has the greater density only if their volumes are equal. Newton writes nothing about equality of volume as a precondition, and what is more, he represents the two objects in the diagram on folio 13 94^r with different shapes, suggesting that their volumes are different. It is possible that at this point in the *Questiones*, as perhaps also in the definition of quantity of matter in the *Principia*, the notions of mass and density are the same and that they are both measures of the solidity of a body (i.e., how much solid matter there is in it, which is the same as the number of atoms in the body).

Bodies do not fall equally because gravity or weight is "as their density" in our sense of mass divided by volume. They fall equally because gravity or weight is as their solidity, in the sense that all atoms are pulled (or have tendency to move toward the center) equally. But because all atoms will have the same gravity, the total gravity of each object will be as the number of atoms it contains, and thus all objects will fall at the same rate. Each atom in a conglomerate falls individually, with each of the other atoms – the whole falls as each of its parts falls.

That mass, as a notion of resistance to projection, should be identified with weight or gravity at this point in Newton's thought is not surprising. In his reading of Charleton's *Physiologica* he would have found the following:

To these 4 Essential Attributes of Atoms, *Empiricus* hath superadded a Fifth, viz. ἀντιτυπία Renitency, or Resistence. But, by his good leave, we cannot understand this to be any distinct Propriety; but as τί ὑποκεί-μενον, something resilient from and dependent on their *solidity*, which is the formal reason of Resistence: besides, we may confound their Renitency with their Gravity, insomuch as we commonly measure the Gravity of any thing, by the renitency of it to our arms in the act of Elevation. Which may be the reason, why *Aphrodisaus* (*lib*.I. *Quest.cap*.2) enumerating the proprieties of Atoms, takes no notice at all of their Gravity; but blends it under the most sensible effect thereof, *Resistence*.[6]

Further evidence for this identification of mass and gravity is to be found in Newton's essay "Of Violent Motion," which was discussed in detail earlier.[7] On folio 21 98^r Newton writes that "violent motion is continued either by the air or by a force impressed, or by the natural gravity in the body moved," and on folio 52 113^v, after rejecting the other alternatives, he concludes, "therefore, it must be moved after its separation from the mover by its own gravity."

That the same thing that offers resistance to projection should also function as the property of bodies that accounts for the continuation of motion after projection appears later in Newton's conception of inertial mass and in the first and second laws of the *Principia*, but the unfolding of this dual role in the *Questiones* should be clear from the foregoing.

Newton is concerned not only with the explanation of gravity but also with clarification of the phenomenon itself. On folio 68 121^v he considers the implications of the definition of uniform acceleration – equal increments of speed in equal increments of time. As a consequence of this definition, he asserts that "in the descension of a body there is to

[6] Charleton's *Physiologia*, p. 112. Sextus Empiricus writes in *Against the Physicists*, Book I, line 437, "But it is possible, also, to repeat our former argument which deduces our thesis in a convincing way: if a body exists, it is either sensible or intelligible. And it is not sensible; for it is 'a complex quality perceived through the combination of form, size, and solidity'; and a quality perceived through a combination of things is not sensible; therefore the body also, conceived as body, is not sensible." Loeb Classical Library, *Sextus Empiricus*, translated by R. G. Bury, 4 vols., Vol. III, pp. 207–9. Gassendi, in his *Syntagma*, Sectio I, Lib. 3, Cap. 6, takes resistance to be an essential property of bodies.

[7] See Chapter 3, Section 5.

be considered the force which it receives every moment from its gravity – which must be least in a swiftest body. . . ." Consider two falling bodies; if one has a greater initial speed, it will cover a fixed distance in less time and thus will receive fewer increments of speed than the other. In this case, Newton seems to think that the one with less initial speed will reach a given point at the same time as the other! In the diagram on folio 68 121^v there is a ball b suspended in such a way that when another ball a is dropped, it will in its descent release a blade that will free b at the moment that a and b are abreast. Newton supposes that a and b will come to rest together at h, in accordance with the foregoing 'argument.' The next entry on the folio records what Newton considers to be a 'fact' from Galileo's *Dialogue,* that is, that 100 pounds Florentine falls 100 braces in 5 seconds.[8] The next diagram shows a version of what is essentially a balance-beam scale, where the degree of displacement of the 'beam' indicates the weight of the object, on a scale using a plumb line as pointer.

2. Some later effluxial theories

That Newton did not abandon the effluxial theory entirely is made evident by his "An Hypothesis explaining the Properties of Light discoursed of in my severall Papers," communicated to Oldenburg in a letter of 7 December 1675, ten years after he wrote the passages in *Questiones.* The "Hypothesis" is an elaboration of his account of the role of the ether in his dispute with Hooke of 1672.[9] Among the various roles of the ether, he once again considers gravitation and elaborates on the effluxial theory. He writes:

so may the gravitating attraction of the Earth be caused by the continuall condensation of some other such like aethereall Spirit . . . as to cause it from above to descend with great celerity for a supply. In wch descent it may beare downe with it the bodyes it pervades with force proportionall to the superficies of all their parts it acts upon; nature makeing a circulation by the slow ascent of as much matter out of the bowells of the Earth

---

[8] Galileo is off by a factor of 2. See Martin Tamny, "Newton and Galileo's Dialogue on the Great World Systems," *Isis,* 68(1977):288–9.

[9] See Westfall, *Never At Rest,* Chap. 7, pp. 238–74, for an account of the dispute.

in an aereall forme wch for a time constitutes the Atmosphere, but being continually boyed up by the new Air, Exhalations, & Vapours riseing underneath, at length (Some parts of the vapours wch returns in rain excepted) vanishes againe into the aethereall Spaces, & there perhaps in time relents & is attenuated into its first principle.[10]

On 27 July 1686 Newton wrote Halley concerning the independence of his discovery of the inverse-square (duplicate proportion) character of gravity from the writings or comments of Hooke. He argues that a passage in the previously quoted letter of 1675 presupposes the inverse-square relation. He writes:

For I there suppose that ye descending spirit acts upon bodies here on ye superficies of ye earth wth force proportional to the superficies of their parts, wch cannot be unless ye diminution of its velocity in acting upon ye first parts of any body it meets with be recompensed by the increase of its density arising from that retardation. Whether this be true is not material. It suffices that 'twas ye Hypothesis. Now if this spirit descend from above wth uniform velocity, its density & consequently its force will be reciprocally proportionall to ye square of its distance from ye center. But if it descend with accelerated motion, its density will every where diminish as much as its velocity increases, & so its force (according to ye Hypothesis) will be ye same as before, that is still reciprocally as ye square of its distance from ye center.[11]

It is now well known that prior to 1675 Newton was indeed aware that the duplicate proportion applies to gravity.[12] It would then seem strange that he would propose an account of gravity that was obviously contrary to that relation. As we have seen, Newton argues (in his letter to Halley) that his supposition that "ye descending spirit acts upon bodies here on ye superficies of ye earth with force proportional to the superficies of their parts" shows that he then understood the duplicate proportion. But whether or not that duplicate proportion *is* presupposed in his effluxial theory as given in his letter to Oldenburg must be examined.

In the letter to Halley, Newton equates the force of gravity with the product of the density and velocity of the stream. This is a collision account, in which density plays the

[10] *The Correspondence*, Vol. I, Item 146, pp. 365–6.
[11] *Ibid.*, Vol. II, Item 291, p. 447.
[12] See Westfall, *Never At Rest*, pp. 151ff., and note 35 on p. 152.

role of mass, and the force of gravity is equated with the momentum of the stream. Newton's argument for the presence of an inverse-square relation proceeds by first considering streams descending toward the center of the Earth with uniform velocity. In this case, the density would be inversely proportional to the square of the distance from the center of the Earth, and because the force of gravity is directly proportional to the product of density and velocity, the force would be inversely proportional to the square of the distance. If we imagine the streams passing through concentric spheres along their radii and consider the points of intersection of the streams with the surface of any given sphere, the density of the streams at that distance from the center is the number of points divided by the area of that sphere. Because the area of a sphere is proportional to the square of the radius, it is clear that the density of the streams would satisfy an inverse-square law.

Because acceleration, not constant velocity, is required, the argument continues in Newton's claim that as the velocity increases in an accelerated stream, its density is diminished correspondingly, so that the product of velocity and density remains the same as in the case of the nonaccelerated streams, and thus is still inversely proportional to the distance from the center. The notion of density employed here is that of density along a radius. If we think of the stream as made up of a line (along a radius) of particles and define the density as the number of particles per unit distance, we can see how the density decreases as the velocity increases. Because in accelerated motion, velocity is a function of time, the first particle in the stream (the first one to begin its descent, and thus the one closest to the Earth) will be moving at a greater velocity than the others and thus will be a greater distance from the second than is the second from the third, and so on. In this way the density will be inversely proportional to the velocity, so that the product of density and velocity will remain the same, and so this case will reduce to the previous case, in which the velocity was constant.

The final complication arises from the fact that when the descending stream strikes an object, its velocity must, as a consequence, be decreased, and so subsequent strikes by that stream will be of less force. Newton tells Halley that such strikes increase the density of the stream, apparently by slowing the first particle along the radius and thus allowing the next particle above to come closer. Once again, although here without obvious argument, Newton claims that the increase in density exactly compensates for the decrease in velocity, and thus the account is consistent with an inverse-square relation, as in the previous cases.

It is unlikely that at the time of Newton's letter to Halley he believed his effluxial theory to be adequate. He certainly did not think that he had good arguments for its truth, because he wrote Halley "whether this be true is not material. It suffices that 'twas y^c Hypothesis." Apparently neither Newton nor Halley believed the effluxial hypothesis had any place in the *Principia*, the preparation of which was their main activity at the time of Newton's letter. One can only assume that Newton realized that the same objections that could be raised against the effuxial theory of the *Questiones* could be raised against this later and more sophisticated account. One need only consider again whether a flat plate is heavier "flat ways or edge ways" to see that the problem of the earlier theory persists into the later theory, despite its apparent 'consistency' with the inverse-square relation.

It is likely that the effluxial theory of the Oldenburg and Halley letters owes much to Kenelm Digby's *Two Treatises*. We have speculated elsewhere that Newton read this work of Digby's during the period of the *Questiones*. Although the effluxial theory of the *Questiones* itself owes little more to the reading of Digby than the fact that it provided the occasion on which he thought about the matter, the later theory of the letters owes far more. In Digby's theory, the ascending stream is rarer (i.e., less dense) than the descending stream. The rising streams become denser near the top of their climb as they are joined by the parts of fire added by the Sun. They become still denser as they fall, because they fall along radii, and thus the streams approach each other as they come closer to the Earth's center. It is this notion that is seized on by Newton in the effluxial theory of the letters. Digby does not attempt to give a quantitative account of the increase in density, and so he makes no suggestion regarding an inverse-square relation. Neither does the account make use of pores. It attributes the fall to the net effect of the less dense streams 'knocking' from below and the more dense streams 'knocking' from above. The accelerated character of the fall is a consequence of the 'fact' that the more dense a body is, the more quickly it falls, for it will cut the air more easily. Digby was clearly confused about the subject of his attribution of density, a confusion that perhaps led Newton to his use of the several different senses of density we have examined.[13]

It is well established that Newton's attempts to construct mechanical accounts of gravitation alternated with the conviction that gravity was a power simply imbued in bodies by God.[14] Thus, Newton's attempts to find a mechanical account do not appear to grow out of a deep commitment to the mechanical philosophy. In the *Questiones* it seems to emerge from his reading of the mechanical philosophers, Boyle,[15] Charleton, and others. Although he does not agree with their specific accounts, he advances a theory of his own, about which he likewise entertains doubts. His returns to effluxial theory seem prompted more by a concern that he may be taken to be an invoker of occult qualities than by any conviction that the effort will be successful. Indeed, when he is bold enough to hazard his view that gravity is an innate power, or even when he is *taken* to be doing so, he is rebuked by stauncher mechanists. Gregory writes in one of his memoranda: "He [Christopher Wren] smiles at Mr. Newton's belief that it [gravity] does not occur by mechanical means, but was introduced originally by the Creator."[16] Leibniz's attack on what he takes to be Newton's nonmechanical theory of gravity of the *Principia* is too well known to require exposition here.

[13] Digby, *The First Treatise . . . of Bodies*, Chap. X, pp. 94–106.

[14] See, for instance, Newton's comments in the General Scholium concerning a nonmechanical notion of gravity, *Principia*, edited by Koyré and Cohen, Vol. II, Lib. III, p. 760.

[15] Principally Boyle's *The Spring of the Air*.

[16] From a memorandum by David Gregory dated "20 Feb 1697/8." *The Correspondence*, Vol. IV, Item 584, p. 267. See J. E. McGuire, "Force, Active Principles, and Newton's Invisible Realm," *Ambix*, 15(1968):154–208.

It is clear that the effluxial theories were never happily embraced by Newton. Their attraction lay in their popularity among mechanists, and the suggestion that they might explain not only gravitation but also magnetism. Not only the mechanists but also the Aristotelians (and, indeed, Aristotle himself) had invoked the effluxes in explanation of various phenomena.

Given the great importance that Newton assigned to the ideas of Epicurus, it is interesting to note that Charleton, in his *Physiologia,* suggests that the use of effluvia to explain attraction might be traced back to Epicurus's notion of rebound in the *Letter to Herodotus.*[17] Epicurus argues that the number of things in an infinite space is unlimited: "For if the void were infinite and bodies finite, the bodies would not have stayed anywhere but would have been dispersed in their course through the infinite void, not having any supports or counter checks to send them back on their upward rebound."[18]

This passage may well have suggested what Charleton claims, though one cannot be sure that this is the passage he had in mind. Charleton offers as evidence Galen's supposed claim in his *On the Natural Faculties* that it was the opinion of Epicurus that "Omnes attractiones per resilitiones atque implexiones Atomorum fieri, that all Attractions were caused by the Resilitions and Implexions of Atoms."[19] Charleton claims that these statements, together with Plato's claim in the *Timaeus* that attraction is not involved in cases of rubbed

[17] Charleton, *Physiologia,* p. 124.

[18] Epicurus, *Letter to Herodotus,* in Diogenes Laertius, *Lives of Emminent Philosophers,* Loeb Classical Library, translated by R. D. Hicks (Cambridge, Mass.: Harvard University Press, 1925), 2 vols., Vol. II, Book X.42, p. 567–70.

[19] Charleton, *Physiologia,* p. 124. In fact, what Galen claims is quite otherwise. He asserts that, despite his atomism, Epicurus allows for the existence of attraction. Galen writes, "His [Epicurus's] view is that the atoms which flow from the stone are related in shape to those flowing from the iron, and so they become easily interlocked with one another; thus it is that, after colliding with each of the two compact masses (the stone and the iron) they then rebound into the middle and so become entangled with each other, and draw the iron after them." Galen, *On the Natural Faculties* (Cambridge, Mass.: Harvard University Press, 1952), Loeb Classical Library, translated by A. J. Brock, Book I, Chap. XIV, pp. 71–2. We have no reason to think that Newton read Galen or had any reason to disbelieve Charleton.

amber and lodestones ("that such wonderful phenomena are attributable to the combination of certain conditions – the nonexistence of a vacuum, the fact that objects push one another round, and that they change places, passing severally into their proper positions as they are divided or combined"),[20] gave "the hint" to Descartes, Regius, Sir Kenelm Digby, and Gilbert.

Robert Boyle was to argue at great length for the utility of effluxial theories in his *Essays of the Strange Subtilty, Determinate Nature, and Great Efficacy of Effluviums* (London, 1673). There he wrote that

there are *at least* six ways, by which the the Effluviums of a Body may notably operate upon another; namely, 1. By the *great number* of emitted Corpuscles. 2. By their *penetrating* and pervading nature. 3. By their *celerity*, and other Modifications of their Motion. 4. By the *congruity* and *incongruity* of their Bulk and Shape to the Pores of the Bodies they are to act upon. 5. By the *motions* of one part *upon another*, that they excite or occasion in the Body they work upon according to its Structure. And 6[ly], By the Fitness and Power they have to make themselves be *assisted*, in their Working, by the *more catholic Agents* of the Universe.[21]

Much of Boyle's thought about effluvia was presaged in *The Spring of the Air*, and Newton's reading of that book at the time of the writing of the *Questiones* probably accounts for the similarity between the uses he made of effluvia and Boyle's outline of such uses written nearly ten years later.

When Newton returned to mechanical theories of gravitation after 1706, it was to another theory, an "electric and elastic"[22] spirit that was meant to explain the behavior of light, magnetism, gravity, and capillarity and how the mind directs the action of the body. Here, too, the motivation is external: the apparent influence of Francis Hauksbee's work on electroluminescence, electrical attraction, capillar-

[20] Plato, *Timaeus*, translated by Benjamin Jowett, in *The Collected Dialogues of Plato*, edited by Edith Hamilton and Huntington Cairns, Bollingen Series LXXI (New York: Pantheon Books, 1963), 80c, p. 1201.

[21] *Essays of the Strange Subtilty Determinate Nature, and Great Efficacy of Effluviums*, in Thomas Birch, *The Works of the Honourable Robert Boyle*, 5 vols., Vol. III, p. 321.

[22] See A. R. Hall and M. B. Hall, "Newton's Electric Spirit: Four Oddities," *Isis*, 50(1959):473–6, also Alexandre Koyré and I. B. Cohen, "Newton's 'Electric & Elastic Spirit,'" *Isis*, 51(1960):337.

ity, and so forth.[23] Thus, Newton's wish to be in the 'main-stream' of the scientific thought of his time and his concern to find theories that would subsume the largest amount of apparently disparate phenomena under a single account lead him back to effluvial theory. The Queries to the *Opticks* simply represent these desires, whereas their lack of quantitative parameters reveals that these remain only hopes that continue to resist the precision that characterizes Newton's mature discoveries.

3. Cohesion and adhesion

Another of Newton's concerns in the *Questiones* is the explanation of how the parts of bodies unite to form wholes. He rejects the Cartesian account that repose or rest (more exactly, a balance between the external pressure of the second matter and the internal pressure of the third matter)[24] is what joins the parts of a body into a whole, "for then sand by rest might be united sooner than by a furnace" (folio 6 90^v).[25] He instead accepts an account apparently based on Boyle's findings regarding air pressure (i.e., that bodies are held together by the pressure of all the surrounding matter). Air pressure itself cannot actually account for this pressing together, because it is not strong enough, as Newton notes from Boyle. Newton argues at folio 6 90^v that if two bodies are fitted to contact each other so that nothing can come between them, "those two may move together as one body and so may increase by having others joined to them in the same manner." If, however, by some chance the particles (parts) cannot come into complete contact, as in the diagrams on folio 6 90^v, then they can be separated again. Newton then writes, "but it may be that the particles of compound matter were created bigger than those which serve for other offices." This makes clear that Newton's pre-

[23] See Henry Guerlac, "Newton's Optical Aether: His Draft of a Proposed Addition to His Opticks," *Notes and Records of the Royal Society,* 22(1967):45–57. Also see Joan L. Hawes, "Newton's Revival of the Aether Hypothesis and the Explanation of Gravitational Attraction," *Notes and Records of the Royal Society,* 23(1968):200–12.

[24] Descartes, *Opera,* p. 44; AT.VIII.71, *Principia,* Part II, Article LV.

[25] Newton raises the question again on folio 18 96^v, but without comment.

vious assertions were meant, at least in part, as an account of how the larger particles of compound bodies are formed from the smaller particles of 'simple' bodies (i.e., apparently atoms). That such an account was later published by Newton as *De natura acidorum,* written in 1692 and published in 1710 in the Introduction to Volume II of the *Lexicon Technicum* of John Harris,[26] adds evidence for this interpretation.

Newton's account of cohesion in the *Questiones* rests on an analogy with air pressure. Cohesion is the result of the pressure of the surrounding medium when a vacuum exists between the bodies that cohere, or rather when a vacuum can be achieved between the two bodies by a perfect fit of the bodies to each other, or through the use of a substance between the bodies that will accomplish the fit, such as a liquid. Newton notes Boyle's experiment in which slabs of marble are made to cohere through the use of a film of water spread between them. Once again, this is a case of air pressure, and the force holding the slabs together is not great enough to account for the integrity of solid objects. The ability of liquids 'to make a join' had been misunderstood by some to be a property of the liquid itself. Digby writes in the *Two Treatises:*

> Suppose that such a liquid part is between two dry parts of a dense body and sticking to them both, becometh in the nature of a glew to hold them together: will it not follow that these two dense parts will be as hard to be separated from one another as the small liquid part by which they stick together is to be divided.[27]

The notion that liquids were the key to cohesion was also suggested by the phenomena of capillarity and "filtration," the percolation of a liquid through a solid. Liquids seemed to be able to adhere to bodies, climbing up tubes of glass, and able to permeate apparently solid objects. Newton writes on folio 31 103^r: "But where air cannot enter water will (as appears in that it will get through a bladder, which air cannot do) . . ." He accounts for this phenomenon once again by the presence of a pressure. These phenomena are to be explained by a pressure theory, rather than them-

[26] Reprinted in *Papers and Letters,* edited by I. B. Cohen, pp. 255–8.

[27] (London, 1665), p. 150. Quoted in E. C. M. Millington, "Theories of Cohesion in the Seventeenth Century," *Annals of Science,* 5(1945):253–69, on p. 254.

selves being used as the foundation of a theory of cohesion. Newton's account as it appears on folio 31 103^r runs as follows: In most bodies, air is "pressed" into their pores, though the air "will have some reluctance outward, like a piece of bent whale bone crowded into a hole with its middle part forward." If water is present, it will be drawn in by the striving of the air to get out. As if not totally satisfied with his whalebone model, Newton suggests another, one quite inconsistent with the previous model. He writes: "The air, too, being continually shaken and moved in its smallest parts by vaporous particles everywhere tossed up and down in it (as appears by its heat), must needs strive to get out of all such cavities which hinder its agitation." Newton does not notice that these models are inconsistent, as the word "too" seems to show. The first model is clearly intended to be mechanical, although it is not clear precisely how it works. The second model is a mix of mechanical and non-mechanical notions. The shaking of the air particles by the water (vaporous) particles is clearly mechanical, but it is not this that makes the air exit and the water take its place, but rather a striving of the air to find a place where its agitation will not be hindered. Surely this can be turned into a mechanical account if we allow for the actual shaking and agitation of the air particles within the pores. But how can this occur if they are wedged in like bent pieces of whalebone? Thus, we are left with tendencies and strivings alone.

In bodies with smaller pores, the parts of the bodies are "crushed closer together than their nature will well permit." In "striving" to get free, these parts draw between themselves air or water, and in the cases where air cannot enter, water does. "By striving to get apart they draw in the parts of water between them." Thus, water is drawn in, not as the glue to hold parts of bodies together, but in the attempt to separate parts "unnaturally" pressed together.

There are indications in the *Questiones* that Newton found interesting those explanations of cohesion that rest on the structure of the minute parts of bodies. For one thing, there are the whalebonelike particles of air mentioned earlier. But more pervasive are the references to the "branchy particles" that Descartes uses to explain the behavior of air and oil.

On folio 16 95^v Newton asks "whether hard bodies stick together by branchy particles folded together. Descartes,"[28] and on folio 25 100^r he asks "whether it [air] consists of branchy particles not folded together but lying upon one another. Descartes."[29] Still more clearly, on folio 45 110^r, where Newton considers Descartes's use of the branchlike particles of oil to explain why it "mixes" with most bodies more easily than water, Newton writes that this phenomenon "may proceed from its [oil's] branches taking hold, like briers, on all adjacent bodies. Whereas water dropped is kept round by the air and crowded together, unless the pores of bodies lie open for its particles to drop into them." Descartes may not have been the only source of Newton's thought on these matters, because Newton's whalebone "springs" of air are not very different from those springlike particles invoked by Boyle in his *The Spring of the Air*, with which Newton was familiar.

[28] See footnote 32 in the Transcription.
[29] See footnote 59 in the Transcription.

7

ASTRONOMY

The character of Newton's early interest in astronomy is reflected in the large number of entries on subjects such as comets, the possible influence of the Sun and Moon on the tides, and what moves the planets. There is little that is conceptually exciting in these entries; they appear to represent Newton's introduction to an area of expertise that he was quickly to master and soon to revolutionize. But at the time of the *Questiones*, Newton was only learning his trade.

The astronomical entries fall into two groups: (1) The cometary notes, which consist of notes on the comets of 1585 and 1618, as well as reports of Newton's observations of two comets in 1664–5, and form a single continuous block of notes in the *Questiones* covering folio 12 93^v and folios 54 114^v through 58 116^v; (2) a number of diverse notes scattered throughout the *Questiones* touching on topics such as the tides (47 111^r and 49 112^r), Cartesian astronomy (11 93^r and 49 112^r), and cosmology (27 101^r and 83 129^r). We shall consider these two groups in turn, referring to them as the "cometary notes" and the "miscellaneous astronomical notes."

1. The cometary notes

Newton's notes on comets begin on folio 12 93^v. He initially considers Cartesian astronomy, including some comments on Descartes's theory of cometary motion and his account of the nature of a comet's tail. We shall consider these notes later. Just before Newton's first report of his own observation of a comet, there is an entry on one of the comets of 1618: "On October 26, 1618, in Scorpio there appeared a

296

comet, its tail being extended between Spica virginis and Arcturus toward the North Pole." We have been unable to trace the exact source of this entry. The entry is peculiar in that the date given is certainly wrong. There were three comets observed in 1618. The first was seen in August in the constellation of Ursa Major. Horatio Grassi reports in his *On the Three Comets of the Year MDCXVIII* (Rome, 1619) that "when the Italian and German observations had been compared, it was discovered that on the twenty-ninth day of August the comet had been between two stars – numbers 22 and 39 – of the Great Bear, and by its [swift] motion it had reached the forelegs in the space of four days, so that on the second of September it was observed under the exterior stars numbered 33 and 34, and there it finally vanished after having completed a journey of about 12°."[1] The second comet appeared on November 18, according to Grassi, and was "of a very slight brilliance but of such great magnitude that it formed a visual angle of about 40°."[2] This comet was located close to the constellation Crater when first seen, and by November 30 it was in the constellation Hydra. On the third comet, Grassi comments that "it surpassed the others in magnitude of light and daily continuance, so it was outstanding as long as it remained by reason of its course and life, and it drew to itself the eyes of all."[3] The exact day of its appearance was disputed to be November 14, 26, or 29. "But, finally, whatever of these dates saw the first light of the comet, it is Scorpio which was its true native land. On the twenty-sixth day, it reached the ecliptic nearly $14\frac{1}{2}°$ inside Scorpio, and on the Twenty-ninth this new foetus was established in Scorpio at a longitude of about $11\frac{1}{2}°$, between the two scales of the Balance with a Northerly latitude of almost 7°."[4]

Newton's reference is clearly to the third of these comets and is misdated by a month. His more complete notes on this comet (those on 54 114^r and 55 115^r) are drawn from Willebrord Snell's *Descriptio cometae* (Lugduni Batavorum,

---

[1] Horatio Grassi, *On the Three Comets of the Year MDCXVIII in the Controversy on the Comets of 1618*, translated and edited by Stillman Drake and C. D. O'Malley (Philadelphia: University of Pennsylvania Press, 1960), p. 7.

[2] *Ibid.*, p. 8. [3] *Ibid.* [4] *Ibid.*, p. 10.

1619), which gives the date of initial appearance as November 27, 1618. It would appear, then, that Newton's entry is from another source, and a mistaken one.

Early in December 1664 a comet appeared that drew the attention of observers everywhere. It was reported to have been seen as early as December 4 near the constellation of Crater.[5] The *Questiones* contain a report on this comet dated December 9 on folio 55 115^r. This report appears to us to have been added after a report of December 10 on folio 12 93^v. The December 9 report reads: "On December 9, 1664 (Old Style), at 4 in the morning the latitude southward of the comet was 20°, its longitude 182°. The length of its tail was 20°." This is all very matter-of-fact for a young man interested in science, and seeing his first comet. If he was the observer, it is surprising he did not say so, because he does in all the other entries. Finally, he writes "the comet" as if it is perfectly clear which comet he is writing about; yet according to the date, this would be the first entry regarding that comet.

The first report of an observation made by Newton himself is that of December 10, 1664[6] (folio 12 93^v). Newton gives the position of the comet in relation to the center of the Moon, which is an unsatisfactory way to record a comet's location in the absence of a theory that allows the determination of the Moon's position on a given date and time. Newton was to spend many years discovering just how difficult the construction of such a theory is, but at the time of this comet observation he seems not to have known that such a theory did not exist. Far more odd, however, is Newton's claim that "Her [the moon's] place being Capricorn 26° 2' or else Aquaris 5°," which he cancelled before cancelling the whole of his first comet report. He had good reason to cancel it, because Capricorn and Aquarius are summer constellations and were not visible on the night of December 10, 1664. It is fairly clear that at the time of this entry Newton was almost a complete novice in astronomy. It also

[5] See *The Correspondence of Henry Oldenburg*, edited and translated by A. R. Holland and M. B. Hall (Madison: University of Wisconsin Press, 1966), Vol. II, Item 366, pp. 353–7.

[6] This date is given in the Julian calendar, as are all of Newton's dates unless he specifies "Gregorian style" or "New Style." See footnote 2 in the Introduction.

appears that Newton quickly recognized his deficiencies and immediately moved to correct them.

Before making his next entry on folio 12 93^v, a comet observation of December 17, Newton read Snell's *Descriptio cometae* and made notes from it on folios 54 114^v and 55 115^r. These notes consist of tables and descriptions regarding the comet of 1585 and the third comet of 1618.[7] They are immediately followed by the report of the comet's position on December 9 discussed earlier. This entry was very likely made after the observation reports of December 10 and 17 on folio 12 93^v, but before that of December 23 on folio 55 115^r.

The report of December 17, 1664, is much in the style of Snell. It gives positions relative to specific stars and a full description of the tail, not only its length but also its position, much like Snell's descriptions, such as the one noted by Newton at the bottom of 54 114^v and the top of 55 115^r. There is, however, one significant difference: Snell gives his positions using latitude and longitude; Newton uses declination and right ascension. This difference can perhaps be explained if Newton was reading other astronomical texts at this time. An examination of Newton's fourth report of a comet observation (folio 55 115^r) gives further reason to believe that that was the case; indeed, it provides a clue to what the other astronomical work was.

On the night of Friday, December 23, 1664, Newton observed the comet a fair distance from where he had last seen it. We know from a letter written by John Wallis (probably to Henry Oldenburg), dated Oxford, December 24, 1664, that Wallis had heard about the comet on December 19 and complained that the nights of December 19 through 22 "were all so cloudy & misty with us yt though I did attempt it, I could not see it."[8] It is therefore likely that Newton, sixty miles away at Cambridge, was confronted with the same weather and was unable to see the comet again until the night of December 23, which Wallis reports "was a very bright Moon-shiny night." Newton writes of his observation on December 23 almost as if he were seeing

---

[7] See footnotes 111–15 in the Transcription.

[8] *The Correspondence of Henry Oldenburg*, Vol. II, Item 362, p. 339.

another comet, showing a caution that is not characteristic of the *Questiones,* and perhaps reflecting again his lack of self-confidence in this new area of study. Here, too, Newton's description involves giving the angular distances of the comet from specific stars (Aldebaran and Rigel) and an elaborate description of the tail (and the nucleus), but he returns to giving the positions in longitude and latitude. In addition, there is a description of the comet's position within the constellation Cetus (the whale): "The comet was then entering into the Whale's mouth at the lower jaw."[9] This descriptive phrase and such others as "the middle star in the Whale's mouth," "star below the whale's eye," and "star in the hinder part of the head" (which are used in subsequent reports) are stock phrases used and codified by Tycho Brahe in his star table, which appeared in its full form in Vincent Wing's *Harmonicon celeste* (London, 1651) under the heading "A Catalogue of 1000 *of the* Fixed Stars, *according to the accurate* Observations *of* Tycho Brahe, *and by* Him *rectified to the beginning of the* year 1601."[10] Although this catalog appeared in a number of books of the period, it was rarely complete, containing all the stock phrases used by Newton in his star descriptions. The *Harmonicon celeste* also contains a table of declinations (pp. 213–15) and a table of right ascensions (pp. 217–19). Furthermore, we know that Newton owned a copy, which is now in the Butler Library of Columbia University.[11] The copy is well annotated in Newton's hand. Thus, we think it likely that in addition to Snell's *Descriptio cometae,* Newton was reading Wing's *Harmonicon celeste* during December of 1664.

From Snell, Newton learned the language of comet description and a mixture of ancient and modern terminology ranging from Aristotle's use of κόμην (beard)[12] to Kepler's use of *cauda* (tail).[13] There is no indication in Newton's re-

---

[9] Folio 55 115ʳ.

[10] Vincent Wing, *Harmonicon celeste: or, The celestial harmony of the visible world* (London, 1651), pp. 240–62.   [11] See Harrison 1744.

[12] See Aristotle's *Meteorologica,* 342b27–345a10.

[13] Kepler, *De cometis* (Augsburg, 1619). Pliny the Elder, in his *Natural History,* Book II, Chapter XXII, gives a large number of descriptive terms used with comets. He writes: "The Greek call them 'comet,' in our language 'long-haired stars,' because they have a blood-red shock of what looks like shaggy hair at their

ports of his acceptance or rejection of any particular theory of the nature of the comet's tail, though there is an entry elsewhere in the *Questiones* that asks if Descartes's theory of the tail as an optical phenomenon will suffice. We shall return to this issue in our subsequent discussion of the "miscellaneous astronomical notes." There is also an indication of what might be his surprise in the report of the observation of December 23, 1664 (55 115^r), when he writes, "I observed a comet whose rays were round her, yet her tail extended itself a little toward the east and parallel to the ecliptic." Aristotle had divided comets into two kinds; those that occur below the Moon and those that occur above. In the case of those below the Moon, the comet is an "independent phenomenon" and does not move with the stars but "seems to fall behind the stars, as it follows the movement of the terrestrial sphere." These comets are further divided into two groups by Aristotle: "[when] a comet is produced, its exact form depending on the form taken by the exhalation . . . if it extends equally in all directions it is called a comet or long-haired star, if it extends lengthwise only it is called a bearded star."[14] It would then be unusual, according to the Aristotelian understanding, for a comet to have rays around it (a "long-haired star") *and* to have a tail (a "bearded star"). This perhaps explains Newton's use of the word "yet" in his report. Although this phenomenon is not particularly unusual, probably neither Aristotle nor Newton had seen many comets.

From Wing, Newton learned the standard ways of describing stars, the various systems of designating star locations, and how to move back and forth between those sys-

top. The Greeks also give the name of 'bearded stars' to those from whose lower part spreads a mane resembling a long beard. 'Javelin-stars' quiver like a dart . . . the same stars when shorter and sloping to a point have been called 'Daggers'; these are the palest of all in colour, and have a gleam like the flash of a sword, and no rays, which even the Quoit-star, which resembles its name in appearance but is in colour like amber, emits in scattered form from its edge." To these, Pliny adds the "Tub-star," the "Horned star," the "Torch-star," the "Horse-star," the "Goat Comets," and the "Mane-shaped" comet. Pliny, *Natural History*, translated by H. Rackham, Loeb Classical Library (Cambridge, Mass.: Harvard University Press, 1938), Vol. I, lines 89–90, pp. 231–3.

[14] Aristotle, *Meteorologica*, translated by H. D. P. Lee, Loeb Classical Library (Cambridge, Mass.: Harvard University Press, 1952), Book I, Chapter VII, 344a19–24, p. 51.

tems. By January 1, 1665, Newton was using latitude and longitude, declination and right ascension, and almacantars and azimuth with apparent equal ease. Another sign of Newton's growing competence during this short period of less than a month is the difference between his drawing of a comet on folio 12 93ᵛ (which tells us very little) and the drawing on folio 56 115ᵛ, which shows the comet's position relative to the fixed stars and illustrates the transparency of its tail.

Newton's observations of the comet of 1664 were made on December 9, 10, 17, 23, 24, 27, 28, 29, and 30 and on January 1, 2, 10, and 23. We know from Wallis's letter to Oldenburg that the comet was difficult or impossible to see at Oxford owing to poor weather on December 19–22, 25, and 26. On December 27, Wallis reported that the comet was not visible at Oxford, though Newton reported seeing it at Cambridge. For December 31 through January 3, Wallis reported that it was too cloudly to give an accurate position. Wallis gave reports for January 4, 6, 8, 9, and 10, but nothing about the possibilities of observation on January 5 and 7. He did report that from January 11 to the writing of his letter on January 21 "the nights were either cloudy, or inconvenient by reason of ye Moon-shine."[15]

By and large, the reports of Wallis and Newton on the comet's position agree, apart from the difference in their reports for December 23. Newton reported the comet entering the mouth of the whale (Cetus) that night, whereas Wallis placed it in the "Voyd space between Taurus & Eridanus." On December 28, Wallis saw it in the head of the whale, once again bringing their reports into agreement. What is even more curious is what Wallis wrote in his letter of December 24: "Last night [December 23] was a very bright Moon-shiny night, when wee had hopes of seeing it. And, to be sure not to miss it, I watched it (with divers more) in or on ye Tower of ye publicke Schooles, from eleven of ye clock till all most six in ye morning, but could see nothing of it: though wee had no disadvantage but ye too much light of ye Moon, & a misty obscurity round the

---

[15] See footnote 5.

horizon; wch in this place [Oxford, in the Thames Valley] we are seldom free from."[16] Thus, Wallis's letters of December 24 and January 21 contradict each other as to whether he did or did not see the comet on December 23. It seems most likely that he did not, because it is unlikely that his memory erred one day after the event in question; and this (perhaps) removes the difficulty of the difference between his report and Newton's report.

Newton's last observation of the comet was a sighting on January 23 indicating that the tail was scarcely discernible, whereas in a report of his observation on January 10 he wrote (57 116^r) of the tail as being very weak and the nucleus as "grown very dim." One would surmise that the comet was not visible to the naked eye after this date, although we know that the comet was visible through telescopes all throughout February and part of March.[17]

In any case, it would appear that Newton's interest in comets was fading. When the comet of 1665 appeared, Newton entered only three reports in the *Questiones*. They are dated April 1, 4, and 5 and are very straightforward, stating the date and time of the observation and the position relative to the fixed stars (using the standard phrases drawn from the Tycho star table in Wing), but without description of the tail. There is no mention in the *Questiones* of the dispute over the comets between Hévélius and Auzout, in which the Royal Society played the roles of both instigator (through its secretary, Henry Oldenburg) and mediator, though we know that Newton read the *Philosophical Transactions* of the society for the year 1665 (its first year of publication). He seems to have lost interest in these matters. Although the Royal Society's astronomers concluded that Auzout's observations were the more accurate, history has redressed the balance; the comets are now known as Hévélius 1664 and Hévélius 1665.

Hévélius 1664 is a comet of indeterminate period, with an orbit of eccentricity 1.0, a perihelion distance in astronomical units of 1.026, an argument of perihelion of 310.7°, the longitude of ascending node 85.3°, and an inclination of its

---

[16] See footnote 8.
[17] *The Oldenburg Correspondence*, Vol. II, Item 378, pp. 405–6.

orbit to the plane of the ecliptic of 158.7°. Hévélius 1665 is also of indeterminate period, with orbit of eccentricity 1.0, perihelion distance 0.106, argument of perihelion 156.1°, longitude of ascending node 232.0°, and inclination of orbit 103.9°.[18]

2. Miscellaneous astronomical notes

The *Questiones* contains only a few notes on astronomical matters apart from those that relate directly to the comet observations of 1664 and 1665. The topics covered are aspects of Cartesian vortex theory, the causes of the tides (also connected to vortex theory), a few references to Descartes's comet theory, and a consideration of biblical evidence that God created time.

Newton attacks (as we have seen) some aspects of Descartes's vortex theory on folio 11 93ʳ.[19] This attack seems directed not at the existence of the Cartesian vortices but rather at the multiplicity of functions that Descartes attributes to them. Newton writes: "Whether Descartes's first element can turn about the vortex and yet drive the matter of it continually from the Sun to produce light, and spend most of its motion in filling up the chinks between the globuli." Can the vortices do all this and more? How can the least globules remain next to the Sun and yet always come from it to cause light? Newton clearly believes the various functions of the vortices on the Cartesian account to be inconsistent with one another. In particular, he believes that what he takes to be Descartes's theory of light places impossible constraints on the vortices and is an internally inconsistent theory as well. For example (as we have noted in Chapter 2, Section 3), Newton asks if it is possible to obscure the Sun (folio 11 93ʳ), given that the motion (pressure) should be transferred through the obscuring object to the matter on the other side. Most of Newton's objections here display a misunderstanding of the Cartesian theory of light. Newton argues as if the Cartesian account involves the motion of particles from the thing seen to the eye, which is

[18] P. L. Brown, *Comets, Meteorites & Men* (New York: Taplinger Publications, 1974), App. IV, p. 236.   [19] See Chapter 2, Section 3.

specifically denied by Descartes and replaced with the notion of "pression."

That Newton's attack is not directed at the vortex notion
itself, but rather at its particulars, is made clear by the fact
that Newton accepted the existence of Cartesian vortices
during the 1660s and into the 1680s. William Whiston reported in his *Memoirs* that when Newton attempted to apply
his inverse-square law of gravitation to the Moon's motion
in 1666,

> He was, in some Degree, disappointed, and the Power that restrained the
> Moon in her Orbit, measured by the versed Sines of that Orbit, appeared
> not to be quite the same that was to be expected, had it been the Power of
> Gravity alone, by which the Moon was there influenc'd. Upon this Disap-
> pointment, which made Sir Isaac suspect that this Power was partly that
> of Gravity, and partly that of Cartesius's Vortices, he threw aside the
> Paper of his Calculations and went to other Studies.[20]

In Conduitt's version of the same incident, he writes, "his
computation did not agree with his theory & inclined him
then to entertain a notion that together with the force of
gravity there might be a mixture of that force wch the moon
would have if it was carried along in a vortex."[21] Finally,
Newton's own notes in his copy of Vincent Wing's *Astronomia Britannica* (London, 1669) give support to the accounts
of Whiston and Conduitt.[22]

On folio 12 93^v Newton asks if the Sun's rotation on its
axis causes the motion of its vortex or if the motion of the
vortex causes the rotation of the Sun. But on folio 49 112^r
he seems convinced that in the case of the Earth, the rotation of the Earth on its axis is not "helped by its vortex." If
it were, then the water and air would move more rapidly
toward the east than the Earth does. He suggests on folio
47 111^r that the issue whether or not the Earth is helped by
the vortex in its annual motion around the Sun is not settled. He writes: "Try also whether the water is higher in
mornings or evenings, to know whether the Earth or its
vortex press forward most in its annual motion." Thus, the

[20] William Whiston, *Memoirs of the Life of Mr. William Whiston by himself* (London, 1749), 2 vols., Vol. I, pp. 35–6 (also quoted in Herivel, *Background*). Whiston asserts that this account was given to him by Newton in 1694.

[21] Keynes ms. 130.4, pp. 10–12. Quoted in Westfall, *Never At Rest*, p. 154.

[22] See note 44, p. 155, of Westfall, *Never At Rest*.

tides, he is suggesting, can be used to see if the Sun's vortex causes the Earth's motion or if the Earth's motion helps the vortex around.

These questions are of great interest because they show that Newton is at this point concerned with what causes the planets to move, and at the same time he allows that that cause may be independent of the vortices. Thus, although he believes that the vortices exist and play some causal role in our understanding of planetary motion, there are other important factors that need to be investigated (e.g., the cause of the motion of the Earth). Also, Newton is not happy with the supposed role that vortices play in Cartesian tidal theory. His notes on folios 47 111^r and 49 112^r reflect a skepticism concerning Descartes's account of the tides, and in one place he seems to doubt that the Moon is the cause of the tides at all. He writes: "Tides cannot be from the Moon's influence for then they would be least at new moons" (folio 47 111^r). One does not know quite what to make of this. It is true that tides are higher at full and new moons, but there seems to be no reason why we *should* expect them to be lowest at a new moon. Descartes's theory purports to explain the occurrence of highest tides at new and full moons by the claim that the Moon and Earth are closest at those times, and so the pressure of the vortex is greatest. Because it is most likely that Newton was thinking of the Cartesian theory when he wrote this remark, one can only surmise that he was unaware of the details of the theory. As we have seen (Chapter 2, Section 3), the rest of Newton's comments consist in stating the consequences of Descartes's theory and making suggestions as to how to test the truth of those consequences, and thus the theory.

Descartes asserts that the increased pressure of the vortex on the side of the Earth toward the Moon causes the Earth's center to be displaced from the center of the vortex away from the direction of the Moon. This leads Newton (as we have seen earlier) to suggest on folio 49 112^r that we should see a monthly parallax of a nearby object like Mars, because, in his account, the center of the Earth moves around the center of the vortex each month while staying opposite the Moon. Had Newton made the observation, he would have

noted such a monthly parallax, but not in confirmation of
Descartes's theory. As Newton was later to predict, the
Earth and Moon rotate each month about a point approxi-
mately one-fifth of the way between the Earth and the
Moon, which is the center of gravity of the Earth-Moon
system. The size of this rotation is much greater than that in
Descartes's account, which places the center of the vortex so
close to the center of the Earth as to lie within the planet.
Thus, the amount of parallax observed would be greater
than that predictable from the Cartesian theory.

Newton comments on three other matters in Descartes's
astronomy. The first is to ask if the vortex can carry a comet
toward the poles. The answer for Descartes is clearly no,
because the comets move more or less along lines tangential
to the rotating vortex and thus at right angles to the axis of
the vortex. The comet can change its direction of motion as
it moves from vortex to vortex, but within each vortex it
tends to travel along a tangent to the rotation of the vortex.
Newton's aim is once again to point to a consequence of a
theory as a means of testing that theory. In this light, his
observation of the comet for the night of December 27
(55 115^r and 56 115^v) becomes extremely interesting, for
there Newton writes that "it [the comet] moved northward
against the stream of the vortex cutting it at an angle of
about 45° or 46°." Such an observation is inconsistent with
the Cartesian theory regarding both the motion against the
vortex and the motion across it.

The last notes we shall discuss are marginally astronomical in that they concern what would now be considered a
cosmological issue – the creation of time. Newton notes two
biblical passages, one from the Old Testament and one
from the New Testament, that appear to support the view
that God created time. The first, on folio 12 93^v, is quoted
from Hebrews, Chapter 1, Verse 2: "God made the worlds
by his son τοὺς αἰῶνας." The King James Bible has the
following for Verses 1 and 2 of Chapter 1, Hebrews:

1   God, who at sundry times and in divers manners spake in time past
unto the fathers by the prophets,
2   Hath in these last days spoken unto us by *his* Son, whom he hath
appointed heir of all things, by whom also he made the worlds;

Newton's interpretation is quite literal and very different from the King James version, which treats τοὺς αἰῶνας as meaning heir to the ages.

Newton quotes the verse again on folio 83 129ʳ in connection with another reference to the Bible. He writes: "Whether Moses saying that the evening and the morning were the first day, Genesis, Chapter 1, proves that God created time. As expanded at Colossians, Chapter 1, verse 16, or Hebrews, Chapter 1, verse 2, τοὺς αἰῶνας ἐποίησεν, he made the worlds, proves that God created time."

These entries draw their interest from our knowledge of Newton's later position, expressed in the *Principia,* that "Absolute, true, and mathematical time, of itself, and from its own nature, flows equably without relation to anything external. . . ." This statement does not explicitly state that time is eternal (i.e., uncreated), but from the properties that Newton indicates, that conclusion could be inferred. It is perhaps not surprising that he should change his mind on this matter between mid-1664 and 1687, but what is interesting is that he may perhaps have done so within as little as two years.

At some time between 1666 and 1669 Newton wrote a manuscript entitled *De gravitatione.* As we have noted, it is a polemic against Descartes's views on motion and body in which Newton attempts to show that extension characterizes space primarily, and body derivatively. Newton puts forward the view that whereas the creation of bodies is a product of God's will, space and time are co-eternal with God, and so not the products of a creative act. They exist because God exists.[23] Newton writes:

Space is an affection of being insofar as it is being. No being exists, or can exist, which is not related to space in some way. God is everywhere, created minds are somewhere, and body is in the space that is occupies; and what is neither everywhere nor anywhere does not exist. Hence it follows that space is an emanative effect of the primarily existing being, because when any being is posited, space is posited. The same may be affirmed of duration: both are affections of being or attributes. According to which the quantity of existence of each individual [being] is de-

[23] See Martin Tamny, "Newton, Creation, and Perception," *Isis,* 70(1979):48–58, and J. E. McGuire, "Existence, Actuality and Necessity: Newton on Space and Time."

nominated as regards its amplitude of presence [*amplitudinem praesentiae*] and its perseverance in existence. So the quantity of existence of God is eternal in relation to duration, and infinite in relation to the space in which he is present; and the quantity of the existence of a created thing is as great, in relation to duration, as the duration since the beginning of its existence, and in relation to its amplitude of presence as great as the space in which it is.[24]

Thus, the notion that time is created does not represent a view that was long to last. But the entries are interesting, for the very reason that they show us that Newton was thinking about this issue as early as mid-1664. One notion present in these entries is to continue into Newton's later theological thought, namely, Christ's role in the creation. The idea that God did not make the world, but used an intermediary, was to become a central question in Newton's Unitarian theology.[25] There is no indication in the *Questiones* that he had yet come to consider this particular theological view.

[24] *De gravitatione*, p. 103, in Hall and Hall. Translation here by J. E. McGuire, as in "Existence, Actuality and Necessity: Newton on Space and Time," *Annals of Science*, 35(1978):463–508, p. 466.

[25] See Frank E. Manuel, *The Religion of Isaac Newton* (Oxford: Clarendon Press, 1974), pp. 57ff.

8

THINGS AND SOULS

1. Things

On folio 37 106ʳ Newton considers the phenomenon of resonance.[1] He notes the fact that "in every sound the eighth above it, but not below it seems to be heard." Thus, he is aware that in resonance, the harmonics, not the subharmonics, are produced. He attempts to give an account of this by invoking the fact that air is composed of "some more subtle, some more gross matter." In this conception, he thinks that the tonic produces a vibration in the "grosser part" of the air and that the various harmonics produce vibrations in successively rarefied media, "the subtlest matter [being] prone to quickest vibrations." There are no subharmonics, because the vibrating string or pipe sets in motion the 'grossest matter' that it is able to move, but any grosser matter is merely separated by the string and does not vibrate. Sympathetic vibrations are explained by Newton in terms of the same model. He says that "one string struck, by the mediation of the air moves an unison string of another instrument better than that which is an eighth above or below it. The string is most easily moved by the air when its motion can be most conformable to the motion of the greatest part of the air."

Unlike modern practice, Newton counts his notes consecutively from a tonic; thus, he characterizes the harmonics of a sound as "its concomitant eighth, and perhaps 15th and 22nd." Newton says that "eighths seem to be unisons," as indeed they are, because they are multiples of vibrations that reinforce one another; the 'eighth' is twice the funda-

---

[1] We thank Daniel Greenberger for a short introduction to the physics of music that helped us make sense of this entry.

mental, the 'sixteenth' is four times, and the 'twenty-fourth' is eight times.

Newton then asserts that "more violent breathing raises the sound an . . . eighth or 15th, not a 12th, or but seldom to a 12th." A twelfth represents vibrations three times the fundamental. If the tonic is middle C, the twelfth is the G above the next C. In the case of instruments like clarinets and oboes, the twelfth is vividly present among the harmonics, so that Newton's assertion wrongly characterizes them. But in the case of an open organ pipe, which is analogous to a string fixed only at one end, the eighth, sixteenth, and twenty-fourth strongly dominate the harmonics, with the twelfth being barely discernible. This is most likely the case that Newton had in mind. If the organ pipe is closed, which is analogous to a string fixed at both ends, the eighth and the 'twelfth' are both discernible.[2]

Newton next asks how the intensity of sound falls off with distance. He answers that "if the sound be a at the distance b it shall be abb/xx at the distance x." This expression involves an inverse-square relationship of distance to the intensity of sound. If the intensity of the sound is a, it is inversely proportional to b^2, $a \propto (1/b^2)$. At the distance x, the intensity of sound is equal to $1/x^2$. Hence, the intensity of sound at the distance x is directly proportional to $(ab^2)/(x^2)$.[3] It is interesting that Newton should at this time consider a phenomenon whose behavior can be characterized by an inverse-square relationship with respect to distance.

This entry indicates two facets of Newton's approach to phenomena that should be emphasized. First, he views the nature of sound as analogous to the nature of light insofar as the feature that distinguishes among colors is the same as the feature that distinguishes among pitches. He asks "whether are acute or grave sounds the swifter?" So here, as in his approach to colors, speed is the differentiating parameter; this remains a characteristic of his lifelong search for unify-

[2] A probable source is Marin Mersenne, *Harmonicorum libri* [12] *in quibus agitur de sonorum natura causis et effectibus: de consonantiis, dissonantiis . . . compositione, orbisque totius harmonicis instrumentis* (Paris, 1635), Part II, *Liber primus*, propositions XLI through XLIII, pp. 62–8.

[3] If $a = 1/b^2$, then $ab^2 = 1$. Substituting ab^2 for 1 in $1/x^2$ yields Newton's expression $(ab^2)/x^2$.

ing features in nature.[4] Second, the entry reveals a concern to establish mathematical relationships that reveal the nature of phenomena. Thus, he is concerned not only to establish proportions that apply to those features of the tonic that produce vibrations in a medium but also to argue that these properties, if mathematically expressed, 'capture' the invariant structure of the world itself. After all, this is what one would expect from a writer concerned to marshal a theory of mathematical indivisibles that correspond to the atomic structure of sensible objects. Later, Newton will carry his sensibility concerning 'nice' proportions further in the *Opticks,* where he argues that the spectral colors answer to the mathematical properties inherent in the divisions of a musical scale.[5] Thus, his interest in the relationship between the mathematical description of the world and the world itself is a clear precondition of his later achievements in mathematical science.

Newton's notes on magnetism (folio 29 102^r) presuppose an effluxial theory of magnetic attraction. The character of these entries is nonspecific enough to make the identification of his sources difficult. The first entry mentions the standard way of identifying magnetic north, a procedure found in every treatise on magnetism. However, Newton turns it into a way of determining "the motion of any magnetical ray." Continuing this line of inquiry, he asks "whether magnetic rays will blow a candle" and "whether a lodestone will not turn around a red hot iron fashioned like windmill sails, as the wind does to them."[6] Newton considers the possibility that the gravitational efflux (see Chapter 6, Sections 1 and 2) could be augmented by a magnetic efflux. For example, he asks whether a magnet is heaviest when the north pole or south pole is uppermost. He suggests that if there is a definite answer, a perpetual-

---

[4] It is frequency (or wavelength) that determines color *and* pitch. Thus, Newton was wrong in his choice of parameter, but correct in thinking it would be the same in both cases.

[5] For a discussion of this sensibility, see Alan Shapiro, "Newton's 'Achromatic' Dispersion Law: Theoretical Background and Experimental Evidence," *Archive for the History of Exact Sciences,* 21(1979/80):91–128, Section 25.

[6] Red-hot iron is not attracted by magnets; yet the efflux will pass over it.

motion device might be constructed by mounting two
magnets in opposite orientation on either side of the axle
of a vertically oriented wheel. Because one of the magnets
will be heavier than the other (given Newton's prior supposition), the wheel will be overbalanced. After the wheel
rotates through 180°, the magnets will have switched their
positions, so that the one that was lighter before is now the
heavier, and thus the wheel will continue to turn (see diagram, 29 102^r). Another perpetual-motion device based on
the same principle appears on the same folio: A single
S-shaped magnet is pivoted vertically, the 'heavier' pole
represented by the arrow in the accompanying diagram.
Two other perpetual-motion devices are drawn by Newton.
They appear to be based on the phenomenon of attraction
and are not dependent in any direct way on the effluxial
theory. In the first case, a magnet is fixed to the rim of a
vertically mounted wheel, so that a magnet adjacent to the
wheel attracts the fixed magnet and draws the wheel
around. The momentum of the wheel carries the fixed
magnet past the adjacent one, with the result that the
other pole of the fixed magnet is repelled, and the wheel
turns round. The second device consists in a spring of steel
fixed at one end and drawn toward the magnet until the
tension in the spring snaps it back, at which time the magnet draws it forward again.

On folio 67 121^r Newton introduces a method for finding
the weights of bodies composed of various substances in
different media such as air or water. He asserts that this
method will work in the determination of the absolute
weight of a body (i.e., its weight in a vacuum). We are asked
to consider two bodies, one of gold and one of silver. In air,
the quantity of gold (a) is equal in weight to the quantity of
silver (z), their weight being b. We are also told that in water
the quantity of gold (a) is equal to the quantity of silver ($2z$).
Furthermore, if the quantity of water displaced by the gold
(a) has weight c, and the quantity of water displaced by the
silver (z) has weight d, then $a : c :: z : d$, or $a/c = z/d$, and
$d = (cz/a)$. Newton then invokes a principle that is expressed in Proposition 7 of Book I of Archimedes, *On Float-*

COMMENTARY

ing Bodies, a work that is contained in Newton's 1615 edition of the *Opera quae extant.*[7] The proposition states that "A solid heavier than a fluid will, if placed in it, descend to the bottom of the fluid, and the solid will, when weighed in the fluid, be lighter than its true weight by the weight of the fluid displaced."[8] By this principle, the weight of the gold a equals $b - c$, and that of the silver z equals $b - d$. Because $d = (cz/z)$, $z = b - (cz/a)$. And, because $a = 2z$ and $a = b - c$, we have $b - c = 2[b - (cz/a)]$, or $b - c = 2b - (2cz/a)$. Solving this last equation for z yields $z = (ab + ac)/(2c)$, or $(2c)/(b + c) = a/z$. This last can be represented as the proportion $2c : b + c :: a : z$. Finally, solving for c in $z = (ab + ac)/(2c)$ yields $c = (ab)/(2z - a)$. In order to find the 'absolute' weight of a body, however, one would have to determine the weight of the volume of *air* displaced by that body.[9] But this Newton does not do.

At the bottom of the folio, Newton writes: "Try whether flame will descend in Torricelli's vacuum." Several other entries to the same effect appear in the *Questiones.* It is likely that Newton's interest in this matter was prompted by his knowledge of Galileo's view that all substances have weight, as opposed to the traditional Aristotelian view that fire has the greatest 'levity' among natural substances.

As we noted at the beginning of Chapter 5, Section 3, Newton is interested in color phenomena as indicators of the ways in which natural substances affect perception and the behavior of light. That is particularly so as regards the differences between objects when looked at and when looked through. His interest in this phenomenon stretches from the *Questiones* to the *Opticks.* Indeed, some of the examples that Newton uses in the *Questiones* are found in Proposition X, Problem V, of Book I, Part II, of the later treatise. For example, in the *Questiones* he uses *Lignum Nephriticum* in entry 47 of folio 94 134[v] to draw attention to this important distinction. The same example appears as

---

[7] See Harrison 75.

[8] Archimedes, *On Floating Bodies,* in *The Works of Archimedes with the Method of Archimedes,* edited by T. L. Heath (New York: Dover Publications, 1953), Book I, Proposition 7, p. 258.

[9] In this connection, it should be noted that c and d are the weights of quantities of water *in air.*

entry 2 of the first folio of Add. 3975 and in the problem from the *Opticks* mentioned earlier.

Newton's interest in color phenomena goes beyond their use in gathering information about the nature of light; he also notes their possible use as indicators of chemical changes in substances, a point not emphasized in that work by Boyle that is the source for most of Newton's information in this area. It is difficult to imagine, of course, that Newton could fail to note that chemical changes produce color changes, a fact recorded in many of his entries. The implications of this fact are well developed in his later *De natura acidorum.*[10]

2. Souls

"Of the Creation," on folio 83 129ʳ, is certainly directed toward the question of *creatio ex nihilo,* but the emphasis is on the problem of individuating individuals and natural kinds. Newton observes that matter itself cannot individuate whales, if it exists prior to their creation. Here he refers to matter generically, because he does not deny that a particular organization of matter can function as a principle of individuation. For "few men are of the same temper, which diversity arises from their bodies, for all their souls are alike; so why may not the several tempers or instincts of diverse kinds of beasts arise from the different tempers and modes of their bodies, they differing one from another more than one man's body from another's." But equally, "the soul or form" cannot function as the principle of individuation, because it is the same in each individual of a particular species. Thus, philosophically considered, Newton's position is Aristotelian in spirit. But a principle of economy directs Newton's thought, revealing his true posi-

[10] In *Experiments Touching Colours*, Boyle wrote of his Experiment XL, which is Newton's entry 32 on folio 92 133ᵛ: "of all the Experiments of Colours, I have yet met with, it seems to be fittest to recommend the Doctrine propos'd in this Treatise, and to shew that we need not suppose, that all Colours must necessarily be Inherent Qualities, flowing from the Substantial Forms of the Bodies they are said to belong to, since by a bare Mechanical change of Texture in the Minute parts of Bodies, two Colours may in a moment be Generated quite *De novo,* and utterly Destroy'd" (pp. 302–3). It is likely that Newton's thoughts were led in the same direction by this experiment.

Of ye Creation

The word בראא wch Gen i.i. is interpreted to create
something out of nothing is used Gen g^{1st} v: 2g where
tis said God created greate Whales &but ye matter out
of wch they were created did exist before wither is
it ment of creating ye soule or forme of ye whale, for y^t is
not ye whale alone. as there may be but one kind of irrati
call souls wch joyned wth severall kinds of bodys make
severall kinds of beasts, for setting aside ye different shape
of their body beasts differ from one another but in some
qualitys wch are called instincts of nature. now as
in men whose soules are of one kind some love hate
feare &c one thing some another & few men are of
y^e same temper wch diversity arises from y^e their body
for all their soules are alike, so why may not y^e se=
verall tempers or instincts of divers kinds of beasts
arise from y^e different tempers & modes of their bodys
they differing one from another more yⁿ one mans body
from anothers. then y^e can y^e soule of y^e whale
be called y^e whale since before it be joyned wth y^e
whale tis as much y^e soule of a horse & this creating
y^e of whales & severall other creatures must be
but modifying matter into ye body of a whale & infusi
an irrationall soule into it. Eccles: 33 v:s 10 Adam
was created of ye earth.
Whither Moses hy saying Gen ye 1st y^t ye eveing & ye mor
ning were ye first day we do prove y^t God created time.
Coll 1. 16) or heb 1 u 2 v Τοὺς αἰῶνας ἐποίησεν espou
did he make ye worlds. prove y^t God created time

Plate 3. Folio 83 129^r of the *Questiones*. The arrow indicates the line in
Sheltonian shorthand.

tion: "To suppose then that God did create diverse kinds of souls for diverse kinds of beasts is to suppose God did more than He needed.[11] How then can the soul of the whale be called the whale, since before it is joined with the whale it is as much the soul of a horse." The conception expressed here is akin to Averroism, in which individuals are viewed as participants in separate forms or souls that subsist independent of matter. Ironically enough, if the Aristotelian position denies individual immortality, the Averroists allow that impersonal immortality is alone possible. But however it is considered, Newton's position is an apostasy from the Christian perspective on the soul. But this is not surprising, given his biblical orientation in the Old Testament. And although this particular doctrine does not occur in his later thought, in other matters Newton remained a heretic in the bosom of Christianity.

This entry again reveals Newton's drive for unity and simplicity. Here the concern is the nature of individuation with respect to the creation of species and kinds. But as the *Questiones* amply illustrate, Newton's interest in individuation also involves the identity of points, units, and the parts of extension, an interest that goes forward into his later thought, as is evident in his attempts to individuate the 'parts' of absolute space and time.

3. Conclusion

The *Questiones* are the work of a young, active, and inquiring mind with an impressive range of interests. Certainly there is abundant evidence of an intellect concerned with matters of practical interest and bent on mastering the techniques of effective knowledge (one need only mention the entries on optics and mathematics to underscore this point). But equally clear is an interest in metaphysics and epistemology. The entries that refer to Descartes's *Meditations*, together with his replies to objections, are among the earlier in the notebook. Also, many of Newton's dog-earings in the *Opera* mark pages that contain various of Descartes's philo-

---

[11] This line was written by Newton in his Sheltonian brevigraphy. See footnote 141 in the Transcription, and Plate 3.

sophical arguments. And although he developed a deep-seated animus against Descartes's science, many of the latter's epistemological and metaphysical distinctions become the permanent possessions of Newton's intellectual makeup. A case in point is the claim that the intellect understands that extension is actually infinite. Although Descartes denies this particular claim, Newton uses his manner of formulating the epistemic distinction between the powers of the intellect and the powers of the imagination to formulate the epistemological credentials of his own doctrine concerning the positive infinity of extension. Also, it is clear that Newton (partly under the influence of Hobbes) attempts to develop a physicalist program in physiology with the intention of explaining many of the cognitive functions that were traditionally ascribed to the soul. Thus, in the *Questiones* we have the record of a young mind attentive to the minutest detail concerning perceptible things, yet eager to speculate on the ultimate reality of existing things. Not surprisingly, these are characteristics often combined in minds capable of outstanding results. The *Questiones* reveal clearly that Newton was not some mysterious demigod whose intellectual achievements cannot be measured by standards applicable to other human beings. On the contrary, they indicate a muscular mind capable of extraordinary industry and penetration, but one quite capable of espousing untenable positions and of following unproductive turns. To maintain perspective, however, it is well to remember that Newton was twenty-one until Christmas day 1664.

The *Questiones* establish clearly the important early influences on Newton's development. The notion of an 'important influence' is not one that we restrict to the *Questiones* themselves. It also refers to the adoption of beliefs during the period of the *Questiones* that are carried forward into Newton's later thought under recognizable descriptions. In this connection, let us lay a ghost that has too long haunted Newtonian scholarship. There is no positive evidence to show that Newton had read Gassendi at first hand. In every case in which a line of thought can be reasonably attributed to Gassendi, it is clear that Newton's knowledge of it can be traced to Charleton's *Physiologia*. And in any event, one

should not make too much of the Gassendist origins of Newton's views. The only case in which it seems likely that Gassendi was an original source for Newton (to be sure, through
the meditation of Charleton) concerns his views on projectile
motion (i.e., the claim that a body moves by its own motion).
As for Newton's knowledge of certain details of ancient atomism, it was neither Charleton nor his source Gassendi that
played a decisive role. Newton read Epicurus himself.

The fact that Newton read a well-edited copy of Diogenes's *Lives of the Philosophers* is significant. It indicates that
he was familiar with a major source for the views on matter
and extension that he developed in the *Questiones*. But also,
his reading of Epicurus provided a basic framework of concepts from which Newton fashioned his cosmology and the
conception of actual infinity. It is certainly clear that Newton's espousal of indivisibles, especially atomic times, owes
much to Epicurus as well as to Sextus Empiricus.

As the large number of references in the text indicate,
Charleton's *Physiologia* was an important source for many of
Newton's views. But even where he is indebted to Charleton, Newton soon develops his own line of argument, taking
what is essential to his position from Charleton's more prolix expositions. As to the influence of Henry More, there
are good grounds for reevaluating Newton's indebtedness
to his writings. Certainly, More was not a major influence
on Newton's views concerning infinity and extension; moreover, his conception of the 'indiscerpibility' of matter was
not motivated by the sorts of reasons that originally led
Newton to adopt arguments for conceptual indivisibilism.
This is not to deny Newton's indebtedness to More's theory
of matter but rather to insist that Newton was acquainted
with many of the ancient sources of atomism on which More
himself drew. On topics of a specific nature, the influences
of Galileo, Barrow, Wallis, Euclid, and Digby are certainly
present. But their contributions to Newton's general intellectual outlook are minimal, if judged in comparative terms.
The references to Descartes are numerous and for the most
part significant. Indeed, it can fairly be said that Newton's
familiarity with Descartes's science and philosophy was perhaps greater than his familiarity with any other important

source. Although there are more references to Boyle than to Descartes, this difference is insignificant, for in the case of Descartes, Newton expends considerable effort in analyzing his views. His attitude toward Boyle is otherwise, the latter's writings being used for the most part as sources of information.

One of the most interesting, if surprising, orientations in the *Questiones* is Newton's initial commitment to indivisibilism. As we have made clear, there are two important ancient sources for this – Aristotle and Epicurus. Both were available to Newton, and it is undoubted that he made use of Diogenes Laertius. The importance of this should not be overlooked. Furthermore, Charleton and More do not provide an analysis of indivisibilism, the latter, in fact, arguing that extension is mentally divisible to infinity. Also, it is important to notice that Newton argues initially for a consistent indivisibilism (i.e., that if an indivisible magnitude of matter is posited, atoms of time, distance, and motion must be posited as well). The sources for this picture of the world are few indeed. Aristotle subjects it to devastating criticism; but it is Epicurus who provides the model for Newton's arguments. Although Francis Hall was a seventeenth-century indivisibilist, whom Newton probably read, it is difficult to believe that Newton's indivisibilism had its source in his *Tractatus*.

It is clear from the *Questiones* that Newton gave up his indivisibilist program. Nevertheless, it left its mark on his mathematical development. We have argued that he probably adopted Charleton's version of the distinction between what can be said according to the mathematical imagination and what can be said about the way things are. But in terms of this distinction, Newton attempted to conceive his mathematical indivisibles as standing in correspondence to the indivisibilist structure that he ascribed to physical quantity. On seeing the difficulties in this position, he adopted Wallis's conception of indivisibles as tools of mathematical analysis that, in the context of 'measuring' a finite quantity, can be conceived as tending toward, but never equal to, zero. This shift in Newton's mathematical *gestalt* probably occurred at the same time as his rejection of 'Epicurean' indivisibles and no doubt coincided with his full appreciation of Euclidean proportion theory. Even when he moved to fluxional analysis, Newton was still concerned that his mathematical notions be well founded on features of the perceptible world – in the case of fluxions, kinematic notions that pertain to the phenomenon of motion.

It is clear in the *Questiones* that Newton is concerned with foundational issues in mathematics, by which we mean the ancient problem of how mathematical analysis is grounded in the extramental world. It is evident from the beginning that he inclines toward mathematical realism, an orientation that remains evident in certain branches of his mature mathematics. This should cause no surprise, given Newton's actualism and his anti-Aristotelianism concerning the structure of the world.

Another interesting feature of the *Questiones* is his adherence to certain epistemological commitments. Not only are these commitments present in the *Questiones*, but they remain evident in his subsequent thought. So, too, does an unresolved tension. On the one hand, he is oriented toward Hobbes's conception that the basis of all knowledge rests with patterns of motion that constitute the natural realms of human experience; on the other hand, he is clearly influenced by the Cartesian distinction between natures and natured things. The first orientation is expressed in Newton's physiological program and is evident in his approach to the optics of light. Here he proceeds not only on the assumption that knowledge has its basis in sensory experience but also according to a program that attempts to locate various cognitive functions in the physiology and optics of the sensory modalities. The Cartesian influence is more subtle. It involves the notion that there are various ways of knowing specific to the mind's cognitive powers that bring perceptible individuals under different descriptions. Thus, there are intentional objects appropriate to the senses, the imagination, and the understanding. Although it is clear that Newton does not explicitly adopt these distinctions in the *Questiones* (this occurs in *De gravitatione*), they are nevertheless involved in his conception of the actual infinity of extension as expressed in the earlier notebook.

Consistent with his epistemological position, Hobbes rejects the existence of actual infinities construed as real natures. But it is just these realities that Newton is committed to by virtue of his espousal of the infinity of extension in the *Questiones,* and later in his doctrines of absolute space and time. And it is distinctions consistent with these commitments that he adopts from Descartes's epistemology (as early as the *Questiones*) to provide a basis on which to justify these cognitive claims. It is clear that the epistemic commitments involved in Newton's program of physiology and optics do not need an ontology of natures that the understanding alone can grasp. On the contrary, Newton's approach is based to a large extent on Hobbesian empiricism. This hiatus in Newton's epistemology remains throughout his development. On the one hand, he can claim that he frames no hypotheses and 'induces' his natural philosophy from the phenomena. On the other, he can warn us against confusing sensible measures with real quantities when considering the structure of absolute space and time, stating that "in philosophical disquisitions, we ought to abstract from our senses, and consider things themselves." It is significant that this epistemic tension begins in the *Questiones,* for it is one that Newton never resolves.

The optical entries are among the most advanced of those devoted specifically to scientific topics. Nowhere else in this early period is Newton's budding scientific genius more evident. From the beginning we have an interest in the spectral colors and the bending of light and an interest in the experimental determination of optical properties. From the start Newton shows an uncanny ability to spot weaknesses in traditional views and to isolate the essential questions that need to be answered. And straightway he attempts to develop a physical optics, one that can handle the operation of light by means of explanatory mechanisms. Furthermore, there is a continuous line of development from the *Questiones* into the investigations of Add. 3975, which marks Newton's earliest exposition of his mature theory of light and colors.

Also from the very beginning Newton conceives physiological and optical phenomena as being intimately related.

He believes that an understanding of the nature of light can be established by studying the organs of vision, as much as he believes that the study of the nature of light helps to explain the workings of sight. Thus, one cannot begin to understand the origins of Newton's optical ideas without reference to his physiological investigations, and conversely. These intimate connections are legitimized for Newton by his belief in cosmic teleology. The eye is suited to receive light as light is suited to affect the eye. The fact that the structures and functions of the eye and light conspire teleologically is for Newton evidence of cosmic design in nature. The belief that things are designed for an overall teleological purpose is clearly related to Newton's religious entries in the *Questiones*. Moreover, the relationship between mechanics and design remains prominent in his later thought. His lifelong adherence to the principle that we must argue for design before we can argue from design begins with the *Questiones*.

As one reads through these entries, certain distinct habits of thought emerge. Newton is a fact collector. For the most part, what he takes from his sources is information, and more often than not it is Newton who supplies the theory on the basis of what he culls from his reading. This attitude of mind is evident in his mature work. Even in cases where he is influenced positively on matters of theory, the position he adopts is quickly transformed according to Newton's lights. Always there is a strong tendency to generalize and to reduce cases to tokens of types. Also evident is a mind critical of assumptions, as well as of results. None of his sources escapes scrutiny, and nothing, except the purest of facts, is taken at face value. It is worth noting that the theological entries are few in number; indeed, religious interests do not take pride of place as they are to do in much of his later thought. Absent, too, is any sign of alchemical interests, nor is there much indication of interest in scriptural exegesis and the chronology of Christian belief—matters of considerable concern later.

This brings us to the question whether or not Newton is an advocate of the mechanical philosophy. Certainly there is no reason to suppose that he is advocating a mechanical view of nature on ideological grounds. There is nothing to show that Newton has unquestioned allegiance to Boyle's program concerning matter and motion conceived as the 'catholic principles' in virtue of which all physical phenomena are to be explained; nor is there an unambiguous endorsement of Descartes's conception that interrelations among natural phenomena are to be explained solely in terms of contact causation. So far as it can be determined, Newton's habits of thought are consistent with a mechanical conception of the workings of nature. But having said that, it is unwarranted to conclude that Newton accepts the mechanical philosophy to the extent that he proceeds uncritically within its framework. The term 'mechanical philosophy' carries evaluative connotations, and in the seventeenth century there are as many different versions of that ideology as there are thinkers persuaded that interchanges among physical phenomena are to be explained in virtue of spatial contact. Also, Newton is too resourceful as a thinker to foreclose on his explanatory options at the beginning of his scientific career. As for the suggestion that he proceeds in the *Questiones* by critically examining both the credentials of the Gassendists and the Cartesians, there is no evidence in its support.

Although it cannot be said that the *Questiones* are a unified work in the sense of Descartes's *Principles* or Hobbes's *Elements,* it does possess a unity of outlook. In the first place, Newton makes use of consistent cross-references to the topic headings that he originally set out before composing his entries. But more important is the fact that he is an atomist and an actualist. This preferred picture of the world is one that informs a majority of the entries of the *Questiones.* This commitment to atomism remains throughout Newton's career. Moreover, it is the background against which his later theories of force and the ether (as deployed in both optics and dynamics) are to be seen. It is important to notice that Newton also emphasizes the notion of actual infinity in both his cosmology and mathematics. The cosmological commitment has its roots in ancient atomism and remains a persistent feature of Newton's mature thought.

If we were to suppose that 1666 is a 'miracle year' as regards Newton's important discoveries, how do the *Ques-*

tiones stand in the development of his thought? With the exception of optics and physiology, there is a clear lack of continuity between the level of Newton's thought in the *Questiones* and what we find a short while later in his other manuscripts. Consider two examples: mechanics and mathematics. The essay "Of Violent Motion" is positively Medieval in character. Charleton, his chief source, argues that natural motion in the void is perpetual; yet Newton leaves that point implicit in his analysis. But a few months later, in the *Waste Book,* he has absorbed Descartes's conception of inertial motion and has surpassed him in the analysis of impact and circular motion. This is perhaps surprising, because Newton singularly fails to engage the central principles of Descartes's mechanics in the *Questiones.* As for mathematics, his initial orientation is 'Epicurean.' Yet in the space of a few months he is launched on the beginnings of his original mathematical researches. These observations do not undermine the importance of the *Questiones* for an understanding of the development of Newton's mature lines of thought. Quite the contrary, they reveal a thinker on the threshold of those investigations that will largely provide the foundations of his later work. Whatever is decided ultimately regarding the place of the *Questiones* in the Newtonian corpus, they reveal the truth of the maxim: ἀρχὴ δέ τοι ἥμισυ παντός.

Woolstro House, Lincolnshire!

S. Wright del.

Sir Isaac Newton, Born Dec.ʳ 25.ᵗʰ 1642, Ob.ᵗ 1726.

Woolsthorpe House, Newton's birthplace and family home.
The halo of light indicates his room and
the veneration in which he was held 50 years after his death.

PART II

TRANSCRIPTION AND
EXPANSION OF *QUESTIONES
QUÆDAM PHILOSOPHICÆ*

Edited by J. E. McGUIRE AND
MARTIN TAMNY

WITH DRAWINGS, AFTER NEWTON, BY MYRNA TAMNY

PRINCIPLES OF THE
TRANSCRIPTION AND
EXPANSION

The manuscript Add. 3996 has the characteristics of an English seventeenth-century manuscript: the vagaries of spelling and grammar, the shorthand devices and conventions. These, combined with the deletions and insertions so frequent in Newton's writing, make his early notebook difficult reading for the nonspecialist. We have attempted to meet these difficulties by providing both a 'diplomatic' Transcription and an Expansion, which treats the Transcription almost as if it were in a foreign language. In the following pages, the Transcription will be found on the left-hand pages, with the corresponding parts of the Expansion on the right-hand pages. The Expansion regularizes the spelling, punctuation, and grammar and also replaces the shorthand devices and conventions employed by Newton with unabbreviated equivalents. Only those of Newton's deletions are indicated in the Expansion that we believe will be of interest to the general reader. We have been guided chiefly by our desire to produce a flowing text that is true to the original. Although we have in many places modernized Newton's English, we have attempted to retain the flavor of seventeenth-century prose. We ask the reader to bear in mind that the Transcription on the facing pages will answer any questions as to what liberties we may have taken with the text.

Although the meanings and uses of most of Newton's devices and conventions should be clear from the Expansion, they are given explicitly following the Glossary. The meanings of technical terms, archaic words, and familiar terms used in unusual ways can be found in the Glossary.

A Table of yᵉ things following[1]

Aer 25.
Antipathy. 44.
Asperity. 17.
Attoms. 3. 4. 63. 64. 65.

Attraction $\begin{cases} \text{Magnet. 29} \\ \text{Electrical. 31.} \end{cases}$

Bodys Conjunction 6. 7.

Comets 12. 54. 55. 56. 57.
Cold. 18
Colours. 36. 69. 70. 71. 72. 73. 74. 91. 92. 93. 94. 95.
Corruption. 40.
Creation 83
Condensation. 13

Density 13
Ductility. 16
Dreames 89

Earth 27
Eternity 9.
fluidity 15
flexility 16
Figure 17
Fier 24
Flux & Reflux 47 49 26
Filtr̄acon. 31.
Fantacy. 43. 75. 76

God 81.
Gravity 19. 67.

---

[1] Following the Glossary there is a discussion of Newton's use of symbols aı
shorthand devices.

Ha⌞r⌟dnesse 16[2]
Humidity 15
Hebetude 17
Heate 18

Imaginācon ⎫
Invention ⎭ 43. 75. 76

Levity 19. 67.
Light 32 82

1ˢᵗ Matter. 1. 2.
Motion. 10. 21. 22. 51. 52. 53. 59.
Memory 41.
Meteors 46
Mineralls 48.

Oyles 45
Odors. 38.
Opacity 14.
Orbes. 11

Place. 8. 67
Planets 12. 54
Perspicuity 14
Phylosophy

Quantity. 5. 67
Quality 28

Rarefaction. 13.
Reflection 23
Refraction 23

Soule 85 86
Sleepe 89

87ᵛ Starres & Sunnes. 12. 54. 55. 56. 57.
Stability & Siccity. 15.
Softness. 16.
Subtility. 17.
Smoothnesse. 17.
Salt. 26. 47. 49.

---

[2] The letters, words, and sentences between corner brackets (⌞ ⌟) are insertions or additions made by Newton at the time of writing the immediately preceding or succeeding material.

[3] Question marks within parentheses indicate illegible letters or words. A horizontal line drawn through a symbol, a letter, or a word indicates that Newton cancelled the material. Occasionally Newton's cancellations are so thorough as to make the material cancelled illegible, thus requiring us to indicate (?).

Amicus Plato amicus Aristoteles magis amica veritas.[5]

Questiònes quædam Philosophcæ[6]

OF yᵉ first mater

Whither it be mathematicall points: or Mathematicall points & parts: or a simple entity before division indistinct: or individualls i.e. Attomes.

1 Not of Mathematicall points since wᵗ wants dimensions cannot constitute a body in theire conjunction because they will sinke into yᵉ same point. An infinite number of mathematicall points sink into one being added together,& yᵗ being still a mathematicall point is indivisible but a body is divisible. ~~In fine a Mathematicall point is Nothing, since it is but an immaginary entity. &c~~

Not of parts & Math:points: for such a point is either something or nothing. if something tis a ₚte & so added betweene 2 ₚs will make a line of 3 ₚts. if nothing, then added betwixt two parts there is still nothing betwixt yᵉ 2 ₚts & consequently yᵉ line consists of nothing still but 2 ₚs. &c

Not of simple entity before division indistinct. for this must be an union of yᵉ parts into wᶜʰ a body is divisible since those parts may againe bee united & become one body as they were before ⌊at the creation.⌋ Now yᵉ nature of union (being but a modall ens) is to depend on its pts (wᶜʰ are

---

[4] The first number indicates Newton's own numbering of the pages, and these are the numbers referred to in his "Table of yᵉ things following." The second number with superscript r (recto) or v (verso) gives the folio number in the manuscript.

[5] A line appearing in various forms in many seventeenth-century authors. Here it is quoted by Newton from page 3 of Walter Charleton, *Physiologia Epicuro-Gassendo-Charltoniana* (London, 1654).

[6] The title should read "Questiones quædam Philosophicæ." Newton has omitted the second *i*.

Amicus Plato, amicus Aristoteles, magis amica veritas

CERTAIN PHILOSOPHICAL QUESTIONS

Whether it be mathematical points, or mathematical points and parts, or a simple entity before division indistinct, or individuals, i.e., atoms.

1. Not mathematical points, since what wants dimensions cannot constitute a body in their conjunction because they will sink into the same point. An infinite number of mathematical points sink into one, being added together, and that, being still a mathematical point, is indivisible. But a body is divisible. ~~In fine a mathematical point is nothing, since it is but an imaginary entity.~~

Not of parts and mathematical points, for such a point is either something or nothing. If something, it is a part, and so added between two parts will make a line of three parts. If nothing, then added between two parts there is still nothing between the two parts, and consequently the line consists of nothing still but two parts.

Not of a simple entity before division indistinct. For this must be a union of the parts into which a body is divisible, since those parts may again be united and become one body as they were before at the creation. Now the nature of union (being but a modal ens) is to depend on its parts absolute entities) therefore it cannot be y^c terms of creation, or first matter for tis a contradiction to say y^c first matter depends on some other subject ~~(except God)~~ since y^t implys some former matter on wch it must depende.[7]

~~homogeneous~~

2[8]   The first matter must be homogeneous & so either all hard or soft or of a midle temper.~~if hard y^n all y^c pts into wch it is divisible will be hard~~ & of y^c same constitution will all y^c pts be into wch it is divisible. so y^t where there is plenitude of mater it will be of y^c same temper y^c first was & there will be no change in nature unles you will allow it either to arise from vacuities interspersed or from y^c severall proportions y^t quantity hath to its substance: Matter acquiring a harder nature by less quantity & a softer by more. likewise there will be no other way for rarefaction & ~~explo~~ condensation.[9]

If y^c first way then y^c body will be no such ~~p~~ continuū as to be wthout distinct ˌparts˩ since it will be every where divided by interspersed inanitys.

The latter will in its due place be proved impossible.

3 Those things wch can exist being actually seperate are really distinct, but such are y^c parts of mater.[10]

4 Suppose y^c first matter one uniforme mass wthout pts how should that body be divided into parts as we se now it is wthout admission of a vacuum. Suppose it be divided into two w^{h}at will be betweene those two parts.not body since it is all in y^c two halfes. but if it be said y^t it was first divided into smaller parts wee ask how came it so wthout less parts y^n those into wch it was ˌat first.˩ divided or els vacuum to succed in theire rome as they came a peices. ~~Else~~ Suppose y^c ~~world~~ first matter were ˌdivided˩ as small as sand y^n divide one of those sands a .third sand cannot succed twixt 'em before they be at some distance unles ~~they be at some dis~~

---

[7] The foregoing, from "whither it be mathematicall points . . . ," is a close paraphrase of Charleton, *Physiologia*, Book II, Chapter III, Section II, pp. 107–10, and Book III, Chapter X, Section I, pp. 264–5.

[8] The double lines in the margin indicate that the material to the right of them was cancelled by Newton.

[9] Charleton, *Physiologia*, Book III, Chapter X, Section I, pp. 263–4.

[10] *Ibid.*, Book II, Chapter III, Section II, p. 108.

(which are absolute entities). Therefore it cannot be the terms of creation or first matter. For it is a contradiction to say the first matter depends on some other subject (except God), since that implies some former matter on which it must depend.

2. The first matter must be homogeneous and so either all hard or soft or of a middle temper. If hard then all the parts into which it is divisible will be hard, and of the same constitution will all the parts be into which it is divisible. So that where there is plenitude of matter, it will be of the same temper the first was, and there will be no change in nature unless you will allow it either to arise from vacuities interspersed or from the several proportions that quantity has to its substance—matter acquiring a harder nature by less quantity and a softer by more. Likewise there will be no other way for rarefaction and condensation to be explained. If the first way, then the body will be no such continuum as to be without distinct parts, since it will be everywhere divided by interspersed inanities. The latter will in its due place be proved impossible.

3. Those things which can exist being actually separate are really distinct, but such are the parts of matter.

4. Suppose the first matter one uniform mass without parts; how should that body be divided into parts, as we see now it is, without admission of a vacuum? Suppose it be divided into two. What will be between those two parts? Not body, since it is all in the two halves. But, if it be said that it was first divided into smaller parts, we ask how came it so without less parts than those into which it was at first divided; or else vacuum to succeed in their room as they came to pieces. Suppose the first matter was divided as small as sand. Then divide one of those sands. A third sand cannot succeed between them before they be at some distance, un-

2 88ᵛ

there might be some smaler matter to run in & keepe out vacuum but to affirme this is to say ye first matter had very little parts be in it before it was divided. But again ⌐now matter is divided⌐ if two ⟨?⟩ parts of matter of ye least size. were seperateing & distant one from another ye space of halfe

pag 4.

Of Attomes.

It remaines therefore y^t ye first matter must be attoms. And y^t Matter may be so small as to be indiscerpible The excellent Dr Moore in his booke of ye soules imortality hath proved beyond all controversie[11] yet I shall use one argument to shew y^t it cannot be divisible in infinitum & y^t is this: Nothing can be divided into ~~infinite~~ ⌐more⌐ parts y^n it can possibly be constituted of. But matter (i.e finite) cannot be constituted of infinite parts. The Major is true for looke into how many parts ~~of~~ a thing is divided those parts added agane make ye same whole thate they were before, & so if any finite quantity were divided into infinite parts (& certainely it may if it be so far divisible) those infinite parts added would make ye same finite quantity they were before wch is against ye Minor; & It is plaine from hence y^t an infinite number of extended parts ⌐(& the least parts of quantity must be extended)⌐ make a thing infinitely extended[12] ~~& all ye points y^t qua~~ this ~~you~~ cannot ⌐be⌐ denyed if I can prove y^t things infinitely extended have fine parts Now vacuum is infinitely extended & so ⌐may⌐ matter be fansied to be. but if ye world were removed & vacuum came in ye roome of it y^t very vacuum would not be infinite we can conceive of interspersed vacuities ~~in~~ ⌐amongst⌐ matter but they are not infinite (though an infi-

---

[11] Henry More, *The Immortality of the Soul* (London, 1659), Part 3 of the Preface and Book I.

[12] The argument is found in Charleton, *Physiologia*, Book II, Chapter II, Section I, pp. 91–2. Newton's argument may also have its origin in two other sources that he knew and owned: Epicurus's *Letter to Herodotus*, which is in Diogenes Laërtius, *De vitis dogmatis et apophthegmatis . . . libri X* (London, 1664), p. 280; More, *Immortality*, Book I, Chapter VI, Axiomes XIV and XV, and Book II, Chapter I, Axiome XXIV.

less there might be some smaller matter to run in and keep out vacuum, but to affirm this is to say the first matter had very little parts in it before it was divided. But again, now matter is divided, if two parts of matter of the least size were separating and distant one from another, the space of half

continued on page 4[1]

Of atoms

It remains, therefore, that the first matter must be atoms. That matter may be so small as to be indiscerpible the excellent Dr. More in his book of the soul's immortality has proved beyond all controversy. Yet, I shall use one argument to show that it cannot be divisible *in infinitum,* and it is this: nothing can be divided into more parts than it can possibly be constituted of, but finite matter cannot be constituted of infinite parts. The major is true, for look into how many parts a thing is divided, those parts added again make the same whole that they were before. And so if any finite quantity were divided into infinite parts (and certainly it may if it be so far divisible), those infinite parts added would make the same finite quantity they were before, which is against the minor. It is plain from hence that an infinite number of extended parts (and the least parts of quantity must be extended) make a thing infinitely extended. This cannot be denied if I can prove that things infinitely extended have fine parts. Now vacuum is infinitely extended and so may matter be fancied to be. But if the world were removed and vacuum came in the room of it, that very vacuum would not be infinite. We can conceive of interspersed vacuities among matter, but they are not infinite (though an infinite number of them would be so).

---

[1] Page 343

nite number of y^m would be so) we see ye parts of matter are finite.& an infinite number of finite unites cannot be finite. To helpe ye conception of ye nature of these leasts, how they are indivisible how extended of w^t figure &c I shall all along draw a similitude from numbers, compareing ~~ciphers~~ ˻Math:points˼ to ciphers, indivisible extension, to unites: divisibility, or compound quantity, to number: i.e. a multitude of attomes, to a multitude of unites. Suppose y^n a number of Mathematicall points were indued wth such a power as y^t they could not touch nor be in one place (for if they touch they will touch all over, & bee in one place) Then ad thees as close in a line as they can stand together every point added must make some extension to ye lenght because it cannot sinke into ye formers place or touch it so here will be a line wch hath partes extra partes; ˻(?)˼ another of these points cannot bee added into ye midst of this line, for y^t implys y^t ye former points did not lie so close but y^t they might lye closer.[13] The distan˻ce˼ y^n twixt each point is ye least y^t can be & so little may an attome be & no lesse: now y^t this distance is ˻(?)˼ indivisible (& therefore ye matter conteined in it) is thus made plaine: Wherever

pag 63

4 89^v Of a Vacuum & Attomes.   pag 2

halfe theire diameter (they will not then touch, for y^n their semidiameter will bee but as a ˻Math:˼ point & theire diameter a two Mathematicall points together i.e. as nothing (for 2 nothings put together make a third nothing.) & so ye least parts of matter mathematicall points wch is absurd). Vacuum will y^n come betwixt if nothing else can, & no matter will come betwixt since ye diameter of ye least particle will be bee as big againe as y^t space.

---

[13] The terms used in the foregoing are similar to those used by Francis Hall in his *Tractatus de corporum inseparabilitate* (London, 1661), pp. 170–82, and recounted by Robert Boyle in his *A Defence of the Doctrine touching the Spring of the Air* (London, 1662), pp. 102f. It seems probable that Newton had read at least the latter before the composition of this passage.

We see the parts of matter are finite and an infinite number of finite units cannot be finite.

To help the conception of the nature of these leasts, how they are indivisible, how extended, of what figure, etc., I shall all along draw a similitude from numbers, comparing mathematical points to ciphers, indivisible extension to units, divisibility or compound quantity to number, i.e., a multitude of atoms to a multitude of units. Suppose then a number of mathematical points were imbued with such a power as that they could not touch or be in one place, for if they touch they will touch all over and be in one place. Then add these as close in a line as they can stand together. Every point added must make some extension to the length, because it cannot sink into the former's place or touch it. So here will be a line which has *partes extra partes;* another of these points cannot be added into the midst of this line, for that implies that the former points did not lie so close but that they might lie closer. The distance then between each point is the least that can be, and so little may an atom be and no less. Now, that this distance is indivisible (and therefore the matter contained in it) is thus made plain: wherever

continued on page 63[2]

Of a vacuum and atoms   4 89ᵛ

continued from page 2[3]

~~their diameter, they will not then touch, for if they did then their semidiameter will be but a mathematical point, and their diameter as two mathematical points together, i.e., as nothing, for two nothings put together make a third nothing, and so the least parts of matter would be mathematical points, which is absurd.~~ Vacuum will then come between if nothing else can, and no matter will come between, since the diameter of the least particle will be as big again as that

---

[2] Page 421   [3] Page 341

If it be said y^t matter ~~at y^t~~ may move ~~y^t~~ ˌover so littleˌ space in an instant & other matter succeed in an instant & so there needes no vacuum) I answer it may as well be moved throuˌghˌ ye Universe in an instant for instantaneous motion is infinitly swift & will carry ye thing in wch it is as soone ~~&~~ through an infinite space in an instant as well as though a finite space of ye breadth of an attome & I cannot conceive how so violent a motion should be stopt wthout some violent effect though in so little an agent what is saide of so little bodys may be said of greater. As thus suppose two globes were to ~~be seperated~~ come together they must pas through all ye intermediate degrees of distance before they can be joyned suppose they then be distant but halfe ye breadth of ye least particle of matter there can be no matter betwixt y^m since all matter is too big to interpose it sefe. Neither can ye two globes touch for y^t implyes y^t semidiameter of ye least attome hath no breadth but had it not breadth ye diameter could have none & so ye least particles of matter would be Mathematicall points. therefore A vacuum must interpose. ~~But you may say~~ ˌUnles y^u say those attomes are as far divided as they ar divisibleˌ y^t there are least parts of matter* ~~was~~ is proved in ye chapter of attomes.

*i.e. so little y^t theire canˌnotˌ be a place too little for y^m to creepe into & yⁿ you will grant w^t I pleade for i.e. indivisible particles. & you must grant too y^t Attomes were either created so or divided by meanes of a Vacuū.

Of Quantity

As ~~a~~ finite lines added in an infinite number to finite lines, make an infinite line: so points added twixt points infinitely, are equivalent to a finite line.

All superficies beare the same proportion to a line yet one superficies may bee greater yⁿ another (ye same may be said of bodys in respect of surfaces) wch happens by reasons y^t a surface is infinit~~ly~~ in respect of a line, soe though ˌallˌ infinite extensions beare ye same proportion to a finite one yet one infinite extension may be greateˌrˌ ˌyⁿ anotherˌ soe one angle of contact may exceed another, yet they are all

~~space. If it be said that matter may move over so little space in an instant, and other matter succeed in an instant and so there need be no vacuum, I answer it may as well be moved through the universe in an instant. For instantaneous motion is infinitely swift and will carry the thing in which it is as soon through an infinite space in an instant as well as through a finite space of the breadth of an atom. I cannot conceive how so violent a motion should be stopped without some violent effect, though in so little an agent.~~ What is said of so little bodies may be said of greater. As thus, suppose two globes were to come together. They must pass through all the intermediate degrees of distance before they can be joined. Suppose they then be distant but half the breadth of the least particle of matter. There can be no matter between them, since all matter is too big to interpose itself; neither can the two globes touch for that implies that the semidiameter of the least atom has no breadth, but had it not breadth the diameter could have none, and so the least particles of matter would be mathematical points. Therefore a vacuum must interpose, unless you say those atoms are as far divided as they are divisible. That there are least parts of matter* is proved in the chapter "Of Atoms."

*I.e., so little that there cannot be a place too little for them to creep into, and then you will grant what I plead for, namely, indivisible particles. You must grant, too, that atoms were either created so or divided by means of a vacuum.

Of quantity

As finite lines added in an infinite number to finite lines make an infinite line, so points added between points infinitely are equivalent to a finite line.

All superficies bear the same proportion to a line, yet one superficies may be greater than another (the same may be said of bodies in respect of surfaces, which happens by reason that a surface is infinite in respect of a line). So, though all infinite extensions bear the same proportion to a finite one, yet one infinite extension may be greater than another. So one angle of contact may exceed another, yet they are all equal when compared to a rectilinear angle viz wch is infi-nitely greater. Thus $\cancel{(?)}\frac{2}{0}\cancel{(?)}$ is double to $\frac{1}{0}$ & $\frac{0}{1}$ is double to $\frac{0}{2}$, for multiply y^e 2 first & divide y^e 2^{ds} by 0, & there results $\frac{2}{1}$: $\frac{1}{1}$ & $\frac{1}{0}$: $\frac{1}{2}$. yet if $\frac{2}{0}$ & $\frac{1}{0}$ have respect to 1 they beare y^e same relation to it y^t is 1 : $\frac{2}{0}$ :: 1 : $\frac{1}{0}$. & ought therefore to bee considered equall in respect of an unite.[14]

The angle of contact is to another angle, as a point to a line, for y^e crookednes in one circle amonts to 4 right angles & y^t crookednesse may bee conceived to consist of an infi-nite number of angles of contact, as a line doth of infinite points.[15]

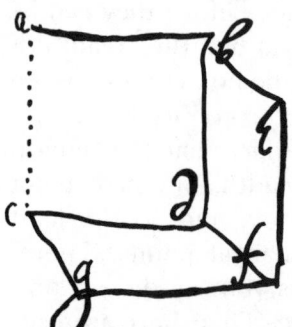

As y^e point a to y^e line ab so y^e line ac to y^e pgr abcd :: pgr dbef : y^e paralelipipedon bg.

Tis indefinite ⌐(y^t is undeteer-mined)⌐ how greate a sphære may be made,how greate a number may be recconed, how far mattar is divis-ible, how much time or extension wee can fansy but all y^e Extension y^t is, Eternity, $\frac{a}{0}$ are infinite.$\frac{a}{0}$ exceeds all number & is soe greate y^t there can bee noe greater, but (finite) number is called indefinite in respect of a greater.[16]

---

[14] Newton's treatment of infinities is influenced by John Wallis, *Opera Mathe-matica*, *Operum Mathematicorum Pars Altera* (Oxford, 1656), which contains the tracts *De Anglo Contactus* and *Arithmetica Infinitorum*. See D. T. Whiteside, *The Mathematical Papers of Isaac Newton*, Vol. I (Cambridge, 1967), pp. 89–91. New-ton's 'numerical calculus' has its source in Wallis's *Mathesis Universalis*, which is found in the *Operum Mathematicorum Pars Prima* (Oxford, 1657). The relevant chapter is 41, which is entitled "De Fractionibus sive Minutiis," pp. 363–9. Spe-cifically, Newton's conception of the unit in his numerical model appears to be influenced by Wallis's argument on p. 364.

[15] *Ibid.*, *Pars Altera, De Anglo Contactus*, Cap. XII, arguments 1–4, pp. 41–5, and Cap. XV, p. 51. Newton's argument finds its context in Wallis's account of the 'horn angle.' As will be made clear in the Commentary, however, his position differs from that of Wallis. The use of indivisibles relates to Newton's more elaborate development of the "method of indivisibles" in ULC, Add 4000, folios 82^r and 82^r–84^r, a pocket book that dates, like much of the *Questiones*, from 1664.

[16] This again seems to bear the influence of Wallis's *Arithmetica Infinitorum* in *Opera mathematicorum, Pars Altera;* Prop. XLIII, pp. 33–35, Prop. CLXXXII, pp. 148–161, CLXXXVIII, p. 167–169, are probable sources. It is clear also that Newton is opposing Descartes's views on infinity and the indefiniteness of extension.

equal when compared to a rectilinear angle, which is infinitely greater. Thus 2/0 is double to 1/0, and 0/1 is double to 0/2, for multiply the first two by 0 and divide the second two by 0, and there results 2/1 : 1/1 and 1/1 : 1/2. Yet, if 2/0 and 1/0 have respect to 1, they bear the same relation to it, that is, 1 : 2/0 :: 1 : 1/0, and ought therefore to be considered equal in respect of a unit.

The angle of contact is to another angle as a point to a line, for the crookedness in one circle amounts to four right angles and that crookedness may be conceived to consist of an infinite number of angles of contact, as a line does of infinite points.

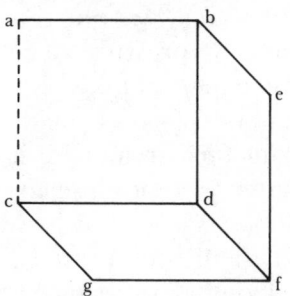

As the point a to the line ab so the line ac to the parallelogram $abcd$ :: parallelogram $dbef$: the parallelepiped bg.

It is indefinite, that is, undetermined, how great a sphere may be made, how great a number may be reckoned, how far matter is divisible, how much time or extension we can fancy; but all the extension that is, eternity, and $a/0$ are infinite. $a/0$ exceeds all number and is so great that there can be no greater, but finite number is called indefinite in respect of a greater.

6 90ᵛ

Conjunction of bodys.

Whither y^e conjunction of bodys ~~depend upon~~ ⌊be from⌋ rest:[17] Neg: For y^n sand by rest might ~~make a sollid body~~ be united sooner y^n by a furnace &c ~~Whither~~

Whither it be from ye close ~~conjuncture~~ ⌊crouding⌋ of all ye matter in ye world affirmated. For y^e aire (though its pressure bee but little in respect of y^t, performed by y^e purer matter of y^e vortex (twixt ☉ & us) receding from y^e center.) ~~F~~ by its pressure to y^e center & consequently crouding ~~(?)~~ all thing close together betwixt wch there is not aire to keepe $\overline{y}$ asunder it maketh $\overline{y}$ stick together, as y^e 2 pollished sides of 2 marbles ⌊y^e ₚts of water⌋ &c but this juncture cañot be very firme by reason y^t y^e pressure of y^e aire is not verry strong as appeares by y^e experiments of Eq^r Boyle.[18] but y^e pressure of all y^e matter twixt ☉ & us ⌊made by reason of its indevor from ☉.⌋ being farr greater (& it may be some other power by wch matter is kept close together &c) when 2 or 3 or more littell bodys once touch so as to admitt noe other matter betwixt them they must be held very fast together al y^e matter about $\overline{y}$ pressing y^m together but nothing striving to ₚte y^m. And when 2 of y^e least particles ~~touch wth very broad sides~~ meete whos sides wth wch they touch one another are pretty broade & fitted to ~~one anto~~ touch close every where, those two may move together as one body & so may increase by haveing others joyned to y^m in y^e same manner. but if y^e circumstant particles chanch to be held of from pressing y^m together ~~as~~ by some accident as those about (a) or be variously pressed as at (b) by y^e bodys c & d they may be againe severed.

[17] Rene Descartes, *Opera philosophica, Principia philosophiæ*, 3rd ed. (Amsterdam, 1656), Part II, Article LV, p. 44.

[18] The experiments referred to are numbers 31 and 32 in Robert Boyle, *New Experiments Physico-mechanicall, touching the Spring of the Air* (Oxford, 1660), pp. 229–36. Boyle discusses the limited power of the air, etc., on pp. 254–5.

Conjunction of bodies

Whether the conjunction of bodies be from rest? No, for then sand by rest might be united sooner than by a furnace, etc.

Whether it be from the close crowding of all the matter in the world affirmed, for the air (though its pressure be but little in respect of that performed by the purer matter of the vortex, between the Sun and us, receding from the center) by its pressure to the center, and consequently crowding all things close together between which there is not air to keep them apart, it makes them stick together; as the two polished sides of two marbles, the parts of water, etc. But this juncture cannot be very firm by reason that the pressure of the air is not very strong, as appears by the experiments of Robert Boyle, Esquire. But the pressure of all the matter between the Sun and us, made by reason of its endeavor from the Sun, being far greater (and it may be some other power by which matter is kept close together, etc.), when two or three or more little bodies once touch, so as to admit no other matter between them, they must be held very fast together; all the matter about them pressing them together but nothing striving to part them. And when two of the least particles meet whose sides, with which they touch one another, are pretty broad and fitted to touch close everywhere, those two may move together as one body and so may increase by having others joined to them in the same manner. But if the surrounding particles chance to be held off from pressing them together by some accident, as those about *a* or be variously pressed, as at *b* by the bodies *c* and *d*, they may be again severed.

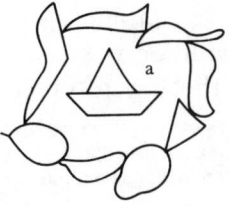

Yet in more compound bodys there is no danger ~~as~~ for the least particles are so wedged toge: y^t neither of these two chanches could undo y^m as for y^e firs it cannot to a whole body but onely in some littell pte of it as some attome of a mans hand may chanch not to be touched by ye water into wch hee puts it. but this cannot happ to his

7 91^r ~~Of Place~~

whole hand. so y^t those particles wch are pressed together may holde a particle wedged as it were amogst y^m so y^t it cannot fall away from y^m though it it chanch not to be pressed to y^m. ye second can~~ot~~ ⌐scarce⌐ happen at all for y^e ~~an~~ neighbouring matter can onely press two touching particles toward ye center or it may be a little awry but not from ye body to wch they adhere but let ye wors hapen y^t can ye particle may be wedged in amongst ye rest. But it may be y^t ye particles of compound matter were created bigger y^n those wch serve for other offices.

8 91^v Of Place

Extension is related to places, as time to days yeares &c. Place is y^e principium individuationis of streight lines & of equall & like figures y^e surfaces of two bodys becomeing but one when they are contiguous becaus but in one place.[19]

9 92^r

Of time & Eternity

The representation of a Clock to goe by water or sand.

Probleme

1. By a line of tangents upon a Suare ruler & a plumet to know ⌐at one view⌐ whither y^e stile of a diall bee true & thereby to erect a stile.

[19] This may have been drawn from the lectures of Isaac Barrow given in 1665, which Newton attended and which were later published as *Lectiones mathematicae XXIII . . . habitae Cantabrigiae A.D. 1664, 1665, 1666* (London, 1684). See Lecture I of 1665, p. 12.

Yet in more compound bodies there is no danger, for the least particles are so wedged together that neither of these two chances could undo them; as for the first it cannot happen to a whole body but only in some little part of it, as some atom of a man's hand may chance not to be touched by the water into which he puts it, but this cannot happen to his

whole hand. So that those particles which are pressed to- 7 91ʳ gether may hold a particle wedged, as it were, among them, so that it cannot fall away from them, though it chance not to be pressed to them. The second can scarcely happen at all, for the neighboring matter can only press two touching particles toward the center, or it may be a little awry, but not from the body to which they adhere; but, let the worst happen that can, the particle may be wedged in among the rest. But it may be that the particles of compound matter were created bigger than those which serve for other offices.

Of place 8 91ᵛ

Extension is related to places, as time to days, years, etc. Place is the *principium individuationis* of straight lines and of equal and like figures; the surfaces of two bodies becoming but one when they are contiguous, because but in one place.

Of time and eternity 9 92ʳ

The representation of a clock to go by water or sand.

Problem

1. By a line of tangents upon a square-ruler and a plummet to know at one view whether the style of a dial be true, and thereby to erect a style.

2. y^c stile erected, by a plumb line ⌊let fall⌋ from y^c stile to find y^c meridian line.

3. By y^c s^d ruler to find y^c substilar & draw y^c other hower lines. Note y^t this may be done though y^c wall bee not eaven &c.[20]

To make metalline Globular dust for ye said clock instead of sand. Daube y^c hollow cone B wth pitch &c on y^c inside, fire it. through wch fire (by y^c helpe of y^e tunnell A) cast y^c filings of brasse or pewter &c wch molten into a globular forme may fall into the bason of water C.

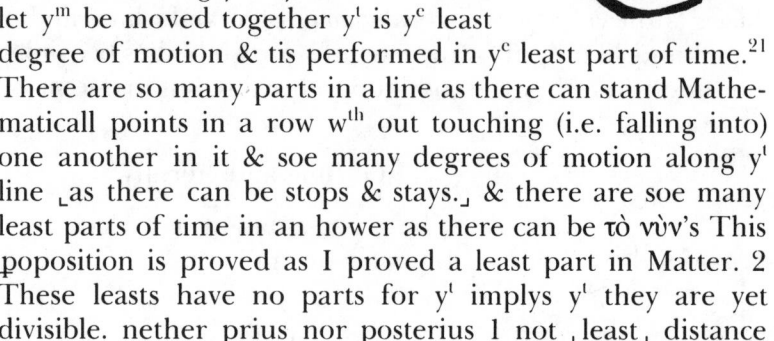

10 92^v

Of Motion

That it may be knowne how motion is swifter or slower consider 1 That there is a least distance, a least progression in motion & a least degree of time. as lay two globes together y^t so close y^t they cannot come any nigher wth out touching y^t is y^c least distance, let y^m be moved together y^t is y^c least degree of motion & tis performed in y^c least part of time.[21] There are so many parts in a line as there can stand Mathematicall points in a row wth out touching (i.e. falling into) one another in it & soe many degrees of motion along y^t line ⌊as there can be stops & stays.⌋ & there are soe many least parts of time in an hower as there can be τὸ νῦν's This poposition is proved as I proved a least part in Matter. 2 These leasts have no parts for y^t implys y^t they are yet divisible. nether prius nor posterius 1 not ⌊least⌋ distance

---

[20] Newton's early interest in the construction of sundials is well known. An even earlier reference by Newton on the construction of dials is to be found in a notebook kept by him during the years 1659–61 and now in the Morgan Library in New York. On folios 18^v–19^r there is an entry headed "The use of y^c table on a Ruler whereby to make a dyall for any latitude."

[21] Newton's discussion is not indebted to Charleton. A probable source for this conception of atomic motion is Epicurus's *Letter to Herodotus*, as found in Newton's 1664 edition of Diogenes Laertius, p. 281.

2. The style erected, by a plumb line let fall from the style to find the meridian line.

3. By the square-ruler to find the substyler and draw the other hour lines. Note that this may be done though the wall be not even, etc.

To make metallic globular dust for the said clock instead of sand: daub the hollow cone *B* with pitch, etc., on the inside and fire it. Through which fire (by the help of the tunnel *A*) cast the filings of brass or pewter, etc., which, melted into a globular form, may fall into the basin of water *C*.

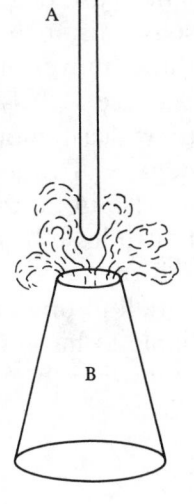

10·92ᵛ

Of motion

That it may be known how motion is swifter or slower consider:

1. That there is a least distance, a least progression in motion, and a least degree of time. Lay two globes together so close that they cannot come any nearer without touching – that is the least distance. Let them be moved together – that is the least degree of motion and it is performed in the least part of time. There are so many parts in a line as there can stand mathematical points in a row without touching (that is, falling into) one another in it, and so many degrees of motion along that line as there can be stops and stays, and there are so many least parts of time in an hour as there can be τὸ νῦν's. This proposition is proved as I proved a least part in matter.

2. These leasts have no parts, for that implies that they are yet divisible, but they are divisible neither *prius* nor *posterius*. (1) Not least distance, since it is passed over in an

since it is passed over in an indivisible part of time & ther cannot be a different time ascribed to ye entrance of a thing into y^t part of space & ye leaving of it. 2 not ye least degree of motion because too y^t is performed in an indivisible pte of time & is no sooner begun y^n done 3 Not y^e least moment of time because first & last imply severall parts of time. 3 The least degree of motion is equal to y^e least ~~degree of space~~ distance & time. 1 distance & not to more: because a thing moves in passing over but one of $\overline{y}$. Nor to less, because y^e least motion is over some distance. 2 Tis æquall to y^e least moment in time, not to more, because in each degree of time wherein a thing moves there will be motion or else in all those degrees put together there will be none: not to lesse becaus no motion is done in an instant or intervall of time. But againe should two parts in $\overline{motio}$ be but equall to one in space or time &c: e contra: y^t one would be liable to have a prius accor to y^e first of y^e two parts a midle according to y^e ~~(?)~~

pag 59

11 93^r Of y^e Celestiall matter & orbes.

Whither Cartes his first element can turne about y^e vortex & yet drive y^e matter of it continually from y^e ☉ to produce light. ⌐& spend most of its motion in filling up ye chinkes betwix ye Globuli.⌐[22] Whither ye least globuli can continue always next y^e ☉ & yet ~~ye globuli of y^e ☉ vortex (?)~~ come always from it to cause light & whither when y^e ☉ is obscured ye motion of ye first Element must cease (& so whither by his hypothesis y^e ☉ can be obscured) & whither upon ye ceasing of ye first elements motion ye Vortex must move slower.[23] Whither ⌐some of⌐ ye first Element comeing (as he confesseth ~~from~~ y^t hee might find out a way to turn ye Globuli about theire one axes to grate ye 3^d El: into wrathes like screws or cockle shells)[24] immediately from ye poles & other vortexes into all ye parts of o^r vortex woul⌐d⌐ not

[22] Drawn from Descartes, *Principia*, Part III, Articles LIV and LV, p. 69, and Articles LXIV–LXXVIII, pp. 74–89.

[23] *Ibid.*, Part III, Articles LXXII–LXXVI, pp. 81–7, and Articles LXXXII, LXXXIV, and LXXXV, pp. 91–5.

[24] *Ibid.*, Part III, Articles LXXXVI–XCII, pp. 95–8.

indivisible part of time, and there cannot be a different time ascribed to the entrance of a thing into that part of space and the leaving of it. (2) Not the least degree of motion, because that too is performed in an indivisible part of time and is no sooner begun than done. (3) Not the least moment of time, because first and last imply several parts of time.

3. The least degree of motion is equal to the least distance and time. (1) Least distance and not to more, because a thing moves in passing over but one of them. Nor to less, because the least motion is over some distance. (2) It is equal to the least moment in time, not to more, because in each degree of time wherein a thing moves there will be motion, or else in all those degrees put together there will be none. Not to less, because no motion is done in an instant or interval of time. But again should two parts in motion be but equal to one in space or time, and conversely, that one would be liable to have a *prius* according to the first of the two parts, a middle according to the

continued on page 59[4]

Of the celestial matter and orbs   11 93[r]

Whether Descartes's first element can turn about the vortex and yet drive the matter of it continually from the Sun to produce light, and spend most of its motion in filling up the chinks between the globuli. Whether the least globuli can continue always next to the Sun and yet come always from it to cause light. Whether when the Sun is obscured, the motion of the first element must cease (and so whether by his hypothesis the Sun can be obscured). Whether upon the ceasing of the first element's motion the vortex must move slower. Whether some of the first element coming (as he confesses that he might find out a way to turn the globuli about their own axes to grate the third element into coils, like screws or cockle shells) immediately from the poles and other vortices into all the parts of our vortex would not

[4] Page 419

impèl yᵉ Globuli so as to cause a light from the poles & those places from whence they come.[25]

Of yᵉ Sunn Starrs & Plannets & Comets

Whither ☉ move yᵉ vortex about, (as Des-Cartes will) by his beames. pag 54 Princip Philos: partis 3ᵃᵉ.[26]

Whither yᵉ vortex can carry a Comet towars yᵉ poles &c[27] Whence tis yᵗ yᵉ ☉ is turned about upon his axis.[28] Whither Cartes his reflexion will will unriddle yᵉ mistery of a Comets bird.[29]

Heb 1 chap: vers 2 by God made yᵉ worlds by his son τȣς αἰῶνας.[30] ~~Carte~~ The ☉s spots are coloured sometimes like yᵉ rainebow.[31]

October 26 1618 in Scorpio appeared a comet yᵉ tayle being extended twixt Spica virginis & Arcturus toward yᵉ North pole it ~~moved went~~ ⌞passed⌟ into libra moveing from yᵉ yᵉ Ecliptick to yᵉ Tropic of cancer from east to west or Northerly

---

On Saturday, Decembʳ 10ᵗʰ, 1664. By a sleighty observation I found yᵉ distance of a Comet from yᵉ ⌞center of yᵉ⌟ Moone to be 9ᵈ, 48, min. Its altitude 3ᵈ, 40ᵐ; or 4ᵈ.

The moons altitude 8ᵈ, 40ᵐ. ~~Her place being Capric 26ᵈ, 2ᵐ or else ♒ 5ᵈ.~~

The longitude of yᵉ Moone.

On Satturday at 30ᵐⁱⁿ: past 4 of yᵉ clock in yᵉ morning Decembʳ 17ᵗʰ 1664 A Comet appeared Whose distance from Sirius was 30ᵈ, 0'. from procion 38ᵈ, 45'. There was little or noe difference twixt yᵉ time of its & Sirius his setting, ⌞it setting about 2' after him.⌟ Soe yᵗ its Right ascention was about 126ᵈ, 32'. & its declina͞con southward 31ᵈ. The length

[25] *Ibid.*, Part III, Article LXIV, pp. 74–5; Articles LXXVI–LXXVIII, pp. 87–9; Article LXXX, p. 89.   [26] *Ibid.*, Part III, Article XXI, p. 54.

[27] *Ibid.*, Part III, Articles CXXVI and CXXVII, pp. 118–19.

[28] *Ibid.*, Part III, Article LXXII, pp. 81–3.

[29] *Ibid.*, Part III, Articles CXXXIII–CXXXIX, pp. 125–9.

[30] This appeares to have been copied from Newton's copy of 'Η Καινὴ Διαθήκη, *Novum Testamentum* (London, 1653), purchased by Newton in April of 1661.

[31] Descartes, *Principia*, Part III, Articles XCVII and XCVIII, p. 100.

impel the globuli so as to cause a light from the poles, and those places from whence they come.

Of the Sun, stars, planets, and comets   12 93ʳ

Whether the Sun moves the vortex about (as Descartes's will) by his beams, page 54, *Principia Philosophia*, Part III. Whether the vortex can carry a comet toward the poles. How is it that the Sun is turned about upon his axis. Whether Descartes's notion of reflection will unriddle the mystery of the comet's tail.

Hebrews, Chapter 1, verse 2, God made the worlds by his son Τοὺς αἰῶνας.

The Sun's spots are colored sometimes like the rainbow.

On October 26, 1618, in Scorpio there appeared a comet, its tail being extended between Spica virginis and Arcturus toward the North Pole. The comet passed into Libra moving from the ecliptic to the Tropic of Cancer from east to west or northerly.

On Saturday, December 10, 1664, by a subtle observation I found the distance of a comet from the center of the Moon to be 9° 48'. Its altitude 3° 40' or 4°. The Moon's altitude, 8° 40'. ~~Her place being Capricorn 26° 2' or else Aquarius 5°.~~ The longitude of the Moon.

On Saturday at 4:30 in the morning, December 17, 1664, a comet appeared whose distance from Sirius was 30°, from Procyon 38° 45'. There was little or no difference between the time of its and Sirius's setting, the comet setting about 2 minutes after Sirius; so that its right ascension was about 126° 32' and its declination southward 31°. The length of its

of its tayle was about 34^d or 35^d & pointed $^{below}_{toward}$ procion or almost to y^e ~~horizon~~ North pole cutting y^e horizon at an angle of about 35^d or 40^d & y^e Ecliptick at 47^d.

vide pag 54

13 94^r

Of Rarity & Density. Rarefaction & Condensa⌊tion.⌋

Corke may be pressed into 40 times less roome y^n it naturally requireth & yet swim in water. By my tryall 48 times. two bodys given to find wch is more dense. Upon y^e ~~body~~

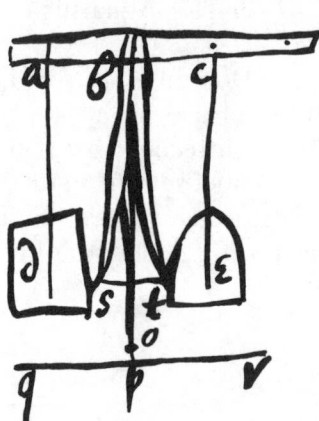

Threds da & ce hang y^e bodys d & e. & exactly twixt y^m hang y^e spring sbt ⌊by a thred⌋ soe y^t it have liberty to move to move any way. then compresse y^e sping bs to bt by y^e thred st. Then $\left\{ {clipping \atop cutting} \right\}$ y^e thred st y^e spring shall cast both y^e body d & e from it & they receve alike swiftnes from y^e spring if there be y^e same quantity of body in both otherwise y^e body bo (being fastened to y^e spring) will move towarsds y^e body wch hath less body in it. wch motion may be observed by comparein ye motion of y^e point (o) to y^e point p & other points in y^e ⌊resting⌋ body qr.

tail was about 34° or 35° and pointed toward-below Procyon, or almost to the North Pole, cutting the horizon at an angle of about 35° or 40°, and the ecliptic at 47°.

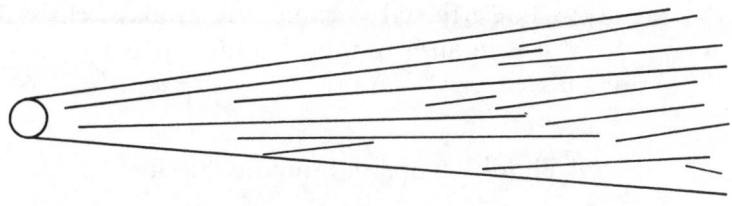

vide page 54[5]

Of rarity, density, rarefaction, and condensation   13 94[r]

Cork may be pressed into 40 times less room than it naturally requires and yet swim in water. By my trial 48 times.

Given two bodies to find which is more dense.

Upon the threads *ad* and *ce* hang the bodies *d* and *e*. Exactly

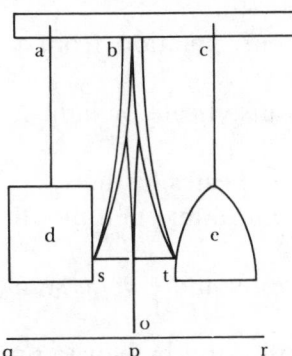

between them hangs the spring *sbt* by a thread so that it is free to move in any way. Then, compress the spring *bs* to *bt* by the thread *st*. Then clipping the thread *st* the spring shall cast both the bodies *d* and *e* from it, and receive alike swiftness from the spring if there be the same quantity of body in both. Otherwise, the body *bo*, being fastened to the string, will move toward the body which has less body in it—which motion may be observed by comparing the motion of the pointer *o* to the point *p* and other points in the resting body *qr*.

[5] Page 411

14 94ᵛ   Of Perspicuity & Opacity

Why though both a dry bladder & water ~~reflect~~ are perspicuous yet a wet blader is not. though oyle bee les diaphaneous y^n water yet it makes a paper more diaphaneous y^n it.
    Perspicuity is not effected ye same way in glass, christall,
water   &c y^t it is in aire, Aether, bladders,paper
    Why water is clearer y^n Vapors.

15 95ʳ   Of Fluidity Stability humidity Siccity

16 95ᵛ   Of Softnesse hardnes Flexility Ductility Tractility

    Why flints doe breake upon a soft thing sooner y^n a hard
one.
    Whither hard bodys stick together by branchy particles
foulded together. Cartes.[32]
    Why ye Adamant dust is harder y^n adamant.

17 96ʳ   Of Figure Subtility hebetude smothnes asperity

18 96ᵛ   Of heate & cold

    Whither things congeale for want of agitation from y^e
ethereall maters ~~agitation.~~ Cartes.[33]
    Whither doth air moved by light cause heate or light it-
selfe &c.
    Why is coale hoter y^n flame but flame heates farther.
    Why doth warme breath proceede out of an open mouth
but cold out of a contracted mouth.[34]
    Why is breath & sweate seene in winter more y^n in sum-
mer.~~& why vapors~~
    Fire heates by heating ye aire because wind by blowing y^e
aire blows heate to a man.
    To make a crucible w~~ᶜʰ shall~~ endure halfe a yeare a very
strong fire lute it in ye outside (thick at ye bottom & thiner
towards ye top) wth a mixture of tobaccopipe clay & salt of

---

[32] *Ibid.*, Part II, Article LXIII, p. 48.
[33] Rene Descartes, *Opera philosophica, Meteora,* 3rd ed. (Amsterdam, 1656),
Caput I, Article VII, p. 156.   [34] *Ibid.*, Caput II, Article V, p. 162.

Of perspicuity and opacity

14 94ᵛ

Why though both a dry bladder and water are perspicuous, yet a wet bladder is not. Though oil be less diaphanous than water, yet it makes a paper more diaphanous than it. Perspicuity is not effected the same way in glass, crystal, water, etc., than it is in air, ether, bladders, paper.
Why water is clearer than vapors.

Of fluidity, stability, humidity, and siccity

15 95ʳ

Of softness, hardness, flexibility, ductility, and tractility

16 95ᵛ

Why flints break upon a soft thing sooner than a hard one.
Whether hard bodies stick together by branchy particles folded together. Descartes.
Why diamond dust is harder than the diamond.

Of figure, subtility, hebetude, smoothness, and asperity

17 96ʳ

Of heat and cold

18 96ᵛ

Whether things congeal for want of agitation from the ethereal matter. Descartes
Why does air moved by light cause heat or why does light itself cause heat?
Why is a coal hotter than flame but flame heats farther?
Why does warm breath proceed out of an open mouth but cold out of a contracted mouth?
Why is breath and sweat seen in winter more than in summer?
Fire heats by heating the air, because wind, by blowing the air, blows heat to a man.
To make a crucible endure a very strong fire half a year lute it on the outside (thick at the bottom and thinner toward the top) with a mixture of tobacco-pipe clay and salt of

Tartar (or Tartar may bee made use of, but not wth so good successe) I think there must bee an equall quantity of each.

Whither may not water bee frozen by drawing ye warme aire from it out of Mr Boyls Receiver.

Snow put in a glasse & Salt or any quick dissolvent put into it ⌊& well mixed wth it⌋ will cause vapors to settle on ye outside of ye glase & to freeze.nay warme water or heated sand powered into ye snow & well shaked together will ~~bee~~ condense vapors on ye outside & perhaps congeale y^m. Mr Boyle.[35]

Cold (~~(?)~~ because bodys condensed therewth move down wards) tends farthest downwards as heat upwards.[36]

Tis best to freze liquors at ye bottome for feare of breaking ye Glasse.[37]

An frozen Egg will thaw much faster when immersed in water y^n w^n in ye ayre, & will freze ye water by its thawing.[38] soe will frozen men,⌊chese,meate,[39]Glasse.⌋[40]

Why does water freeze first & most next ye Aire.

19 97^r Of Gravity & Levity

The matter causing gravity must pass through all y^e pores of a body. it must ascend againe. 1 for else y^e bowells of y^e earth must have had large cavitys & inantys to conteine it in ⌊2⌋ or else y^e matter must swell it. 3 y^e matter y^t hath so forcibly borne do$\overline{w}$ y^e earth & all other bodys to ~~for~~ y^e center (unles you will have it growne to as gross a consistance as y^e Earth is, & hardly y^n) cannot if added to ~~y^e Earth~~ gether be of a bulke so little as y^e Earth, For it must descend exceding fast & swift as appeares by y^e falling of bodys, & exceeding weighty press⌊ure⌋ to ye Earth. It must ascend in another forme y^n it descendeth or else it would have a like force to beare bodys up y^t it hath to press y^m downe & so there would bee no gravity. It must ascend in a grosser consistence y^n it descends 1 because it may be slower & not strike boddys wth so greate a force to impell y^m upward 2 y^t

---

[35] A summary of pp. 115–36 of Robert Boyle, *New Experiments, and observations touching Cold, or An experimental history of cold* (London, 1665).
[36] *Ibid.*, p. 175.   [37] *Ibid.*, pp. 180–1.   [38] *Ibid.*, p. 187.
[39] *Ibid.*, pp. 196–8.   [40] *Ibid.*, p. 193.

tartar (or tartar may be used but not with so good success). I think there must be an equal quantity of each.

Whether water may not be frozen by drawing the warm air from it, out of Mr. Boyle's receiver.

Snow put in a glass and salt or any quick dissolvent put into it, and well mixed with it, will cause vapors to settle on the outside of the glass, and to freeze. Warm water or heated sand powdered into the snow and well shaken will condense vapors on the outside, and perhaps congeal them. Mr. Boyle.

Cold (because bodies condensed therewith move downward) tends farthest downward as heat upward.

It is best to freeze liquids at the bottom for fear of breaking the glass.

A frozen egg will thaw much faster when immersed in water than when in the air, and will freeze the water by its thawing. So will frozen men, cheese, meat, and glass.

Why does water freeze first and most when next to the air?

Of gravity and levity

19 97'

The matter causing gravity must pass through all of the pores of a body. It must ascend again, (1) for either the bowels of the Earth must have had large cavities and inanities to contain it, (2) or else the matter must swell the Earth, or (3) the matter that has so forcibly borne down the Earth and all other bodies to the center (unless you will have it grown to as gross a consistence as the Earth is, and hardly then) cannot if added together be of a bulk so little as is the Earth. For it must descend very fast and swift as appears by the falling of bodies and by the great pressure toward the Earth. It must ascend in another form than it descends, or else it would have a like force to bear bodies up as it has to press them down, and so there would be no gravity. It must ascend in a grosser consistency than it descends, (1) because it may be slower and not strike bodies with so great a force as to impel them upward, (2) that it may only force the

it may onely force ye outside of a body & not sinke into every pore & y^n its densness will little availe it because it will yeild from ye superficies of a body wth ease to run in an easier channell as though it never strove against y^m. if it should ascend thinner it can have onely this advantage y^t it would not hit bodys wth so weighty a force but y^n it would hit more ⱳts of ye body & would have more ⱳts to hit wth & hit wth a ~~swifter~~ ˌsmarterˌ force: & so cause ascension wth more force y^n ye others could do descension. Wee know no body that not sinke into ye pores of bodys finer y^n aire & it will sink into most if it be forcibly crouded in. ye stream descending will lay some hould on ye streame ascending & so press it closer & make it denser & therefore twill rise ye slower. ye streame descending will grow thicker as it comes nigher to ye earth but will not loose its swiftnesse untill it find a much opposition as it hath helpe from ye following flood behind it. but when ye streames meete on all sides in ye midst of ye Earth they must needs be coarcted into a narrow roome & closely press together & find very much opposition one from another so as either to turne back ye same way y^t they came or croud through one

pag 67

Of Heate & Cold.

Apples, Eggs, Cheeses, Men &c: frozen are ~~(?)~~ vitiated by freezing but not soe much when thawed by water or snow as by fire.[41] Frost will breake stones, crack trees,[42] make ye Humor chrystall looke white.[43]

A man cannot feele where hee is frozen & though frozen all over feeles onely a prickling in his recovery, hee may bee recovered being dipped in water or rubbed over wth snow, but not by a ~~hot~~ Stove. Nay any frozen part is lost wch is thawed in a Stove. & ye fier pains us in warming o^r cold fingers.[44] Frozen meate layd to thaw & roast by ye fire will bee raw in ye midst after many Howers.[45]

[41] *Ibid.*, pp. 199ff.   [42] *Ibid.*, pp. 213–15.
[43] *Ibid.*, pp. 207–8. The humor chrystal is the lens of the eye.
[44] *Ibid.*, pp. 219–21.   [45] *Ibid.*, p. 219.

outside of a body and not sink into every pore, and then its denseness will little avail it, because it will yield from the superficies of a body with ease so as to run in an easier channel as though it never strove against them. If it should ascend thinner it can have only this advantage: that it would not hit bodies with so weighty a force, but then it would hit more parts of the body and would have more parts with which to hit with a smarter force, and so cause ascension with more force than the others could do descension. We know of no body that will not sink into the pores of bodies better than air, and it will sink into most if it be forcibly crowded in. The stream descending will lay some hold on the stream ascending, and so press it closer and make it denser. Therefore it will rise the slower. The stream descending will grow thicker as it comes nearer to the Earth; but it will not lose its swiftness until it finds as much opposition as it has help from the flood following behind it. But when the streams meet on all sides in the midst of the Earth, they must needs be compressed into a narrow room, closely pressed together, and thus very much opposing one another either turn back the same way that they came, or crowd through one

continued on page 67[6]

Of heat and cold   20 97ᵛ

When apples, eggs, cheeses, men, etc., are frozen, they are impaired by that freezing, but not so much when thawed by water or snow as when by fire. Frost will break stones, crack trees, and make the humor crystal look white.

A man cannot feel where he is frozen, and though frozen all over, feels only a prickling in his recovery. He may be recovered by being dipped in water or rubbed over with snow, but not by a stove. Nay, any frozen part is lost which is thawed in a stove. The fire pains us in warming our cold fingers. Frozen meat laid to thaw and roast by the fire will be raw in the middle after many hours.

---

[6] Page 427

Though frost change & destroy bodys (espetialy in) y^c thawing) yet cold preserves them.[46]

Ice (to wch noe fresh aire in y^c Freezing could come y^c botome of y^c water being first frozen &c) is full of bubbles wee ⌊greate as⌋ sands, shott, & pease wch bubbles are fewer & lesse if y^c water bee first purged of aire by y^c receiver. And those bubbes in thawing shrink againe perhaps into as little rome as at first.[47]

Cold shrinkes liquors, Oyle shrinke in frezing, water scarce shrinkes before frezing, but swells before & in frezing;[48] & Ice is about one ninth or tenth parte greater y^n water.[49]

Water & aire shut up in a glasse egg wth a shank, y^c aire being in y^c shanke was crouded into 19 times lesse roome by y^c freezing of y^c water till it broke y^c glasse.[50]

Cold will penetrate through Boyles Vacuum,[51] Oyle of Turpentine,[52] & a little through ~~spirit of~~ strong bryan,[53] & perhaps through hot mediums to freze water.[54]

21 98^r ~~Of Reflection Refraction & Undulation~~

Of violent motion

Violent motion is ~~made~~ ⌊continued⌋ either by y^c aire or by ~~motion~~ a force imprest. or by the naturall gravity in y^c body moved. Not by y^c aire since y^c aire ~~since y^c air~~ crowds more uppon y^c thing projected before, y^n behind, & must therefore rather hinder it ~~but if y^c aire helps it why doth y^c thing at last fall~~ for you may observe in water y^t a thing moved in it doth carry ⌊y^c⌋ same water behind it along wth it as in a cone or at least y^c water is moved from behind it wth but a small force as you may observe by y^c motes in y^c water supose

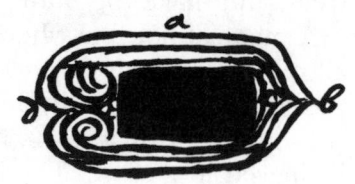

[46] *Ibid.*, p. 201.   [47] *Ibid.*, pp. 245ff.   [48] *Ibid.*, pp. 294–5.
[49] *Ibid.*, pp. 281–2.   [50] *Ibid.*, pp. 323–5.   [51] *Ibid.*, pp. 353–4.
[52] *Ibid.*, p. 358.   [53] *Ibid.*, p. 361.   [54] *Ibid.*, pp. 354–6.

Though frost changes and destroys bodies (especially in the thawing), yet cold preserves them.

Ice (to which no fresh air in the freezing could come, the bottom of the water being first frozen, etc.) is full of bubbles as great as sands, shot, and peas; which bubbles are fewer and less if the water be first purged of air by the receiver. And those bubbles in thawing shrink again, perhaps into as little room as at first.

Cold shrinks liquids. Oil shrinks in freezing. Water scarce shrinks before freezing, but swells before and during freezing. Ice is about one ninth or tenth part greater than water.

Water and air shut up in a glass egg with a shank, the air, being in the shank, was crowded into 19 times less room by the freezing of the water – till it broke the glass.

Cold will penetrate through Boyle's vacuum, oil of turpentine, and a little through strong brine – perhaps through hot mediums – to freeze water.

Of violent motion   21 98ʳ

Violent motion is continued either by the air or by a force impressed, or by the natural gravity in the body moved. Not by the air, since the air crowds more upon the thing projected before than behind, and must therefore rather hinder it, ~~but if the air helps it why does the thing at last fall.~~ For you may observe in water that a thing moved in it does carry along with it the water behind it, as in a cone, or at least the water is moved from behind it with but a small force as you may observe by the motes in the water.

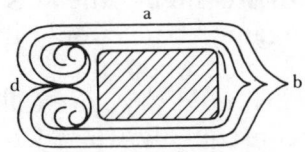

348 QUESTIONES QUÆDAM PHILOSOPHICÆ

(a) to be y^e boddy moved: (b. d. e.
f.) to be y^e water moving behind (a)
to give it place. (c) y^e water behind
(a) following it & going along wth it.
The if y^e water at (f) ran so violently
against y^e backsid of (a) it would
beate away y^e water at (c) wth vio-

lence but y^t water is moved very slowly from behind a. ~~as you~~
if it be moved away: as you may perceive by y^e motes in y^e
water.[55] ~~againe this wil~~ the like must hapen in aire if you say
no I answer must y^n move (a) forwards in water. So if hot
leade drop into water y^t ̵pte wch is behind ~~(?)~~ will be pointed.
ye fore ̵pte round wch would be otherwise if ye aire pressed as
much on it ~~before~~ hind⌐ as ~~behind~~fore. thirdly how can y^e
aire continue ye motion of a globe on it axis. Fourthly

22 98^v in ye former figure ye aire is supposed to have ye same
propensity to motion wch ye ball (a) is supposed to have that
is will move no longer y^n it is propelled on. y^n I say ye water
at (c) cannot move ye ball unles ye ball do at ye same time
move (b) y^t (b) may (g) & (h) & (g) may move (d) & (d) move
(i) & (i) move (f) & f move c & force it to rush uppon ye ball
& consequently at ye same instant (c) must ye ball, & ye ball
move (c) wch cannot be. But suppose ye aire & ye ball were
detained from motion by some outward agent, & yet kept y^e
same respect to one-another in situation as they did in
theire flight: then as soone as they were both let loose
againe ye aire would have as much power to move ye ball as
it had when they were in theire former flight. If it be
answered y^t ye aire will be more compressed at (f) y^n at (b) &
consequenly when let loose againe it will dilate it self & so
begin a new motion. I answer how comes ye aire to be more
crouded behind ye ball then before it since since (a) will
communicate as much force on (b) as it receives from c & y^e
fore ̵pte of ye aire will croud no more on ye latter parte y^n y^e
ball will croude on ~~(?)~~ it. Againe whence is it y^t a ~~of~~ peice of
leade will move ~~faster then a peice~~ farther & wth more force

[55] This attack on antiperistasis was probably directed against Descartes's use of
a partial antiperistasis in his account of motion in a fluid as presented in *Principia*, Part II, Articles LVII–LIX, pp. 46–7, and Article LXI, p. 48.

Suppose *a* to be the body moved; *b, d, e,*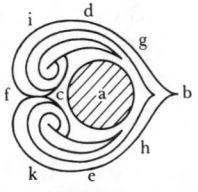
and *f* to be the water moving behind *a* to
give it place; *c* the water behind *a* following it and going along with it. Then, if the
water at *f* ran so violently against the
backside of *a*, it would beat away the
water at *c* with violence. But that water is moved very slowly
from behind *a*, if it be moved away, as you may perceive by
the motes in the water. The like must happen in air. If you
say no, I answer that then *a* must move forward in water. If
hot lead drop into water that part which is behind will be
pointed and the forepart round, which would be otherwise
if the water pressed as much on it behind as before.
Thirdly, how can the air continue the motion of a globe on
its axis.
Fourthly,

in the former figure the air is supposed to have the same   22 98ᵛ
propensity to motion which the ball *a* is supposed to have,
that is, it will move no longer than it is propelled onward.
Then I say, the water as *c* cannot move the ball unless the
ball does at the same time move *b*, that *b* may move *g* and *h*,
and *g* may move *d*, and *d* move *i*, and *i* move *f*, and *f* move *c*
and force it to rush upon the ball, and consequently at the
same instant *c* must move the ball and the ball move *c*,
which cannot be. But suppose that air and the ball were
detained from motion by some outward agent, and kept the
same respect to one another in situation as they did in their
flight, then as soon as they were both let loose again the air
would have as much power to move the ball as it had when
they were in their former flight. If it be answered that the
air will be more compressed at *f* and at *b* and consequently
when let loose again it will dilate itself and so begin a new
motion, I answer, how comes the air to be more crowded
behind the ball than before it, since *a* will communicate as
much force on *b* as it receives from *c*, and the forepart of
the air will crowd no more on the latter part than the ball
will crowd on it. Again, why is it that a piece of lead will
move farther and with more force than a piece of wood of

y^n a peice of wood of ye same bignesse since ye aire will have ye same influence on both.

vide pag 51

23 99^r

Of Reflection undulation & refraction.

Why refraction is less in hot water y^n cold.

If a peice of silver be boyled (that is bee first brushed & y^n decocted wth salt & tartar ⌐& perhaps other ingredients⌐) it will looke very white, but burnish it wth a peice of steele & it will be a perfect speculum.[56]

Whither ye backsid of a cleare glas reflect light in vacu.

Since there is refraction ⌐in vacuo⌐ as in ye aire it follows y^t ye same subtile matter in ye aire & in vacuo causeth refraction.

Try whither ~~ye aire~~ Glasse hath ye same refraction in Mr Boyles Receiver, ye aire being drawn out, wch it hath in ye open aire.

How long a pendulum will undulate in Mr Boyles Receiver? &c.[57]

24 99^v

Of Fier

Whither flame will descend in Tor: vacuo or not.[58] & what other Phænomena (as dilata̅co̅n & transparency) hath it.

25 100^r

Of Aer

Whither ye parts of air be les y^n y^m of light or no.

Whither it consist of branchy bodys not foulded together but lying upon one another. Cartes.[59]

~~Whither~~

[56] Robert Boyle, *Experiments and Considerations Touching Colours* (London, 1664), p. 115.

[57] Boyle, *Spring of the Air*, Experiment 26, pp. 202–4.

[58] Boyle considers in which direction a flame will travel in Experiment 14 of *Spring of the Air*, p. 101. Although he reports that the flame traveled upward, Boyle speculates that this may have been due to some still-present air.

[59] Descartes, *Meteora*, Caput I, Article III, p. 154.

the same bigness since the air will have the same influence on both.

vide page 51[7]

Of reflection, undulation, and refraction   23 99ʳ

Why refraction is less in hot water than cold? If a piece of silver be boiled (that is, be first brushed and then decocted with salt and tartar and perhaps other ingredients), it will look very white, but burnish it with a piece of steel and it will be a perfect speculum.

Whether the backside of a clear glass reflects light in a vacuum. Since there is refraction *in vacuo* as in the air, it follows that the same subtle matter in the air and *in vacuo* causes refraction. Try whether glass has the same refraction in Mr. Boyle's receiver, the air being drawn out, which it has in the open air.

How long a pendulum will undulate in Mr. Boyle's receiver?

Of fire   24 99ᵛ

Whether flame will descend in Torricelli's vacuum or not, and what other phenomena (such as dilation and transparency) it has.

Of air   25 100ʳ

Whether the parts of air be less than those of light or no.

Whether it consists of branchy bodies not folded together but lying upon one another. Descartes.

---

[7] Page 407

The height of ye Atmosphere may bee known from Torricellius his experiment.[60]

What is y^e utmost ⌞naturall⌟ delatacon of y^e aire may be know̅ by Torricellius his Experiment.

The velocity of air, wind, or water may bee know̅ by y^e resistance wch a moveing body hath in standing air or water.

What ~~quantity of~~ angle ought a Wind mill saile to make wth y^e wind. &c. in y^e resolution of this must be considered y^e ordinary velocity of y^e wind & of y^e saile (y^e quantity of ye wind hitting ye saile i:e:) ye perpendicular breadth of ye saile to ye wind, & ye obliquity of ye saile to ye wind.

26 100^v   Whither is salt or fresh water easlier moved & more pellucid. refracts more & easlier frozen.

Of Water & Salt.

Whither fresh water consist of long bending ~~attomes~~ ⌞parts⌟ & salt of stiffe & long ones.[61] ye first is false because it could never bee frozen. 2 they would twist ~~(?)~~ about one another so as they would not be fluid but onely soft. 3 they would ly too close together to admit light through them for being pliable they would fill up every corner & hence they would bee exceeding heavy. 4 they would not refract light so well, for ~~being~~ they would bee soft & so not firmly resist y^e ~~(?)~~ pure matter as Glasse doth & Cartes would have y^t matter to passe swiftlier where it findes strongest resistance & refraction to be from hence y^t ye matter passeth swiftliest wch therefore should bee in water.[62] 5 ~~twould scarce seperate salt so soone.~~ & ~~when~~ if it seperate it by laping about it

---

[60] Boyle, *Spring of the Air,* pp. 297–9.

[61] Descartes, *Meteora,* Caput III, Article II, p. 164.

[62] Rene Descartes, *Opera philosophica, Dioptrices,* 3rd ed. (Amsterdam, 1656), Caput II, Articles IV and V, pp. 58–9.

The height of the atmosphere may be known from Torricelli's experiment.

What the utmost natural dilation of the air is may be known by Torricelli's experiment.

The velocity of air, wind, or water may be known by the resistance which a moving body has in standing air or water.

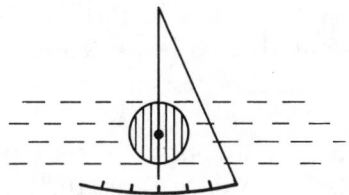

What angle ought a windmill sail make with the wind. The resolution of this must consider the ordinary velocity of the wind and of the sail (that is, the quantity of wind hitting the sail), the perpendicular breadth of the sail to the wind, and the obliquity of the sail to the wind.

Of water and salt   26 100ʳ

Whether salt or fresh water is more easily moved and more pellucid, refracts more, and is more easily frozen.

Whether fresh water consists of long bending parts and salt of stiff and long ones. The first is false, because it could never be frozen. (2) They would twist about one another so as not to be fluid but only soft. (3) They would lie too close together to admit light through them, for being pliable they would fill up every corner, and hence they would be exceedingly heavy. (4) They would not refract light so well, for they would be soft and so not firmly resist the pure matter as glass does, and Descartes would have that matter to pass more swiftly where it finds strongest resistance, and refraction to be from hence, that the matter passes swiftest which therefore should be in water. (5) If water separates salt by lapping about it, then when it has separated as much

y^n when it hath seperated as much as it can of one kind of Salt it could seperate no more of another wch is false. 6 The aire being a stubborne body (because of branchy parts) would instantly quell there circular motion when they are rarefied. Why water is clearer y^n vapors. Whither burning waters & hot spirits be of small spericall or ovall figud parts & have many such globuli as fire is of they are 1 because such are (?) easliest seperated in distilllations[63] 2 because they are easliest agitated & so heate & enliven men 3 they must have many small & sollid attomes in y^m because so easly fired.

Why doth hot water first contract it selfe (viz in cooleing) & y^n dilate it selfe before & as it freeseth.[64]

Why doth salt & snow freese other water.[65] & why is heated water sooner frozen y^n other raw water.[66]

Whither be ther more vapors when y^c aire is clearest.[67]

How salt hinders corruption. but fresh water helpes it.[68]

Why (though salt bee heavier yet will mix wth water. & gather into graines at y^c top of it.[69]

Whither water be salter at y^c poles y^n equater y^n y^c poles Becaus tis there exhaled but may fall againe at y^c poles. Cartes.[70]

Why soft water sea water is not so apt to quench fier & why it will sparkle in y^c night but not if kept long in a vassell.[71] why y^c superficies of water is lesse divisible y^n tis within

vid pag 47

27 101^r Of Earth

Its conflagration testified 2 peter 3^d, vers 6, 7, 10, 11, 12. The wiked probably to be punished thereby 2 Pet: 3^{chap}: vers 7.

[63] Descartes, *Meteora*, Caput II, Article VIII, p. 163.
[64] Ibid., Caput I, Article VIII, p. 157.
[65] Ibid., Caput III, Article V, pp. 165–6.
[66] Ibid., Caput I, Article VIII, p. 157.
[67] Ibid., Caput III, Article X, p. 168.
[68] Ibid., Caput III, Article II, p. 164.
[69] Ibid., Caput III, Article X, p. 168.
[70] Ibid., Caput III, Article VIII, p. 167.
[71] Ibid., Caput III, Article IX, pp. 167–8.

as it can of one kind of salt it could separate no more of another, which is false. (6) The air being a stubborn body (because of branchy parts) would instantly quell their circular motion when they are rarefied.

Why water is clearer than vapors?

Whether burning waters and hot spirits be of small spherical or oval figured parts, and have many such globuli as fire is of. They are of such parts (1) because such are most easily separated in distillations, (2) because they are most easily agitated and so heat and enliven men, and (3) they must have many small and solid atoms in them because they are so easily fired.

Why does hot water first contract itself (viz., in cooling), and then dilate itself before and as it freezes?

Why does salt and snow freeze other water? Why is heated water sooner frozen than raw water?

Whether there be more vapors when air is clearest? How salt hinders corruption, but fresh water helps it. Why, though salt be heavier, yet it will mix with water and gather into grains at the top of it?

Whether water be saltier at the equator than the poles because it is there exhaled, but may fall again at the poles. Descartes.

Why sea water is not so apt to quench fire, and why it will sparkle in the night but not if kept long in a vessel? Why the superficies of water is less divisible than it is within,

vide page 47[8]

Of Earth 27 101ʳ

Its conflagration testified Peter 2, Chapter 3, verses 6, 7, 10, 11, and 12. The wicked probably to be punished thereby, Peter 2, Chapter 3, verse 7. The succession of worlds is

---

[8] Page 399

The succession of worlds, probable from Pet 3^c. 13^v. in wch text an emphasis upon y^e word WEE is not countenanced by y^e Originall. Rev 21^c. 1^v. Isa: 65^c, 17^v. 66^c, 22^v. Days & nights after y^e Judgm Rev 20^c, 10^v.

28 101^v

Philosophy ~~Occult Qualityes~~

The nature of things is more securely & naturally deduced from their oper$\overline{\text{acons}}$ one upon another $\overline{y}$ upon o^r senses. And when by y^e former Experiments we have found y^e nature of bodys, by y^e latter wee may more clearly find y^e nature of o^r senses. But so long as wee are ignorant of y^e nature of both soule & body wee cannot clearly distinguish how far an act of sensation proceeds from y^e soule & how far from ye body &c.[72]

29 102^r

Atraction Magneticall

1 The motion of any magneticall ray may bee knowne by attracting a needle in a corke on water.

2 Whither a magneticall pendulum is perpendicular to y^e Horizon or not, & whither iron is heaviest w^n impregnated, or when y^e north pole or south pole is upmost. Coroll: A perpetuall motion[73]

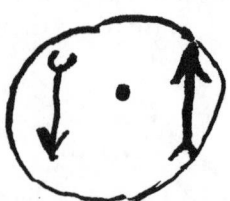

3 Whither magneticall rays will blow a candle move a red hot copper ⌊or iron⌋ needle, or passe through a red hot plate of copper or iron.

---

[72] This entry is drawn from pp. 52–4 of Thomas Hobbes, *Elements of Philosophy. The First Section, Concerning Body* (London, 1656).

[73] Newton's early interest in perpetual motion is also reflected in a set of notes in the Morgan Notebook (see footnote 20) headed "Of a Perpetuall Motion," "Of Drebles Motion," and "Of a perpetuall Lamp" on folios 17^r–18^r. The notes are drawn from John Wilkins, *Mathematicall Magick or, The Wonders that may be Performed by Mechanical Geometry* (London, 1648), pp. 228–30 and pp. 246–52.

probable from Peter 2, Chapter 3, verse 13, in which text an emphasis upon the word "we" is not countenanced by the original. Revelations, Chapter 21, verse 1; Isaiah, Chapter 65, verse 17, Chapter 66, verse 22. Days and nights after the Judgment, Revelations, Chapter 20, verse 10.

Philosophy ~~Occult Qualities~~   28 101ᵛ

The nature of things is more securely and naturally deduced from their operations one upon another than upon our senses. And when by the former experiments we have found the nature of bodies, by the latter we may more clearly find the nature of our senses. But so long as we are ignorant of the nature of both soul and body we cannot clearly distinguish how far an act of sensation proceeds from the soul and how far from the body.

Magnetic attraction   29 102ʳ

1. The motion of any magnetic ray may be known by attracting a needle in a cork on water.
2. Whether a magnetic pendulum is perpendicular to the horizon or not, and whether iron is heaviest when impregnated, or when the north pole or south pole is uppermost. Corollary: a perpetual motion

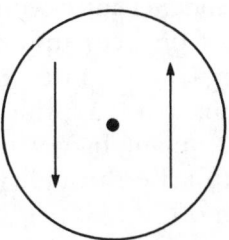

3. Whether magnetic rays will blow a candle, move a red hot copper or iron needle, or pass through a red hot plate of copper or iron.

4 A perpetuall motion

5   Whither a loadestone will not turne around a red hot irón fashioned like a wind=mill=sailes, as y^c wind doth y^m. Perhaps cold iron may reflect y^c magn: rays ~~6~~ wth y^t pole wch shuns y^c lodestone.
6.   A perpet: motion

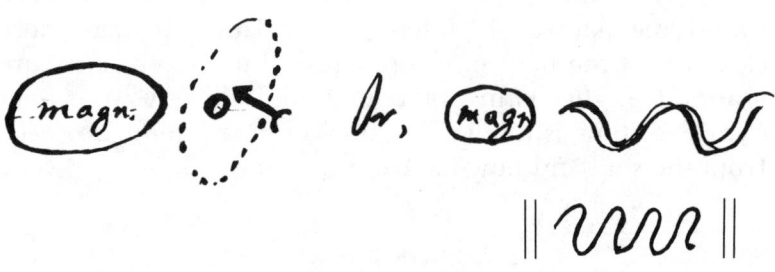

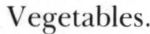

30 102^v Vegetables.

Suppose ab y^c pore of a Vegitable filled wth fluid mater & y^t y^c Globule c doth hitt away y^c particle b, y^n y^c rest of subtile matter in y^c pore riseth from a towards b. & by this meanes juices continually arise from y^c rootes of trees upward: wch juices leaving dreggs in y^c pores & y^n wanting passage stretch y^c pores to make y^m as wide as before they were clogged. wch makes y^c plant bigger untill y^c pores are too narrow for y^c juice to arise through y^c pores & y^n y^c plant ceaseth to grow any more.

31 103^r Attraction Electricall & Filtration.

Whither filtration be thus caused. The aire beiing a stubborne body if it be next little pores into wch it can enter it will be pressed into $\overline{y}$ (unles theye be filled by something else) ~~(?) yet ther it will have some reluctance out wards like a~~

4. A perpetual motion

5. Whether a lodestone will not turn around a red hot iron fashioned like windmill sails, as the wind does to them. Perhaps cold iron may reflect the magnetic rays with that pole which shuns the lodestone.

6. A perpetual motion

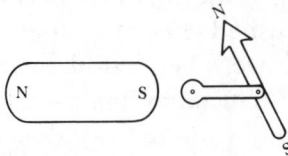   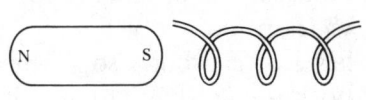

Vegetables   30 102ᵛ

Suppose *ab* the pore of a vegetable filled with fluid matter and that the globule *c* does hit away the particle *b*, then the rest of the subtle matter in the pore rises from *a* toward *b* and by this means juices continually arise from the roots of trees upward. These juices, leaving dregs in the pores, and then wanting passage stretch the pores to make them as wide as before they were clogged. This makes the plant bigger until the pores are too narrow for the juice to rise through the pores, and then the plant ceases to grow any more.

Filtration and electrical attraction   31 103ʳ

Whether filtration be thus caused: the air being a stubborn body, if it be next to little pores into which it can enter, it will be pressed into them (unless they be filled by something else). Yet it will have some reluctance outward, like a piece

peice of bended whale bone crouded into a hole wth its middle ͬpte forwards. if y^n water whose (ͬpts are loose & pliable) have opportunity to enter y^t hole ye aire will draw it in by strivei it selfe to get out. The aire too being continually shake̅ & moved in its smallest parts by vaporous particles every where tossed up & downe in it as appeares by its heate, it must needs strive to get out of all such cavity wch doe hinder its agitation: & this may be ye cheife reason ~~of~~ sponges draw up water. But in paper ropes hempe theds fiddle-strings betwixt whose particles there is noe aire or but a little & it so pend up y^t it can scarce get out the cause may be this. y^t ye parts of those bodys are crushed closer together y^n there nature will well permit & as it were bended like ye laths of crosbows so y^t they have some reluctancy against y^t position & striv̅ to get liberty wch they cannot fully doe unless some oth^r bodys come betweene y^m as aire or water but where aire cannot enter water will (as appeares in y^t it will get through a bladder wch aire cannot doe &c) wherefor when opportunity offers it selfe by striveing to get assunder they draw in y^e ͬpts of water betwixt y̅

32 103^v

see pag 82

Of light.

Why light passeth easlier through white y^n black paper. & yet more efficaciously reflected from it.

How light is conveyed from ye Sun or fire wthout stops. Light is easlier admitted into black y^n reflected from it, for hold a paper twixt you & light wth a black spot in it & it is blacker when towards you, y^n when to ye light. Light cañot be by pressio̅ &c for y^n wee should see in ye night a wel or better y^n in ye day we should se a bright light above us becaus we are pressed downewards ~~we should se when we shut o^r eyes because o^r eye lids presse y^m or when we put o^r hands on o^r eys.~~ The sun would be long ~~so long as we looke at ye same thing ye pression is ye same~~ nay far greater. ther could be no refraction since ~~matter~~ ye same matter cannot presse 2 ways. y^e *sun could not be quite eclipsed.y^e

of bent whale bone crowded into a hole with its middle part forward. If then water, whose parts are loose and pliable, has opportunity to enter that hole, the air will draw it in by itself striving to get out. The air, too, being continually shaken and moved in its smallest parts by vaporous particles everywhere tossed up and down in it (as appears by its heat), must needs strive to get out of all such cavities which hinder its agitation. This may be the chief reason that sponges draw up water. But in paper ropes, hemp threads, and fiddle strings between whose particles there is no air – or but a little – and so pent up that it can scarcely get out, the cause may be this: that the parts of those bodies are crushed closer together than their nature will well permit. It is as if the parts were bent like the laths of crossbows, so that they have some reluctance against that position, and striving to get liberty which they cannot fully do unless some other bodies come between them, such as air or water. But where air cannot enter water will (as appears in that it will get through a bladder, which air cannot do) wherefore, when opportunity offers itself, by striving to get apart they draw in the parts of water between them.

Of light   32 103ᵛ

Why light passes more easily through white than black paper and yet is more efficaciously reflected from it.

How light is conveyed from the Sun or a fire without stops.

Light is more easily admitted into black than reflected from it, for hold a paper between you and light with a black spot in it and it is blacker when toward you than when to the light.

Light cannot be by pression, for then we should see in the night as well, or better, than in the day. We should see a bright light above us, because we are pressed downward. ~~We should see when we shut our eyes, because our lids press them, or when we put our hands on our eyes.~~ The Sun would be long. ~~So long as we look at the same thing the pression is the same,~~ nay far greater. There could be no refraction since the same matter cannot press two ways. The Sun could not be quite eclipsed. The Moon and planets

Mo⌊o⌋ne & planetts would shine like sunns. A man goeing or running would see in y^c night. When a fire or candle is extinguish we lookeing another way should see a light. The whole East would shine in y^c day time & y^c west in y^c night by reason of y^c flood wch carrys o^r Vortex a light would shine from y^c Earth since y^c subtill matter ~~flows~~ ⌊tends⌋ from y^c center. a *little body interposed could not hinder us from seing pression could not render shapes so distinct. There is y^c greatest pression on y^t side of y^c earth from y^c ☉ or else it would not move about ⌊in equilibrio⌋ but from y^c ☉, therefore y^c nights should be lighest. ⌊Also y^c Vortex is Ellipticall therefore light cannot alway come from the same directon &c.⌋[74]

Whither y^c rays of light may not move a body as wind doth a mill saile.

To know how swift light is. Set a broade well pollised looking glasse on a high steeple soe y^t wth a Telescope 1, 2, 3, 10, or 20 miles of you may see yor selfe in it & having by you a great candle in the night cover it & uncover it & observe how long tis before you see the

33 104^r Of Sensation

The senses of divers men are diversly affected by y^c same objects according to y^c diversity of theire constitution.

To them of Java pepper is cold.

If y^c orifice of y^c stomach is wounded it sooner dispatches a ma̅ y^n if y^c head: y^c former having greate sympathy wth y^c heart deads it & stops it motion & so sence ceaseth: y^c latter though it take away sence yet y^c hearts motion is not impeded thereby.[75]

The Common Sensorium is either 1 y^c Whole body 2 y^c Orifice of y^c Stomack 3 y^c heart 4 y^c braine 5 y^c membranes 6 y^c septum lucidum 7 Some very small & perfectly sollid particle in ye body 8 y^c Conarion 9 y^c Concurse of nerves about y^c 4th ventrickle of y^c braine. 10 The animall spirits in y^t 4th ventricle.[76]

[74] Descartes, *Principia*, Part III, Article XXX, p. 58, Articles LV–LXIV, pp. 69–74. Also see Article CLIII, pp. 134–5, for Descartes's description of the Earth's vortex as an ellipse.

[75] More, *Immortality*, p. 187.   [76] *Ibid.*, pp. 155–6.

would shine like suns. A man going or running would see in the night. When a fire or candle is extinguished we, looking another way, should see a light. The whole East would shine in the day time and the West in the night by reason of the flood which carries our vortex. A light would shine from the Earth, since the subtle matter tends from the center. A little body interposed could not hinder us from seeing. Pression could not render shapes so distinct. There is the greatest pression on that side of the Earth from the Sun, or else it would not move about *in equilibrio,* but from the Sun, therefore the nights should be lightest. Also the vortex is elliptical, therefore light cannot always come from the same direction, etc.

Whether the rays of light may not move a body as wind does a mill sail.

To know how swift light is: set a broad, well polished, looking-glass on a high steeple so that, with a telescope 1, 2, 3, 10, or 20 miles off, you may see yourself in it. Having by you a great candle, in the night, cover it and uncover it and observe how long it is before you see the

see page 82[9]

Of sensation   33 104ʳ

The senses of diverse men are diversely affected by the same objects according to the diversity of their constitution.

To them of Java pepper is cold.

If the orifice of the stomach is wounded it sooner dispatches a man than if the head be wounded: the former having greater sympathy with the heart deads it and stops its motion, and so sense ceases. The latter though it take away sense, yet the heart's motion is not impeded thereby.

The common sensorium is either: (1) the whole body, (2) the orifice of the stomach, (3) the heart, (4) the brain, (5) the membranes, (6) the *septum lucidum,* (7) some very small and perfectly solid particle in the body, (8) the conarion, (9) the concourse of nerves about the 4th ventricle of the brain, or (10) the animal spirits in that 4th ventricle.

[9] Page 447

A ligature being tied sence & motion will be twixt ye ligature & ye head but not downwards.[77] A frogs braine being
peirced it looseth both sence & motion but it will leape &
have its sence though its bowells bee taken out.[78]

Phisitians find ye causes of lethargies Apoplexies Epilepsies
&c diseases y^t seiz on ye Animall functions in ye head.[79]

Unles ye braine be peirced so deepe as to reach ye ventricles ye wound will not take away sence & motion.[80]

A man cannot see through ye hole wch a trepan makes in
his head.[81] Stones have beene found in ye glandula pinealis
& it is invironed with a net of veines & arteries.[82]

A Vertigo must be from y^e turning round of y^e spirits.[83]

The least weight upon a mans braine when hee is trepanned maketh him wholly devoyd of sensation & motion.

34 104^v

Of Species visible.

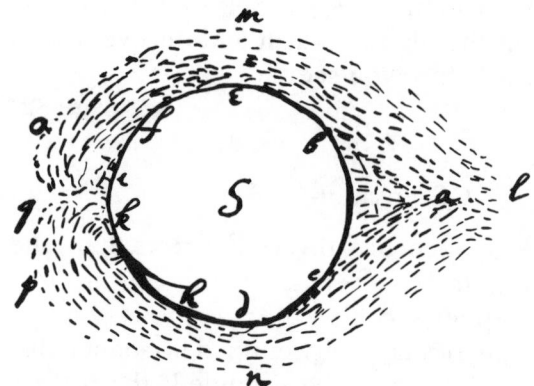

S y^e Globulus of
light. abc a cone
of subtile matter
wch it carrys be-
fore it the better
to cut y^e ether, w^{ch}
serves also to re-
flect it from other
bodys. mln y^e part-
ing of y^e matter
wch lieth so much

ye closer by how much it is nigher ye globulus. (?)

f, h, ye matter pressing on ye backsid & consequently helping it forwards.

pqk, oqi, ye matter returning to communicate ye motion to
ye globulus wch it had before received of it. When it is reflected from i, k, i.e. when it hath given ye ball its motion
againe tis either reflected toward i, k, by ye matter comeing
from o & p towards q &c or else it serves to swell ye matter at
oqp & so is left in ye same condition y^t ye globulus found it in.

[77] *Ibid.*, pp. 190–1.   [78] *Ibid.*, p. 191.   [79] *Ibid.*   [80] *Ibid.*, p. 192.
[81] *Ibid.*, p. 193.   [82] *Ibid.*, p. 197.   [83] *Ibid.*, pp. 203–4.

A ligature being tied, sense and motion will be between the ligature and the head, but not downward. A frog's brain being pierced it loses both sense and motion, but it will leap and have sense, though its bowels be taken out.

Physicians find the cause of lethargies, apoplexies, epilepsies, etc., diseases that seize on the animal functions in the head.

Unless the brain be pierced so deep as to reach the ventricles, the wound will not take away sense and motion.

A man cannot see through the hole which a trepan makes in his head. Stones have been found in the *glandula pinealis,* and it is environed with a net of veins and arteries.

A vertigo must be from the turning round of the spirits.

The least weight upon a man's brain when he is trepanned makes him wholly devoid of sensation and motion.

Of species visible

34 104ᵛ

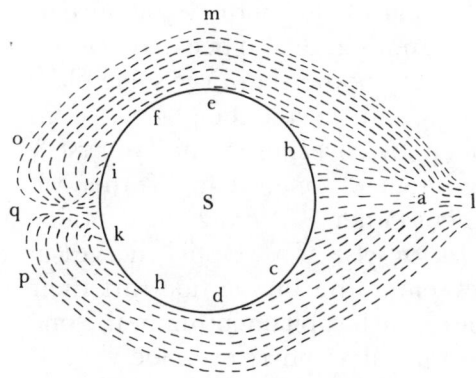

S, the globulus of light. *abc,* a cone of subtle matter which it carries before it the better to cut the ether, which serves also to reflect it from other bodies. *mln,* the parting of the matter which lies so much the closer by how much it is nearer the globulus. *f, h,* the matter pressing on the back side and, consequently, helping it forward. *pqk* and *oqi,* the matter returning to communicate the motion to the globulus which it had before received of it. When it is reflected from *i* and *k,* i.e. when it has given the ball its motion again, it is either reflected toward *i* and *k* by the matter coming from *o* and *p* toward *q,* or else it serves to swell the matter at *oqp,* and so is left in the same condition that the globulus found it in.

Of Vision

There is required some permanency in ye object to perfect vison thus a coale whirled round is not like a coale but fiery circle or who can se a bullet pass by him. yet if ye eye keep motion ~~g~~ wth ye thing moved it sees it right. One eye shutting or being perst ye pupill of ye other open, both ye eyes being opened together, dilate, & cotract, & dilate.

Uppon passion these spirits ebb & flow. Dymnesse may come from ye deficiency of these spirits, & ye Optick nerve ob⌐structed⌐.

Rays from ye same point in an object strikeing upon several respective places in ~~either~~ ⌐both y^e⌐ eys do ~~seeme~~ make ye object seeme two as when an eye is deprest viz: y^n theire two axes respect not ye same point, in ye object. ~~Blen~~ Things of ye darker colo^r are easliest burned i.e. have loosest parts. A mans Eye struck sparkeleth.[84]

The Eys being distorted a man may see ye same parte of ye same object, in two divers places ⌐wth both eys⌐ at once or successively wth one ey after another. & also hee may see two divers objects ~~wth one eye after another but~~ in ye same place sucessively ~~but not at once~~ (wthout varying the posture of his eys or of ye object, but onely by covering ye one eye first & then ye other, or by being intent wth one eye first & then wth the other. wch is an argument either y^t ye image in ye braine is painted on a superficies, or else y^t ye Optick nerves at their meeting before they enter the braine doe sort their capilamenta ~~(?)~~ uniting there, each capilamentum of the one eye to its fellow capilamentum of ye other eye. Soe y^t w^n ye eyes conspire not towards ye same object but have divers pictures painted upon y^e ~~same capil~~ correspondent capillamenta ye stronger picture at their meeting of those capillament̄ drownes the weaker &c:[85]

Quære: whither fantasy as well as sight is not done in plano? That whither wee can imagine two like things to bee one behind another as we can do them one beside another. Resp: Noe but wee can understand them so.

Quære. why doth ye forcible turning of ye eye on way wth ye finger make ye object seeme to move the other way but not ye voluntary turning of it.

---

[84] Boyle, *Colours*, p. 12.   [85] See ULC, Add. 3975 (Appendix, p. 485).

Of vision

There is required some permanency in the object to perfect vision. Thus, a coal whirled round is not like a coal but like a fiery circle, or who can see a bullet pass by him. Yet, if the eye keep motion with the thing moved, it sees it right. One eye shutting or being pressed, the pupil of the other being open, both the eyes then being opened together, dilate, and contract, and dilate. Upon being affected these spirits ebb and flow. Dimness may come from the deficiency of these spirits, and the optic nerve obstructed.[10]

Rays from the same point in an object striking upon several respective places in both the eyes do make the object seem two, as when an eye is depressed, since then their two axes respect not the same point in the object. Things of the darker colors are most easily burned, i.e., have the loosest parts.

A man's eye struck sparkles.

The eyes being distorted, a man may see the same part of the same object in two diverse places with both eyes at once, or successively with one eye after another. He may also see two diverse objects in the same place successively without varying the posture of his eyes or of the object, but only by covering the one eye first and then the other, or by being intent with one eye first and then with the other. Which is an argument either that the image in the brain is painted on a superficies, or else that the optic nerves, at their meeting before they enter the brain, do sort their capillamenta by uniting there each capillamentum of the one eye to its fellow capillamentum of the other eye. So that when the eyes conspire not toward the same object but have diverse pictures painted upon the correspondent capillamenta, the stronger picture at the meeting of those capillamenta drowns the weaker.

Quære: Whether fantasy as well as sight is not done in a plane? Whether we can imagine two like things to be one behind another as we can do them one beside another? Resp: No, but we can understand them so.

Quære: Whey does the forcible turning of the eye one way with the finger make the object seem to move the other way, but not the voluntary turning of it?

---

[10] Another reading may be "Dimness may come from the deficiency of these spirits or from an obstruction of the optic nerve."

Of Colours.

That darke colours seeme further of y^n light ones may be
from hence ~~bee~~ y^t ~~they~~ beames loose little of theire force in
reflecting from a white body because they are powerfully
resisted thereby but a darke body by reason of ye loosenes
of its parts give some admission to ye light & reflects it but
weakly & so ye reflection from whitenes will be sooner at y^e
eye. or else because ye whit sends beams wth more force to y^e
eye & givs it a feircer knock.

Coulors arise either from shaddows intermixed wth light,
or or stronger & weaker reflection. or parts of ye body
mixed wth & carried away by light.
From some of these ariseth splendor & dullnesse.

A shining colour though black reflecteth an Objects rayes
perfecter y^n dull white one as black horne, black pollished
leather &c better y^n white paper. & contrary wise. But
pollished black shineth best.

A window lying open to ye south will bee tincted wth y^e
colour of ye curtane. A paper written on put twixt ye eye &
ye light ye letters towards ye light looke dim ye light being
refracted in ye paper after its past ye inke: but ye letters on
this side looke perfect ye light comeing streight to ye eye
wthout any refraction.
Why are coles black & ashes white.
No colour will arise out of ye mixture of pure black & white
for y^n pictures drawne wth inke would be coloured or
printed would seeme coloured ˪at a distance˩ & ye verges of
shadows would be coloured. ~~& therefore~~ & lamb black &
spanish whiteing would produce colours whence they cannot arise from more or lesse reflection of light or shadows
mixed wth light.

                                                    vide pag 69

Of Sounds

A man may heare ye beatings of his owne pulse.

In every sound ye eight above it but not below it seemes to
bee heard. (for there ~~being~~ ˪is˩ some more Subtile, some
more grosse matter in ye aire, & ye subtilest matter is prone

Of colors

That dark colors seem farther off than light ones may be from hence: that the beams lose little of their force in reflecting from a white body because they are powerfully resisted thereby, but a dark body, by reason of the looseness of its parts, gives some admission to the light and reflects it but weakly. And so the reflections from whiteness will be sooner at the eye. Or else, because the white sends beams with some force to the eye and gives it a fiercer knock.

Colors arise either from shadows intermixed with light, or stronger and weaker reflections. Or, parts of the body mixed with and carried away by light.

From some of these arise splendor and dullness: A shining color, though black, reflects an object's rays more perfectly than dull white. One, such as black horn, polished leather, etc., better than white paper. And contrary wise. But polished black shines best.

A window lying open to the South will be tinctured with the color of the curtain. A paper written on put between the eye and the light, the letters toward the light look dim, the light being refracted in the paper after it's past the ink; but the letters on this side look perfect, the light coming straight to the eye without any refraction.

Why are coals black and ashes white?

No color will arise out of the mixture of pure black and white, for then pictures, drawn with ink would be colored, or printed would seem colored at a distance, and the verges of shadows would be colored, and lamp-black and Spanish whiting would produce colors. Whence they cannot arise from more or less reflection of light or shadows mixed with light.

vide page 69[11]

Of sounds

A man may hear the beatings of his own pulse.

In every sound the eighth above it, but not below it seems to be heard. For there is some more subtle, some more gross matter in the air, and the subtlest matter is prone to

[11] Page 431

to quickest vibrations, though y^c motion of both proceede from y^c same cause, as y^c vibration of a string or pipe ⌐thus twiggs vibrate after y^c branches⌐. Also these motions doe least check one y^c other & are most congruous to y^c string or pipes motion when y^c ones vibrations are double in number to y^c others. Hence a sound & its eight are never seperate. The greatest & grossest ͺpt of y^c ~~aire~~ matter in y^c aire ~~will~~ ⌐doth⌐ comply wth y^c strings motion ~~for were it slower it could not give way to y^c motion of y^c string & it will scarce bee quicker by reason of its tenacious nature~~ for one string struck, ⌐by y^c mediation of y^c air⌐ moves an unison string of another instrument better y^n y^t wch is an Eight above or below it & y^c ~~(?)~~ string is easliest moved by y^c air when its ~~(?)~~ motion can be most conformable to y^c ⌐motion of y^c⌐ greatest ͺpte of y^c aire.[86] Nor can any considerable quantity of matter ~~(?)~~ move slower y^n y^c string because it gives way to y^c strings ⌐motion⌐ ~~Hence y^c~~ & were there, yet it motion ⌐being⌐ 4 times slower y^n y^t ⌐of⌐ y^c subtilest matter can scarce bee perceptible. Hence each sound hath its concomitant 8th, & ͺphaps 15th & 22th to a good eare, above but not below it. Hence 8^{ths} seeme to bee unisons. ~~And~~ And violenter breathing raiseth y^c sound & eight or 15th, not a 12th, or but seldome to a twelft.

Quære, In w^t proportion y^c sound decreaseth in its progresse from y^c fountaine. viz: If y^c sound bee (a) at y^c distance (b) it shall be $(\frac{abb}{xx})$ at y^c distance x.

Why doth y^c sound of a Bell quaver or shake like a mans voyce? Because y^c Bell vibrates somtimes directly somtimes obliquly towars a man.

How swiftly doe sounds move, & whither are acute or grave sounds the swifter?

Sounds are much fainter in y^c exhausted receiver then in the open Aire. Boyle Exper 27.[87]

                   ## Of Odoars. & Sapors

---

[86] This essay was probably suggested by Boyle's discussion of a proposed experiment in *Spring of the Air*, pp. 210–11.   [87] *Ibid.*, pp. 205–14.

quickest vibrations, though the motion of both proceed from the same cause (as the vibration of a string or pipe), thus twigs vibrate after the branches. Also, these motions do least check one the other and are most congruous to the string or pipe's motion when the one's vibrations are double in number to the other's. Hence, a sound and its eighth are never separate. The greatest and grossest part of the matter in the air does comply with the string's motion ~~for were it slower, it could not give way to the motion of the string, and it will scarcely be quicker by reason of its tenacious nature~~ for one string struck, by the mediation of the air moves an unison string of another instrument better than that which is an eighth above or below it. The string is most easily moved by the air when its motion can be most conformable to the motion of the greatest part of the air. Nor can any considerable quantity of matter move slower than the string, because it gives way to the string's motion. And were there any considerable quantity of matter moving slower yet, its motion, being four times slower than that of the subtlest matter, can scarcely be perceptible. Hence, each sound has its concomitant eighth, and perhaps 15th and 22nd to a good ear, above, but not below, it. Hence eighths seem to be unisons. And more violent breathing raises the sound an eighth or 15th, not a 12th, or but seldom to a 12th.

Quære: In what proportion the sound decreases in its progress from the fountain. Viz: If the sound be a at the distance b, it should be abb/xx at the distance x.

Why does the sound of a bell quaver or shake like a man's voice? Because the bell vibrates sometimes directly, sometimes obliquely, toward a man.

How swiftly do sounds move, and whether are acute or grave sounds the swifter?

Sounds are much fainter in the exhausted receiver than in the open air. Boyle, Experiment 27.

Of odors and sapors   38 106ᵛ

39 107ʳ

Of Touching.

A man hath beene deprived of his feeling. Sʳ K: Digb.

40 107ᵛ

Of Generation & Coruption.

In winter expose yᶜ liquor of decocted hearbes to yᶜ cold aire & in yᶜ morning under yᶜ ice, there will appeare yᶜ figure & colour of yᶜ plant ~~from~~ wᶜʰ ~~it~~ was taken from it. it may be yᶜ ice keep those attomes from avolition.

There is an artificiall resurrection of plants from theire ashes. dissolved salt uppon its fixation returnes to its affected cubes. Figures of mineralls are regular: as Chistall Hexagonall: yᶜ Fairy stone hemisphæricall. yᶜ stone Asteria of a stellar figure. yᶜ Misselto, & mosse grows upon other trees. A worme may turne to a Butterfly Tadpoles grow to frogs flys eggs to be wormes & yⁿ flys againe.

41 108ʳ

Of Memory

Messala Corvinus forgot his owne name. One by a blow wᵗʰ a stone forgot all his learning. Another by a fall from a horse forgot his mothers name & kinsfolkes. A young student of Montpelier by a wound lost his memory so yᵗ he was faine to be taught yᶜ letters of yᶜ Alphabet againe. The like befell a Franciscan frier after a fever. Thucidides writes of some who after theire recovery from yᶜ greate pestilence at Athens ~~who~~ forgot yᶜ names & persons of theire freinds & themselves too not knowing who they wer or by wᵗ names they were called.

Atque etiam quosdam cepisse oblivia rerum
Cunctarum, neque se possent cognoscere ut ipsi. } lucretiu

Dʳ Mores immort:[88]

Things out of mind are remembred sometimes by meeting wᵗʰ other things of like nature: as dreames never thought uppon in yᶜ morning at yᶜ time of awakeing are

---

[88] More, *Immortality*, p. 255. *De rerum natura*, VI, 1213–14.

Of touching

A man has been deprived of his feelings. Sir Kenelm Digby.

Of generation and corruption

In winter expose the liquor of decocted herbs to the cold air, and, in the morning under the ice, there will appear the figure and color of the plant from which it was taken. It may be that the ice keeps those atoms from escaping.

There is an artificial resurrection of plants from their ashes. Dissolved salt, upon its fixation, returns to its affected cubes. Figures of minerals are regular: as crystal hexagonal, the fairystone hemispherical, and the stone asteria of a stellar figure. The mistletoe and moss grow upon other trees. A worm may turn to a butterfly. Tadpoles grow to frogs. Fly's eggs grow to be worms and then flies again.

Of memory

Messala Corvinus forgot his own name. One, by a blow with a stone, forgot all his learning. Another, by a fall from a horse, forgot his mother's name and kinfolk. A young student of Montpellier, by a wound, lost his memory, so that he was fain to be taught the letters of the alphabet again. The like befell a Franciscan friar after a fever. Thucydides writes of some who, after their recovery from the great pestilence at Athens, forgot the names and persons of their friends and themselves too; not knowing who they were or by what names they were called.

And certain others fell into [cepere] forget-
fulness of all things, so that they could not   } Lucretius
even know themselves.

Dr. More's *Immortality of the Soul*

Things out of mind are remembered sometimes by meeting with other things of like nature, as dreams* never thought upon in the morning at the time of awaking are remembred by some actions of y^c like nature met wth all in y^c day time.

Forgetfullnesse ariseth sometimes out of y^c want of thinking of things. Things seene & words heard at y^c same distance are distinctly remembred. So are distance & widenes or extesion & bignesse. So are things wch enter not y^c sences as meditations, thoughts, dreames, & y^t a man hath remembred.

Meditations reminde a man of actions, & actions of meditatio.

Colours, actions, sounds loud softly, high & low, Time as y^t 2 things were done together or so long after one another reconing how long since such a thing done by counting y^c time from one action to another untill y^c præsent time.

A man cannot remember what hee never thought uppon as a blow or prick ⌊or noise⌋ in his sleepe y^c things & sounds wch hee heares & sees but minds not.

Objects from either eye or eare affect y^c memory alike. The same thing seene or heard from divers places or distances acte alike on y^c memor.

Things done in y^c same time helpe y^c memory of one another.

If memory bee done by characters in y^c braine yet y^c soule remembers too, for shee must remember those characters.

42 108^v [Blank folio]

43 109^r vid: pag: 75

Immagination. & Phantasie & invention

We can fancie y^c thing wee see in a right posture wth y^c heeles upward. Phantasie is helped by good aire fasting moderate wine

but spoiled by ~~not (?)~~ drunkenesse, Gluttony, too much study, (whence & from extreame passion cometh madnesse) dizzinesse, commotions of y^c spirits

Meditation heates a y^c~~oun~~ braine in some to distraction in others to an akeing & dizzinesse.

The boyling blood of youth puts y^c spirits upon too much

remembered by some actions of the like nature met with in the day time.

Forgetfulness arises sometimes out of the want of thinking of things. Things seen and words heard at the same distance are distinctly remembered. So are distance and wideness or extension and bigness. So are things which enter not the senses: as meditations, thoughts, dreams, and rememberings that a man has.

Meditations remind a man of actions, and actions of meditations. Colors, actions, sound (loud and soft, high and low), time (as that two things were done together, or so long after one another, reckoning how long since such a thing by counting the time from one action to another until the present time). A man cannot remember what he never thought upon, as a blow or prick or noise in his sleep, the things and sounds which he hears and sees but minds not. Objects from either eye or ear affect the memory alike. The same thing seen or heard from diverse places or distances acts alike on the memory. Things done in the same time help the memory of one another.

If memory be done by characters in the brain yet the soul remembers too, for she must remember those characters.

[Blank folio]   42 108ᵛ

Imagination and fantasy and invention   43 109ʳ

We can fancy the things we see in a right posture with the heels upward. Fantasy is helped by good air, fasting, and moderate wine. But spoiled by drunkenness, gluttony, too much study (whence, and from extreme passion, comes madness), dizziness, and commotions of the spirits.

Meditation heats the brain, in some, to distraction; in others, to an aching and dizziness.

The boiling blood of youth puts the spirits upon too

motion or else causet too many spirits. but could age makes y^c brain either two dry to move roundly through or else is defective of spirits yet theire memory is bad.

A man by heitning his fansie & immagination may bind anothers to thinke what hee thinks as in y^c story of y^c Oxford Scollar in Glanvill Va̅n̅ of Dogmatizing.[89]

When I had looked upon y^c Sun ~~I shut my eyes & there appeared nothing untill I strongly fancied y^c ⊙ to be befo~~ all light co~~u~~loured bodys appeared red & darke coloured bodys appeared ~~red~~ blew. 2 If I looked on white paper wth my bare eye it looked red, but if I looked on it through a very little hole so y^t but a little light could come to my eye fm y^c paper it looked greene. ~~hence I guess y^t~~ 3 after y^c motion of y^c spirits in my eye were almost decayed ⌞that I could see all thing wth theire natu̅r̅ colours⌟ I shut it & could see noe colour or image till I heightned my fantasie of seeing ⊙ & y^n began to appeare a blew spot wch grew ligter by degrees in y^c midst untill it ⌞was⌟ white & bright in y^c midst next to wch were cicles of red, yellow, grene, blew, purple, all wch were sometimes encompassed wth a darke greene or red. Sometimes y^c whole spot would turne very blew sometimes most of it red. After I opened my eye againe, white bodys looked red & darke ones blew as if I ~~was~~ had newly looked on y^c Sunne whence

44 109^v Sympathy & Antipathie

To one pallate y^t is sweete wch is bitter to another.
The same thing smells gratefully to one displeasinly to another.
Objects of sight move not some but cast others into an extasie.
Musicall aires are not heard by all wth alike pleasure.
The like of touching.

45 110^r Of Oyly bodys

Whither they ~~be oyly be~~ consist of branchlike particles onely touching superficially & foulded together. Cartes.[90]

[89] Joseph Glanvill, *The Vanity of Dogmatizing* (London, 1661), pp. 196–8.
[90] Descartes, *Meteora*, Caput I, Article III, p. 154.

much motion, or else causes too many spirits. But old age makes the brain either too dry to move roundly through, or else is defective of spirits – yet their memory is bad.

A man by heightening his fancy and imagination may bind another's to think what he thinks, as in the story of the Oxford scholar in Glanvill's *Vanity of Dogmatizing*.

When I had looked upon the Sun ~~I shut my eyes and there appeared nothing until I strongly fancied the Sun to be before~~ all light colored bodies appeared red, and dark colored bodies appeared blue. (2) If I looked on white paper with my bare eye, it looked red, but if I looked on it through a very little hole, so that but a little light could come to my eye from the paper, it looked green. (3) After the motion of the spirits in my eye were almost decayed, so that I could see all things with their natural colors, I shut it, and could see no color or image till I heightened my fantasy of seeing the Sun. Then began to appear a blue spot which grew lighter by degrees in the midst until it was white and bright in the midst; next to which were circles of red, yellow, green, blue, and purple. All which were sometimes encompassed with a dark green or red. Sometimes the whole spot would turn very blue, sometimes most of it red. After I opened my eye again, white bodies looked red and dark ones blue, as if I had newly looked on the Sun. Whence

vide page 75[12]

Sympathy and antipathy

To one palate that is sweet which is bitter to another.

The same thing smells gratefully to one, displeasingly to another.

Objects of sight move not some, but cast others into an ecstasy.

Musical airs are not heard by all with alike pleasure.

The like of touching.

Of oily bodies

Whether they consist of branchlike particles only, touching superficially and folded together. Descartes.

[12] Page 443

That Oyle (though thicker yet) mixeth sooner wth most bodys y^n water & spreades quicklier when dropt upon bodys, may proceede from its branches taking hold like briers on all adjacent bodys whereas water ⌐dropt⌐ is kept round by y^e aire ~~unlesse y^c~~ & crouded together unless y^e pores of bodys lye oper for its particles to drop into them &c

Of Meteors.

Whither fierce winds dry bodys by beating out y^e moisture from other bodys Cartes Met:[91]

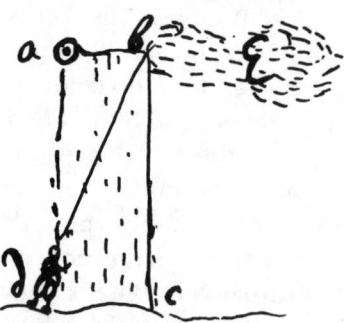

To measure y^e height of y^e clouds. Suppose (E) y^e cloud (b) y^e edge of it casting y^e shaddow c. (d) y^e man measuring it ~~y^c~~ & a y^e Sunne. y^n ∠adb = ∠dbc may bee found & likewise dc its sine, whence y^e radius db or y^e distance of y^e cloude from y^e man is easly found. Feb 19 166$\frac{4}{5}$ at night I observed a Halo about y^e ☽ 22^d 35″ distant from the ☽ it was Ellipticall & its long diameter perpendicular ~~from~~ to y^e Horison verying below farthest from the ☽. neare y^e moone (?) were two rainbow y^e diameter of y^e 1st was 3^d 0′ of y^e 2^d, 5^{degr} 30′. y^e order of y^e colours from ye moone were white, blewish greene, red, yellow; blewish greene, red, yellow.[92]

flux & reflux of ye Sea

Of Water & Salt.

vide page 26 & 49.
insomuch y^t what will swim in its surface will sinke in it.[93]

---

[91] *Ibid.*, Caput II, Article VII, p. 162.

[92] Such halos about the Moon are discussed by Descartes in *Meteora*, Caput IV, pp. 227–34.

[93] Descartes, *Meteora*, Caput III, Article XI, p. 168.

That oil (though thicker yet) mixes sooner, with most bodies, than water, and spreads more quickly when dropped upon bodies, may proceed from its branches taking hold, like briers, on all adjacent bodies. Whereas water dropped is kept round by the air and crowded together, unless the pores of bodies lie open for its particles to drop into them.

Of meteors

Whether fierce winds dry bodies by beating out the moisture from other bodies. Descartes's *Meteorology.*

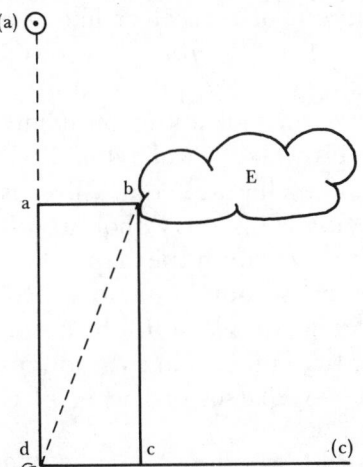

To measure the height of the clouds: Suppose *E* the cloud, *b* the edge of it casting the shadow *c, d* the man measuring it, and *a* the Sun. Then $\angle adb = \angle dbc$ may be found and likewise *dc*, its sine, whence the radius *db*, or the distance of the cloud from the man is easily found.

Feb. 19, $166\frac{1}{5}$ at night I observed a halo about the Moon, 22° 35″ distant from the Moon. It was elliptical and its long diameter perpendicular to the horizon, varying below farthest from the Moon. Near the Moon were two rainbows. The diameter of the first was 3° 0′, of the second, 5° 30′. The order of the colors from the Moon were white, bluish green, red, yellow, bluish green, red, and yellow.

Of water and salt   Flux and reflux of the sea   47 111ʳ

vide page 26[13] and 49[14]

insomuch that what will swim in its surface will sink in it.

[13] Page 373   [14] Page 405

Why is salt of a square figure haveing a hollow & broade top & a narrow base.[94] & why graines of salt will crack in y^c fire but not if they be rubed first asunder.[95]

Whither ye pleasant smell of white salt & y^c colour of black salt proceed from some other mixture. Cartes.[96]

Whither salt is melted by suddaine ˻heate˼ because there is water in it & not by a gentle fire because y^t exhales y^c water out by degrees.˻Cartes.˼[97]

How oyle or spirits of salt (so sharp y^t they will disolve gold) is extracted out of salt. Cartes Met: of Salt.[98]

Tides canot be from y^e ☽s influence for then they would be lest at new moones.[99]

Whither water may be drawne out of a receiver like aire or not. This may bee ye best way [to][100] empty ye receiver.

To try whither ye moone pressing ye Atmosphære cause ye flux & reflux of ye sea.[101] Take a tube of above 30 inches filled w^{th} ~~water~~ quicksilver, or else take a tube filled wth water wch is soe much longer y^n 30 inches as y^c quicksilver is weightier y^n water & ye top being stopped y^e liquor will sinke 3 or 4 inches below it leaving a vacuum (perhaps) then as y^e aire is more or lesse pressed w^{th}out by ☽ so will y^e water rise or fall as it doth in a witherglasse by heate or cold.[102] The same may ˻be done by˼ compareing y^e motions of y^c water of 2 ~~looking~~ ˻weather=˼ glasses one wher of is wth in a vassell of water y^e other not.

Observe if y^e sea water rise not in ~~mornings~~ ˻days˼ & fall at nights by reason of y^e earth pressing from ☉ uppon y^e night water &c. Try also whither y^e water is higher in mornings or evenings to know whither ⊖ or its vortex press forward most in its annuall motion.

---

[94] *Ibid.*, Caput III, Article XIII, pp. 169–70.

[95] *Ibid.*, Caput III, Article XVI, p. 171.

[96] *Ibid.*, Caput III, Article XVII, p. 171.

[97] *Ibid.*, Caput III, Article XVIII, pp. 171–2.

[98] *Ibid.*, Caput III, Article XIX, p. 172.

[99] An argument against Descartes's view as expressed in *Principia*, Part IV, Article LI, p. 160.

[100] In the manuscript, a single inscription of the word "to" serves here and for the first word in the next paragraph.

[101] Descartes, *Principia*, Part IV, Articles XLIX–LVI, pp. 158–61.

[102] The experiment was suggested by Christopher Wren and reported in Boyle's *Spring of the Air*, pp. 132–3. Boyle indicates that the results were inconclusive.

Why is salt of a square figure having a hollow and broad top and a narrow base? Why grains of salt will crack in the fire, but not if they be first rubbed apart.

Whether the pleasant smell of white salt and the color of black salt proceed from some other mixture. Descartes.

Whether salt is melted by sudden heat because there is water in it, and not by gentle fire because that exhales the water out by degrees. Descartes.

How oil or spirits of salt (so sharp that they will dissolve gold) is extracted out of salt. Descartes, *Meteorology,* Of Salt.

Tides cannot be from the Moon's influence for then they would be least at new moons.

Whether water may be drawn out of a receiver like air or not? This may be the best way to empty the receiver.

To try whether the Moon pressing the atmosphere causes the flux and reflux of the sea. Take a tube of above 30 inches filled with quicksilver, or else take a tube filled with water, which is so much longer than 30 inches as the quicksilver is weightier than water. The top being stopped, the liquor will sink three or four inches below it leaving a vacuum (perhaps). Then, as the air is more or less pressed without by the Moon, so will the water rise or fall as it does in a weatherglass by heat or cold. The same may be done by comparing the motions of the water of two weatherglasses, one whereof is within a vessel of water, the other not.

Observe if the sea water rises not in days and falls at nights by reason of the Earth pressing from the Sun upon the night water. Try also whether the water is higher in mornings or evenings, to know whether the Earth or its vortex press forward most in its annual motion.

48 111ᵛ   Of Mineralls

Why doth quicksilver sinke so readily into mettalls & into nothing else.

Foure ounces of copper, & one ounce of Tinglasse melted together compose a body coloured like Gold.

Brasse is Compounded of Copper & lapis Calaminariæ or fire=stone for ☉locks, melted together.

Pewter of tin & brasse.

Bell mettall of Tin one ounce & 3 ounces of Copper.

Mettall for reflections ~~of~~ may bee thus made: Melt throughly 3 pounds of Copper then take 4 ounces of ˌwhiteˌ Arsnick 6 ounces of Tartar & 3 ounces of Saltpeeter finely poudered together & put yᵐ into yᶜ melted copper & stirr yᵐ well together wᵗʰ a rodd of iron until they have done smoaking (but beware of yᶜ pernicious fume for yᶜ Arsenick is poyson). Then after a little blowing yᶜ fire to make it as hot as before put in 6 ounces of Tin=glas 2 ounces of Regulus of Antimony & after another blast or two put in a pound of Tin & stirr it a very little & immediately cast it.

———————

The Tinglasse makes yᶜ mettall tough, & yᶜ Antimony makes it fine & of a steele colour, two much of will make it bleaw. The Saltpeeter opens yᶜ poores of yᶜ mettall to let yᶜ filth evaporate & yᶜ Tartar helpeth to carry it away. If this mettall must bee cast smooth line the sand mold wᵗʰ the smoake of a linke.[103]

If there be mettalls of equall weight there proportions are tin = 10000, Iron = 9250, copper 8222, Silver = 7161, leade = 6435, quicksilver = 5453, gold = 3895. Or if they bee equall the weight of gold is 10000, of quicksilver 7143, of leade 6053, of silver 5438, of copper 4737, of iron 4210, of tin 3895.

---

[103] Compare with Item 35, pp. 82–3, and Item 36, pp. 84–5, of *The Correspondence of Isaac Newton,* edited by H. W. Turnbull (Cambridge University Press, 1959–77), 7 vols., Vol. I. Also compare with Add. 3973, published as Item 37, pp. 85–8, *The Correspondence,* and in David Brewster, *Memoirs of the Life, Writings, and Discoveries of Sir Isaac Newton* (New York: Johnson Reprint Corp., 1965), reprint of the Edinburgh edition of 1855, 2 vols., Vol. II, Appendix XXXI. Turnbull judges Add. 3973 to date from 1665–72.

Of minerals

Why does quicksilver sink so readily into metals and into nothing else?

Four ounces of copper and one ounce of tinglass melted together compose a body colored like gold.

Brass is compounded of copper and lapis calaminaris, or fire-stone for gold locks, melted together.

Pewter of tin and brass.

Bell metal of tin (one ounce) and three ounces of copper.

Metal for reflections may be thus made; melt thoroughly three pounds of copper. Then take four ounces of white arsenic, six ounces of tartar and three ounces of saltpeter finely powdered together, and put them into the melted copper. Stir them well together with a rod of iron until they have done smoking (but beware of the pernicious fumes, for the arsenic is poison). Then, after a little blowing of the fire to make it as hot as before, put in six ounces of tinglass, two ounces of regulus of antimony, and after another blast or two, put in a pound of tin and stir it a very little and immediately cast it.

The tinglass makes the metal tough, and the antimony makes it fine and of a steel color (too much of it will make it blue). The saltpeter opens the pores of the metal to let the filth evaporate, and the tartar helps to carry it away. If this metal must be cast smooth, line the sand mold with the smoke of a link.

If there be metals of equal weight, their proportions are: tin = 10,000; iron = 9,250; copper = 8,222; silver = 7,161; lead = 6,435; quicksilver = 5,453; and gold = 3,895. Or, if they be equal, the weight of gold is 10,000; of quicksilver 7,143; of lead 6,053; of silver 5,438; of copper 4,737; of iron 4,210; and of tin 3,895.

49 112ʳ   Of yᶜ Flux & reflux of yᶜ sea.   vide pag 26 ⌐47⌐

earth water & vortices.

Note yᵗ yᶜ Earths diurnall motion is not helped by its vortex, for by yᶜ same force it would move yᶜ water ⌐& air⌐ along wᵗʰ it, or rather faster.[104]

Try whither yᶜ ~~Earths~~ ⌐Seas⌐ flux & reflux bee greater in Spring or Autume ⌐in winter or Sommer⌐ by reason of yᶜ ⊖s Aphelion & perihelion.[105]

Whither yᶜ Earth moved out of its Vortexes center byᶜ Moones pression cause not a monethly Parallax in Mars &c.[106]

In yᶜ Island Berneray scituated betwixt yᶜ Islands Eust & Herris In yᶜ tract of Islands west to scotland called by yᶜ inhabitants yᶜ long Island (Berneray is 3 miles long from east to west, & more yⁿ one mile broade) foure days before & after yᶜ full moone yᶜ tide flood runs east & its ebb west (yᶜ spring tides riseing 14 or 15 foot upright). but foure days before & after yᶜ quarters (a southerly moone makeing there full sea) One ~~ebb~~ fload & ebb runns east ward from about 9ʰ 30′ to 3ʰ 30′ in yᶜ day & in yᶜ other 12 night howers the flood & ebb run west ward in yᶜ summʳ ½ yeare when yᶜ ⊙ hath northerne declin̄, but in winter it runs westward in yᶜ day eastward in yᶜ night.[107]

The Danube runns swiftest at noone & midnight & slowest at six of yᶜ clock (as is perceived by yᶜ motion & noyse of yᶜ clackers in Mills) & yet there is noe ebb nor flow yᶜ water keeping at a constant height. Also another ⌐turbulent⌐ river[108] wᶜʰ swimms into the Danube mixeth not wᵗʰ it, nor swim they ⌐like water & oyle⌐ one above (but besides) yᶜ other.

50 112ᵛ   [Blank folio]

---

[104]   Descartes, *Principia*, Part III, Articles CXLIX–CLI, pp. 133–4.
[105]   *Ibid.*, Part IV, Articles XLIX–LI, pp. 158–60.
[106]   *Ibid.*, Part III, Articles CLII and CLIII, p. 134.
[107]   From "A Relation of some extraordinary Tydes in the West-Isles of Scotland, as it was communicated by Sr. Robert Moray," *Philosophical Transactions* I:4(1665):53–4. See also Newton, ULC, Add. 3958, folio 9ʳ.
[108]   A space is left as if Newton intended later to add the name of the river.

Of the flux and reflux of the sea

Earth, water, and vortices   *vide* pages 26[15] and 47[16]

Note that the Earth's diurnal motion is not helped by its vortex, for by the same force it would move the water and air along with it, or rather faster.

Try whether the sea's flux and reflux be greater in spring or autumn, in winter or summer, by reason of the Earth's aphelion and perihelion.

Whether the Earth moved out of its vortex's center by the Moon's pression causes not a monthly parallax in Mars.

In the Island Berneray, situated between the islands Uist and Harris in the tract of islands west of Scotland, called the long island (Berneray is three miles long from east to west, and more than one mile broad), four days before and after the full moon the tide flood runs east and its ebb west (the spring tides rising 14 or 15 foot upright). But four days before and after the quarters (a southerly Moon making there full sea), one flood and ebb runs eastward from about 9:30 to 3:30 in the day. In the other 12 night hours, the flood and ebb run westward in the summer ½ year when the Sun has northern declination. But in winter, it runs westward in the day, and eastward in the night.

The Danube runs swiftest at noon and midnight, and slowest at six o'clock (as is perceived by the motion and noise of the clackers in mills); yet there is no ebb nor flow, the water keeping at a constant height. Also, another turbulent river which swims into the Danube does not mix with it, nor do they swim like water and oil, one above the other, but one beside the other.

[Blank folio]

---

[15] Page 373   [16] Page 399

2   This motion is not continued by a force imprest because yᵗ force must be communicated from yᵉ mover ~~withe~~ into yᵉ moved either ˌbyˌ some corporeall efflux or incorporeall one or nothing. if by corporeall attomes we are still at a loss how those attomes must continue theire one motion. if by an incorporeall efflux it must be by either spirit or some quality. if by a spirit how comes yᵉ spirit to be so ~~soone~~ easly united to yᵉ body & not to slip through it & when united to it how comes yᵉ spirits to cease so soone & yᵉ spirits to leave it & hence every little attome must have soules in store to cast away uppon every body they meete wᵗʰ. if a quality then qualitas transmigrat de subjecto in subjectum. & this quality cannot be yᵉ motion of yᵉ mover since it & yᵉ mover are seperated at once from yᵉ thing moved. In a word how can yᵗ give a power of ~~being~~ moveing wᶜʰ it selfe hath not.

     Of Violent motion

Therefore it must be moved after its seperation from yᵉ mover by it one gravity. Which will be cleare by seing whither there can be motion in a vacuum & what that motion is & so compareing it wᵗʰ motion in pleno.

That there may be motion in vacuo let us suppose (ab) to be a body as a peice of ~~earth~~ ˌAireˌ (c. d. e.) to be three globes, (fghi) & all yᵉ space about yᵉ globes & that ~~earth~~ ˌaireˌ to be inane now in yᵉ chapter de vacuo wee have shewed that those three globes would be really seperate & not touch one another. you will grant yᵗ halfe yᵉ globes are

in places ~~& halfe in vacuo~~ & consequently may move, suppose yⁿ yᵗ halfe of (c) in yᵉ ~~Earth~~ ˌaireˌ move towards (d) we

Of violent motion   *vide* page 22[17]

2   This motion is continued by a force impressed because that force must be communicated from the mover into the moved either by some corporeal efflux or incorporeal one or by nothing. If by corporeal atoms, we are still at a loss how those atoms must continue their own motion. If by an incorporeal efflux, it must be by either spirit or some quality. If by a spirit, how comes the spirit to be so easily united to the body, and not to slip through it, and when united to it how comes the spirits to cease so soon and the spirits to leave it. Hence, every little atom must have souls in store to cast away upon every body they meet with. If a quality, then *qualitas transmigrat de subjecto in subjectum.* This quality cannot be the motion of the mover, since it and the mover are separated at once from the thing moved. In a word, how can that give a power of moving which itself has not.

Of violent motion

Therefore, it must be moved after its separation from the mover by its own gravity. Which will be clear by seeing whether there can be motion in a vacuum, and what the motion is, and so comparing it with motion *in pleno.*

That there may be motion *in vacuo:* Let us suppose *AB* to be a body, as a piece of air; *cde* to be three globes; *fghi,* the space about the globes, and the air to be inane. Now, in the chapter *de vacuo,* we have shown that those three globes would be really separate and not touch one another.

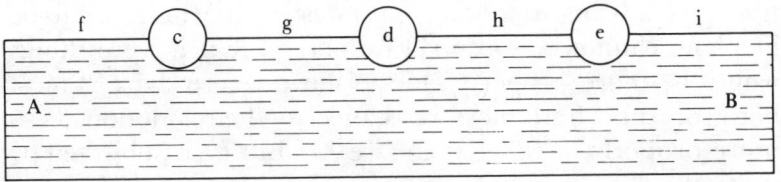

You will grant that half the globes are in places and consequently may move. Suppose then that half of *c* in the air

[17] Page 369

aske whither that part in vacuo would move along wth it or stay behind & ⌞be⌟ seperated frome it if y^c first we have o^r desire if y^c last wee ask what should seperate it from it not y^c vacuum since y^t is accounted nothing. but you may say y^t it is not truly motion for y^c upper ꝑte of c to be carried to (d) we answer y^t where there is action (for such is y^c passing of (c) to (d)) & where there are new respects acquired to ~~other~~ ⌞y^c same⌟ bodys there must be motion, but y^c upper part of ⌞(c)⌟ hath neither y^c same respect to y^c aire ~~w^{ch}~~ nor to (d) wch it had before it began to pass towards d. ~~but this but to quarrell wth termes~~ If this ~~motion~~ ⌞going of c to d⌟ be not motion I aske what it is. But this is onely to strive about termes & if it please you not to call it motion cal

it what you will but it is y^t which we aimed to prove & there is but this difference twixt it & motion in pleno. y^t y^c one is environed with such ~~bodys~~ mater as is impenitrable & consequently y^t mater must be crouded out of y^c moving bodys way before or rather at y^c same time y^t y^c body moves, it must needs impede y^c motion ~~by~~ ⌞to⌟ be continually thrusting against & resiste⌞d⌟ by y^c body before it: but in vacuo it meetes wth nothing ⌞impenitrable⌟ to stay it ~~& certainely then~~ tis true God is as far as vacuum extends but he being a spirit & penetrating all matter can be no obstacle to y^c motiō of matter noe more y^n if nothing were in its way. Let mee aske why one should be motion more y^n another since ~~a vacuum ca~~ in pleno motion is so stopped by one body rubbing uppon another & in vacuum it hath its liberty can y^c same thing (viz ⌞a⌟ being invironed wth bodys) at y^c same time give a being to motion & yet destroy it, wherefore to be in pleno cannot be essentiall to motion. & if it were things would be more properly sade to move where there is most body or they find most resistance to theire motion & so more properly in water y^n in aire &c: But it ⌞is⌟ objected by Aristotle y^t a Vacuum is uniforme & every where alike & a body hath y^c same respects to a vacuum in all places alike but there is no motion with[109] some mutation of circum-

---

[109] Newton clearly meant to write "without." *Phys.*, IV. 8, 215a19–22.

moves toward *d*. We ask whether that part *in vacuo* would move along with it or stay behind and be separated from it. If the first, we have our desire. If the last, we ask what should separate it from it; not the vacuum, since that is accounted nothing. But you may say that it is not truly motion for the upper part of *c* to be carried to *d*. We answer that where there is action (for such is the passing of *c* to *d*) and where there are new respects acquired to the same bodies, there must be motion. But the upper part of *c* has neither the same respect to the air nor to *d* which it had before it began to pass toward *d*. If this going of *c* to *d* be not motion I ask what it is. But this is only to strive about terms, and if it pleases you not to call it motion call

Of motion 53 114ʳ

it what you will, but it is that which we aimed to prove. There is but this difference between it and motion *in pleno*, that the one is environed with such matter as is impenetrable. Consequently, that matter must be crowded out of the moving body's way before, or rather at the same time that, the body moves. It must needs impede the motion to be continually thrusting against, and resisted by, the body before it. But *in vacuo* it meets with nothing impenetrable to stay it. It is true God is as far as vacuum extends, but he, being a spirit and penetrating all matter, can be no obstacle to the motion of matter; no more than if nothing were in its way. Let me ask why one should be motion more than another, since *in pleno* motion is so stopped by one body rubbing upon another and in vacuum it has its liberty. Can the same thing (viz., a being environed with bodies) at the same time give a being to motion and yet destroy it, wherefore to be *in pleno* cannot be essential to motion. If it were essential, things would be more properly said to move where there is most body, or they find most resistance to their motion (so more properly in water than in air, etc.). It is objected by Aristotle that a vacuum is uniform and everywhere alike, and thus a body has the same respects to a vacuum in all places alike. Since there is no motion without some mutation of circumstance, so *in vacuo* there is no mo-

stances And so in Vacuo no motion I answer as to o^r senses ye aire is uniform. And we judge a thing to be moved ~~as~~ ⌞when⌟ we se it come nigher or goe farther from some thing wch o^r senses can perceive & so we judge not a thing to be moved in respect of ye aire but of ye earth or some thing.[110]

54 114^v

Of Comets.

The motion of a Comet[111]

Anno 1585.

Die mensis		longitude			latid Austr		Die		longitud			lat Septen:	
								23	29^d		10′	0^d	40′
Octob.	8	23^d	♓	16′	13^d	51′	Octob.	24	0	♉′	45	1	20
	9	26		17	12	48		25	2		15	1	58
	10	29		13	11	44		26	3		40	2	34
	11	2	♈	4	10	39		27	5		0	3	7
	12	4		50	9	35		28	6		16	3	37
	13	7		31	8	30		29	7		28	4	6
At midnight	14	10		7	7	26		30	8		36	4	33
	15	12		38	6	23		31	9		40½	4	58
	16	15		4	5	22	Nove͞	1	10		42	5	21
	17	17		24½	4	22		2	11		40	5	43
	18	19		39	3	25		3	12		35	6	3
	19	21		47	2	30		4	13		27	6	22
	20	23		48	1	38		5	14		16	6	39
	21	25		42	0	49		6	15		22	6	54
	22	27		29	0	3		7	15		45	7	7
	~~23~~							8	16		25	7	18
	~~24~~							9	17		2	7	28
								10	17	♉	36	7	37

Its beard was round about it.

[110] Newton's essay may owe something to Gassendi. See P. Gassendi, *De motu impresso a motore, epistolae duae* (Paris, 1642), Epistle I, pp. 35–46. For the most part, however, Newton is developing his own views.

[111] The following table is from Willebrordus Snellius, *Descriptio Cometæ* (Lugduni Batavorum, 1619), pp. 88–9. Newton omits Snell's columns headed "motus cometæ diurnus in longitudinem" and "motus cometæ diurnus in latitudinem."

tion. I answer, as to our senses the air is uniform, and we judge a thing to be moved when we see it come nearer or go farther from some thing which our senses *can* perceive. So we judge not a thing to be moved in respect of the air, but of the Earth or some thing.

Of Comets   54 114'

The motion of a comet

Anno 1585

Month and day		Longitude			Latitude south	
October	8	23°	Pisces	16'	13°	51'
	9	26		17	12	48
	10	29		13	11	44
	11	2	Aries	4	10	39
	12	4		50	9	35
	13	7		31	8	30
	14	10		7	7	26
	15	12		38	6	23
	16	15		4	5	22
	17	17		24½	4	22
	18	19		39	3	25
	19	21		47	2	30
	20	23		48	1	38
	21	25		42	0	49
	22	27		29	0	3
	23	29		10	0	40
	24	0	Taurus'	45	1	20
	25	2		15	1	58
	26	3		40	2	34
	27	5		0	3	7
	28	6		16	3	37
	29	7		28	4	6
	30	8		36	4	33
	31	9		40½	4	58
November	1	10		42	5	21
	2	11		40	5	43
	3	12		35	6	3
	4	13		27	6	22
	5	14		16	6	39
	6	15		22	6	54
	7	15		45	7	7
	8	16		25	7	18
	9	17		2	7	28
	10	17	Taurus	36	7	37

(At midnight.)

Its beard was round about it.

Octob 8[th](?) at 11[h] afternoone this comet was distant from the 3[d] star of ♈ 45[d] 16'½ 11[h] 10[m] from scapula Pegasi 33[d] 43'¼. Therfor it longitude was 23[d] ♓ 9' 2". latitud 13[d], 52', 9".[112]

Anno 1618 A Comet appeared Stile Gregor:[113]

	longitudo	lat: Bor:
Decemb 13 hor: 4½ matu	♎ 17[d],42',5"	41[d],47',40".
Dec 14[d].5 ¾[h] mat.	♎ 15[d],41'15"	46 4 24
Dec 24[d] 4½[h] mat:	♍ 28[d],44 2	58 33 37
Dec. 11[d], 6½[h] mane	♎ 23, 21, 0	37, 3, 50
Dec. 2 day.ho 6½ mat	♏ 8[d], 23'43"	14, 12, 0

The motion of y[e] comet

from 2[day] to 11th day 26[d] 31'. From 11th to 13th; 6[d],24.' from 13th to 14th 2[d] 35' 50". from 13th day to 24th it moved 23[degr] 45' 54".[114] The 1st day y[e] tayle was averse from ☉ exactly its ⌊tayle⌋ reached beyond/bolow y[e] ~~skies~~

55 115[r]   lower wheeles in y[e] ⌊Great⌋ Beares belly & was in ⌊a⌋ manner // to y[m]. But it declined a little ⌊afterwards⌋ from ☉ toward ♀ ⌊upwards.⌋ Sometimes y[e] tayle declines from ☉ 20 or 30[degres] east or westward. The tayle of this last Comet was sometimes 25[d] in length. The beard of it was so raire as y[t] starrs might bee seene through it.[115]

1664. Dec 9th old stile at 4 of y[e] clock in y[e] Morning y[e] latitude southward of y[e] Comet was 20[d], its longit 182[d]. The length of its tayle 20[d].

On fryday ⌊before midnight⌋ Decemb[r]. 23[d] 1664 I observed a Comet whose rays were round her, yet her tayle extended

[112] Snellius, *Descriptio Cometæ*, pp. 76–8.

[113] *Ibid.*, p. 14. Newton has changed the order, which in Snell is chronological. For longitude on December 14, Snell has "15gr. 1 scr. 16 sec."

[114] *Ibid.*, p. 17.   [115] *Ibid.*, pp. 5–6.

On October 8 this comet was 45° 16½' distant from the 3rd star of Aries at 11 o'clock in the afternoon, and 33° 43¼' from scapula Pegasi at 11:10. Therefore its longitude was 23° Pisces 9' 2" and its latitude 13° 52'9".

Anno 1618 (Gregorian style) A comet appeared

		Longitude			Latitude north		
December 13, 4:30 a.m.	Libra	17°	42'	5"	41°	47'	40"
December 14, 5:45 a.m.	Libra	15	41	15	46	4	24
December 24, 4:30 a.m.	Virgo	28	44	2	58	33	37
December 11, 6:30 a.m.	Libra	23	21	0	37	3	50
December 2, 6:30 a.m.	Scorpio	8	23	43	14	12	0

The motion of the comet from the 2nd day to the 11th day was 26° 31'; from the 11th day to the 13th, 6° 24'; from the 13th day to the 14th, 2° 35' 50"; and from the 13th day to the 24th it moved 23° 45' 54". The first day the tail was exactly averse from the Sun. Its tail reached beyond/below the

lower wheels in the Great Bear's belly, and was, in a man- 55 115'
ner, parallel to them. But it declined a little afterward from the Sun upward toward Venus. Sometimes the tail declines 20° or 30° east or westward from the Sun. The tail of this last comet was sometimes 25° in length. The beard of it was so rarefied as that stars might be seen through it.

On December 9, 1664 (Old Style), at 4 in the morning the latitude southward of the comet was 20°, its longitude 182°. The length of its tail was 20°.

Before midnight on Friday, December 23, 1664, I observed a comet whose rays were round her, yet her tail

it selfe a little towards y^e east parallell to y^e Ecliptick The $*$ it ˌselfe was notˌ seene onely it looked like a little cloude The altitude of Sirius at y^e time of observation was 16^d, The comet was y^n entering into y^e whales mouth ˌat y^e nether jawˌ being distant frō Aldeboran 23^d $21'$ & as much from Rigel. Therefore y^e longitude of it was (?) 48^d $4'$. its latitude 22^d, $3'$, $44''$. At about 9^h $24'$ at night.

December 24th it appeared as on y^e day before, being distant from Rigel 28^d, $24'$; from Aldeb: 24^d, $12'$. Sirius being 20^d high y^t is it was 10^h $\frac{26'}{28'}$ at night. Whence its longitude was 44^d $7'$ its latitude ~~Northward~~ ˌSouthwardˌ 18^d $23'$.

Dec 27th before midnight Sirius being 16^d high y^e distance of y^e Comet from Aldeboran was 28^d $11'$. from Rigel 38^d $36'½$. Its longitude was 37^d. $41'$. $13''$.

its longitude was 37^d, $4'$, $13''$. its latitude south 10^d, $20'$ $47''$. at 9^h $8'$ at night.

The length of its tayle was about 11^d being extended towards Aldeboran or a little below it parallell to y^e Ecliptick. The tayle now perfectly manifested it selfe on y^t ~~west~~ east side ~~& y^e $*$ began to discover it selfe, y^e cloude being now dissipated.~~ It moved ˌnorthwardˌ against y^e streame of y^e Vortex cutting it at an angle of about 45^d or 46^d. There was stil a very bright $\left\{ \begin{array}{c} \text{haire} \\ \text{beard} \end{array} \right\}$ round about y^e Comet & it seemed to bee nothing but bird raying ~~for I could~~ from y^e center of it for I could not see y^e limits of y^e $*$.

Dec 28th y^e comet was distant from y^e brighter p$*$ in y^e jaw of y^e Whaile 5^d $52'$. from y^e midde star ˌq$*$ˌ in y^e Whailes mouth 3^d $43'$. Covering y^e $*$ twixt $*$e (?) y^e 3^d & 5^t of y^e Goate wth its haire but being rather above (?) it as in y^e figure. Its tayle extending to y^e 3^d

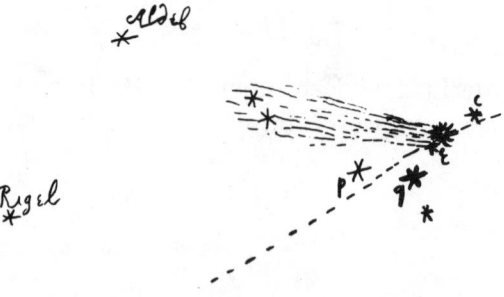

& 4th stars in y^e section of y^e bull.

extended itself a little toward the east and parallel to the ecliptic. The nucleus itself was not seen, but appeared like a little cloud. The altitude of Sirius at the time of observation was 16°. The comet was then entering into the Whale's mouth at the lower jaw, being distant from Aldebaran 23° 21' and as much from Rigel. Therefore the longitude of it was 48° 4' and its latitude 22° 3' 44", at about 9:24 at night.

On December 24, it appeared as on the day before, being 28° 24' distant from Rigel and 24° 12' from Aldebaran; Sirius being 20° high, that is it was 10:26 or 10:28 at night. Whence the comet's longitude was 44° 7' and its latitude southward was 18° 23'.

On December 27 before midnight, Sirius being 16° high, the distance of the comet from Aldebaran was 28° 11' and from Rigel 38° 36½'. Its longitude was 37° 4' 13"

and its latitude south was 10° 20' 47", at 9:08 at night.

The length of its tail was about 11°, being extended toward Aldebaran, or a little below it, parallel to the ecliptic. The tail now perfectly manifested itself on the east side and the nucleus began to reveal itself, the cloud now being dissipated. It moved northward against the stream of the vortex cutting it at an angle of about 45° or 46°.

There was still a very bright hair/beard round about the comet and it seemed to be nothing but bird-raying from the center of it, for I could not see the limits of the nucleus.

On December 28 the comet was 5° 52' distant from the bright star p in the jaw of the Whale, and 3° 43' from the middle star q in the Whale's mouth. Covering the star between the star e and the 3rd and 5th of the Goat with its hair, but being rather above the star e as in the figure. Its tail extending to

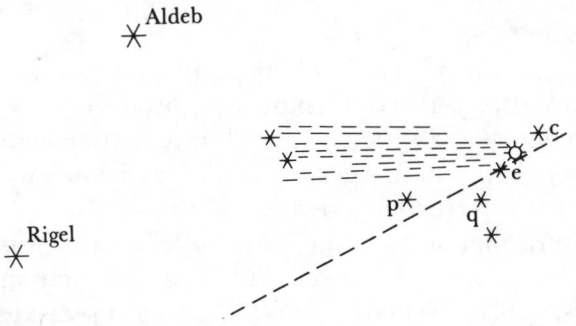

the 3rd and 4th stars in the section of the Bull.

Dec: 29th when 'twas in y^e meridian its altitude was 44^d 13'. Therefore its declin$\overline{acon}$ was 6^d 30' Northward. ~~It passed a little below y^e ✳^{c} wch is below y^e whales eye.~~

Dec: 30th, 15^{min} past 7^h at night. The right ascen: of y^e Comet was y^e same wth y^e ✳s below the whailes eye its de-cin$\overline{acon}$ about 28' more northward.

57 116^r Jan 1st Sirius being 20^d 50' high, or 21^d high. The comet had 4' in Almicanthers more y^n y^e ✳ below y^e whailes eye, 5' in Azimuths westward more y^n y^t in y^e hinder ͜pte of y^e head. Its tayle being 12^d in length & extending towards y^e 3 ✳^{s} below suculæ under Aldeboran.

Jan 2^d ~~Sirius~~ ͺAldebor͜ being Jan 2^d Rigells Altitude be-ing 25^d 30' or lesse The Comet had ye same Azimuth wth y^e ✳ in ye hinder ͜pte in ye Whailes head, being perpendicu-larly over it & distant from it 2^d 41'.

On Tuesday Jan 10th at 10 of ye clock ~~(?)~~ ͺat night͜ The comet was distant from the northerne ✳ following ye 1st ✳ of ♈ 6^d, 10'. from ye midle & brighter ✳ in ye north knot of ye fishes 6^d, 8'. Its tayle poynting towards ye biggest of ye 3 ✳^{s} in ye taile of ♈ but being very weake & ye star it selfe growed very dim.

On Munday Jan 23^d at 8^h at night The Comet was distant from ye middle bright ✳ in ye North knot of y^e ♓ 3^{d}58'. & twixt y^t ✳ & y^e ✳ in ye Rams neck exactly. its tayle being scarse discernable. This comets motion was swiftest ͺ& it nerest to us͜ when it was in ye belly of Syrius its right ascen-tion being 100^d.

58 116^v On Satturday Aprill 1st 1665 There appeared another Comet wch at 20' after 3 in ye morning was distant from ye Head of Andromeda 3^d, 29'. & From y^e ✳ in ye left knee of Pegasus 16^d, 18'. ͺbeing in or very neare ye Tropick ~~wth~~ wth longitude [2(?)]4^d or thereabouts by the Globe.͜

On Tuesday morning at 20' after 3 Apr: 4th ye Comet was distant from Andromeda's head, 8^d, 26'. From ye Star in Cassiopeas Brest ͺcalled Schedir͜ 28^d 22'. Being on y^t side of them towards Perseus, ~~Therefore it is~~

On December 29 it was in the meridian and its altitude was 44° 13'. Therefore its declination was 6° 30' northward. ~~It passed a little below the star *c* which is below the Whale's eye.~~

On December 30 at 7:15 at night, the right ascension of the comet was the same with the stars below the Whale's eye, its declination about 28' more northward.

On January 1, Sirius being 20° 50' or 21° high, the comet had 4' in almacantars more than the star below the Whale's eye, and 5' westward in azimuth more than the star in the hinder part of the head. Its tail being 12° in length and extending toward the three stars below Suculæ under Aldebaran.

On January 2, Rigel's altitude being 25° 30' or less, the comet had the same azimuth as the star in the hinder part in the Whale's head, being perpendicularly over it and 2° 41' distant from it.

On Tuesday, January 10, at 10 o'clock at night the comet was 6° 10' distant from the northern star following the 1st star of Aries, and 6° 8' from the middle and brighter star in the north knot of the Fishes. And its tail was pointing toward the biggest of the three stars in the tail of Aries, but being very weak and the nucleus itself grown very dim.

On Monday, January 23, at 8 o'clock at night the comet was 3° 58' distant from the middle bright star in the north knot of Pisces, and exactly between that star and the star in the Ram's neck, its tail scarcely being discernible. This comet's motion was swiftest and nearest to us when it was in the belly of Sirius its right ascension being 100°.

On Saturday, April 1, 1665, another comet appeared which was 3° 29' distant from the head of Andromeda and 16° 18' from the star in the left knee of Pegasus (being in or very near the tropic with longitude 24° or thereabouts by the globe), at 3:20 in the morning.

On Tuesday, April 4, the comet was 8° 26' distant from Andromeda's head and 28° 22' from the star in Cassiopeia's breast (called Shedir) at 3:20 in the morning – being on that side of them toward Perseus.

57 116'

58 116'

On wednesday apr 5th at 20^{m} after 3^{h} y^c Comet was distant from y^c head of Andromeda 10^d 40'. 28^d. 58'. from Cassiopeas brest.[116]

59 117^r ~~The first matter attomes, mixt wth v~~

                                                     vid pag 10
                              Of Motion.

y^c joyn⌐in⌐g & meeting of y^c two parts & posterius according to y^c latter of y^c two parts & so be ⌐still⌐ liable still to divisibility wch contradicts y^c notion of an indivisible part. But to explaine how theese leasts have no parts

---

An Artist will play a lesson not minding a stroke & sing neither minding nor missing a note a man may walke wthout thinking of it. &c[117]

In a wheele (divided into 24 parts by y^c 24 letters. A cannot move before b nor b before c &c to z, y^n z will not move untill A hath nor A till z y^c reason is becaus a can have no place but b's nor y^t till b hath left it. If they move all together. y^n in y^c instant y^t b leaves its place it is in't or not: if in't y^n a can't move into't in y^c same instant y^t it leaves it if not in't y^n it had left it before. A less & greater ◎ in a wheele move equally swift ~~yet y^c ((?)~~ or els a ⌐straight⌐ line drawne from y^c center to y^c circumference would be inflected i.e if some parts move faster y^n others) yet y^c greater circle passeth over more space. A little wheele on y^c same axis wth 2 large ones will pass over equall space wth equall revolutions. Glanvill.[118] When a snaile creepes a gale of spirits circuit from her head downe her back to her taile & up her belly to her head againe.[119] Cartes defines motion 2^a ⌐pte Pr:Ph: to be The Translantion of one part of matter or one body from y^c vicinity of those bodys wch imediately touch it ~~to y^c vi~~ & seem to rest, to y^c vicinity of others.[120]

[116] The star descriptions appear to come from "A Catalogue of 1000 of the Fixed Stars, according to the Accurate observations of Tycho Brahe, and by Him rectified to the beginning of the Year 1601" as published in Vincent Wing, *Harmonicon Celeste* (London, 1651), pp. 240–62.   [117] Glanvill, *Vanity*, p. 26.
[118] *Ibid.*, pp. 54–61.   [119] More, *Immortality*, p. 203.
[120] Descartes, *Principia*, Part II, Article XXV, pp. 32–3.

On Wednesday, April 5, the comet was 10° 40' distant from the head of Andromeda and 28° 58' from Cassiopeia's breast, at 3:20 in the morning.

59 117^r

Of motion   *vide* page 10[18]

joining and meeting of the two parts, and *posterius* according to the latter of the two parts: so it is still liable to divisibility which contradicts the motion of an indivisible part. But to explain how these leasts have no parts

---

An artist will play a lesson not minding a stroke, sing neither minding nor missing a note, and a man may walk without thinking of it, and so on.

In a wheel divided into 24 parts by the 24 letters, *a* cannot move before *b* nor *b* before *c* and so forth to *z*. The *z* will not move until *a* has, nor *a* until *z*. The reason is because *a* can have no place but *b*'s, nor that till *b* has left it. If they move all together, then in the instant that *b* leaves its place it is in it or not: if in it then *a* can't move into it in the same instant that it leaves it, if not in it then it had left it before. A less and greater wheel in a wheel move equally swift (i.e., if some parts move faster than others, a straight line drawn from the center to the circumference would be inflected), yet the greater circle passes over more space. A little wheel on the same axis with two large ones will pass over equal space with equal revolutions. Glanvill.

When a snail creeps a gale of spirits circuit from her head down her back to her tail and up her belly to her head again.

Descartes defines motion in the second part of the *Principia Philosophiæ* to be the translation of one part of matter or one body from the vicinity of those bodies which immediately touch it and seem to rest, to the vicinity of others.

[18] Page 353

The motion of y^c Stomack in vomiting (though wholly agains our will & therefore merly mechanicall) by y^c touch of a whalebone onely, doth much more illustrate y^c {?} actions of brutes to bee mechanicall & independent of {?} soules, then Chartes his instance of winking at y^c shaking of a freinds hand by y^c eye.

How much longer will a pendulum move in y^c Receiver then in y^c free aire. Hence may bee conjecttured w^t bodys there bee in the receiver to hinder y^c motion of the pendulum.[121]

60 117^v [Blank folio]

61 118^r [Blank folio]

62 118^v [Blank folio]

63 119^r Of Attomes.   vide pag 3.

Division can be made a Mathematicall point or superficies may be betwixt y^c parts divided ⌞divisible⌟ but a point cannot be put or conceived in this little space to divide it. Add y^n a ⌞naked⌟ point {?} to any space in the line I ask {?} whither it bee in y^c space wthout touching y^c ⌞any of y^c⌟ other points y^t make y^c line or not. if you say tis y^n I answer y^t a point wch would have resisted touching might have as well beene added before wthout an absurdity. If you say y^n y^t it must touch one of y^c other points y^t makes y^c line. I say y^n y^t that point is in y^c same place wth y^c point wch it touches, & so not in y^c space wch you would have divided. how can y^n y^t space be divided into wch a mathematicall point cannot enter to seperate its ⌞pts⌟? & so how can an attom be divided wch is no larger y^n to fill up y^t space. It like manner (y^t it may appeare ⌞of⌟ how nigh kindred number & ma Extension in matter is, in so much y^t nothing can be supposed of one, but may be so of y^c other) suppose there were ciphers of such a nature & quality y^t they will resist being y^c same. Lay y^m together they will not turne all to one nothing since theire qualities require some distance or difference amongst y^m

[121] See footnote 57.

The motion of the stomach in vomiting (though wholly against our will, and therefore merely mechanical) by the mere touch of a whalebone, does much more illustrate the actions of brutes to be mechanical and independent of souls, then Descartes's instance of the eye winking at the shaking of a friend's hand.

How much longer will a pendulum move in the receiver than in the free air. Hence it may be conjectured what bodies there be in the receiver to hinder the motion of the pendulum.

[Blank folio]   60 117ˢ

[Blank folio]   61 118ⁱ

[Blank folio]   62 118ˢ

Of atoms   vide page 3[19]   63 119ⁱ

division can be made a mathematical point or a superficies may be between the parts divisible, but a point cannot be put or conceived in this little space to divide it. Add then a naked point to any space in the line, I ask whether it be in the space without touching any of the other points that make the line or not. If you say it is, then I answer that a point which would have resisted touching might have as well been added before without an absurdity. If you say then that it might touch one of the other points that makes the line, I say then that that point is in the same place with the point which it touches, and so not in the space which you would have divided. How can then that space be divided into which a mathematical point cannot enter to separate its parts? And so, how can an atom be divided which is no larger than to fill up that space. In like manner (that it may appear of how nearly alike number and extension in matter is, in so much that nothing can be supposed of one, but may be so of the other) suppose there were ciphers of such a nature and quality that they will resist being the same. Lay them together, they will not turn all to one nothing, since their qualities require some distance or difference

[19] Page 341

betwixt each one of y^m let y^c first be a bare cipher, ad another twill be different from y^t & it can differ no less y^n an unite y^c third y^n must differ from both of, $\overline{y}$ & y^t can be no lesse y^n 2 units &c Every cipher thus qualified being different or distant from all y^c former by y^c quantitie of an unite if y^n 11 such ciphers added make y^c difference of 10 unites it make y^c number of 10. See y^n if you can ad another cipher thus qualified into y^c midst of y^m as betweene 5 & 6 y^c cipher will be neith^r 5 nor 6 for y^n it would bee added too y^m not betweene y^m & so be y^c same wth y^t to wch it is added wch is against its nature. There is no difference betweene 5 & 6 where by it may not be y^c same wth some other number und^r 10 therefore it cannot bee admitted into y^c number 10, so add a naked cipher to y^c number there is nothing twixt 5 & 6 (y^c difference or distance of y^c numbers ⌞or space be-twene y^m⌟) in wch it can be therefore it must be either add to 5 or 6 y^c joynt

(as may say of ~~those~~ ⌞y^c⌟ numbred unites.) By y^c way you may note y^t tis not y^c ciphers haveing power to keepe seper-ated or different from y^t sort of being to wch it is added but y^c actuall resistance of being one wth y^m & therefore y^c first ciper of y^c multitude thus qualified will be still a plaine cipher because there is no former cipher ~~whose nature it~~ wth wch it should be one w^{th}, but y^c 2^d ciper refusing to be what y^c first is makes y^c unite or indivisible basis of number. So a Math: point is not extended by haveing power to resist con-junction wth another unles theire be another point wth wch it refuseth to be seperated, & y^n there is distance betwixt y^m two though indivisible & y^c least y^t can be yet y^c basis of all other extensios & y^c mould of attoms.

I would not be mistaken as if I thought a point or Cipher (which are nothings) were capable of powers or qualities but because I thought it a supposition easie to conceive of & fit for y^c purpose I ventred upon it & though it be impossible y^t y^c thing should be so yet it is not so to conceive it: nay if attomes be so small tis necessary to conceive they are ter-

between each one of them. Let the first be a bare cipher. Add another, it will be different from the first, and it can differ no less than a unit. The third then must differ from both of them, and that can be no less than two units, etc. Every cipher thus qualified being different or distant from all the former by the quantity of a unit. If then all eleven such ciphers added make the difference of ten units, it makes the number of ten. See then if you can add another cipher thus qualified into the midst of them, as between five and six. The cipher will be neither five nor six, for then it would be added to them, not between them, and so be the same with that to which it is added, which is against its nature. There is no difference between five and six whereby it may not be the same with some other number under ten, therefore it cannot be admitted into the number ten. So, add a naked cipher to the number, there is nothing between five and six (the difference or distance of the numbers or space between them) in which it can be, therefore it must be either added to five or six, the joint

(as one may say of the numbered units). By the way, you may note that it is not the ciphers having power to keep separated or different from the sort of being to which it is added, but the actual resistance of being one with them. Therefore the first cipher of the multitude thus qualified will be still a plain cipher, because there is no former cipher with which it should be one with. But the second cipher refusing to be what the first is makes the unit, or indivisible basis of number. So a mathematical point is not extended by having power to resist conjunction with another unless there be another point with which it refuses to be joined, and then there is distance between the two, though indivisible, and the least that can be, yet the basis of all other extensions and the mold of atoms. I would not be mistaken as if I thought a point or cipher (which are nothings) were capable of powers or qualities, but because I thought it a supposition easy to conceive of and fit for the purpose, I ventured upon it. And though it be impossible that the thing should be so, yet it is not so to conceive it; nay, if atoms be so small it is necessary to conceive they are

64 119ᵛ

mined ˻& touch others˼ by Math: points & superficies ~~tho~~ at so small a distance as is described here, though held asund^r by y^c attome & no power of theire owne.

Object: y^c least extension is infinitely ~~(?)~~ larger y^n a point & theirefore can conteine it & be divided by it.

Resp: I confess it is so & therefore can contèine an infinite number of points but they must be all in y^c borders or ~~extreames~~ ˻sides & outward superficies of it˼ & y^l can not ~~point~~ ˻make˼ out a place for division: ~~for~~ y^c least distance in y^c whole attome is from one ~~par~~ extrem to another it hath no inside, no midst, nor center ~~but is it selfe y^c~~ but is it selfe all. center inside & midst to y^c invironing superficies & all it can do is to keep those points on eitherside it from touching You cannot put a point wthin it becaus it hath no inside set a point uppon it & y^n it touches but its superficies How can a point y^n be in y^c Attom & distant from one extream & yet not in y^c other. put a point to one extreame of it & lett it move ~~into y^c place where you conceive it might be wthout touching either~~ side towards y^c

other extreame y^n it is no sooner from one extreame but it is at y^c other (because it is no soner out of one ˻extreame of a˼ place (for it can˻not˼ be in a place) but it is in another & y^c 2 ertreames of it are y^c termini of y^c nighest & joyning parts of place) can it y^n be where tis impossible for it to rest from motion nay if it should get from one extreame & not reach y^c other y^l would not be motion because y^c least motion is over y^c least distance & y^c least distance is from one side to anoth^r in y^c Attome. y^c whole attom is all in y^c same place.

What ever can be objected against indefinite divisibility ~~(?)~~ in bodys may also bee objected against y^c same in quantity & number. but if y^c fraction 10/3 bee reduced to decimall it will be 3,33333333 &c infinitely. & what doth ~~(?)~~ every figure signifie but a ˻pte˼ of y^c fraction 10/3 wch therefore is divisible into infinite ˻pts˼.

[Blank folio]

bounded and touch others by mathematical points and superficies at so small a distance as is described here, though
held apart by the atom and no power of their own.

Objection: The least extension is infinitely larger than a
point and therefore can contain it and be divided by it.

Response: I confess it is so, and therefore can contain an
infinite number of points, but they must be all in the
borders or sides and outward superficies of it, and that
cannot make out a place for division. The least distance in
the whole atom is from one extreme to another. It has no
inside, no midst, nor center, but is itself all (center, inside,
and midst) to the environing superficies, and all it can do is
to keep those points on either side of it from touching. You
cannot put a point within it, because it has no inside. Set a
point upon it and then it touches but its superficies. How
can a point then be in the atom and distant from one extreme and yet not in the other? Put a point to one extreme
of it and let it move toward the

other extreme, then it is no sooner from one extreme but it
is at the other (because it is no sooner out of one extreme of
a place (for it cannot be in a place) but it is in another and
the two extremes of it are the termini of the nearest and
joining parts of place). Can it then be where it is impossible
for it to rest from motion? No, if it should get from one
extreme and not reach the other that would not be motion,
because the least motion is over the least distance and the
least distance is from one side to another in the atom. The
whole atom is all in the same place.

Whatever can be objected against indefinite divisibility in
bodies may also be objected against the same in quantity
and number. But if the fraction $\frac{10}{3}$ be reduced to decimal
form it will be 3.33333333 etc. infinitely, and what does
every figure signify but a part of the fraction $\frac{10}{3}$ which,
therefore, is divisible into infinitely many parts.

[Blank folio]

Gravity & levity

anothers streames wth much difficulty, & pressure & so be compacted & y^e descending streame will keepe y^m so by continually pressing y^m to y^e Earth till they arise to y^e place from whence they came, & there they will attaine theire former liberty.[122]

The gravity of a body in divers places as at ye top & bottome of a hill; in differen latitudes &c: may bee measured by an instrument of this forme

The weight of water is to ye weight of quicsilver as 1 to 14.[123] Water is 400[124] ⌊(perhaps 2000)⌋[125] times heavier y^n aire & gold 19 times heavier y^n water.

Quæst: What proportion ye weights of two bodys ⌊as gold & silver⌋ have have in divers mediums as in vacuo aere aqua &c: wch known ye weight of ye aire or water in vacuo ⌊or the quantity of gold to ye silver⌋ is given &c: As if in aire y^e Gold (a) is equiponderant to y^e silver (z) ~~& in vacuo y^e Gold (3a) is equiponderant to y^e gold (2b.)~~ ⌊y^e weight of it being called b. And in⌋ water y^e gold (a) is equiponderant to y^e silver (2z). let (c) bee y^e weight of so much water ⌊in y^e aire⌋ as is equall to y^e Gold a then is $\frac{cz}{a}$ y^e weight of so much water as is equall to y^e silver z. ~~And b−c:z−$\frac{cz}{a}$::a:2z, for~~ ⌊Then⌋ y^e gold & silver weighed in water their weights are diminished by y^e weight of y^e water whose place they conteine. Therefor ~~2az − ⌊2⌋cz = az−cz. Or a = c~~ b−c is y^e weight of y^e gold a in y^e water & b − $\frac{cz}{a}$ is y^e weight of y^e silver (z) in it, & since a is equiponderant to 2z in it Therer b − c = 2b − $\frac{2cz}{a}$. or $\frac{ab + ac}{2c}$

---

[122] Newton's account bears some resemblance to those of Descartes and Kenelm Digby, but it was probably most influenced by Boyle, *Spring of the Air*, pp. 217–29. However, the theory is certainly Newton's own.

[123] Boyle, *Spring of the Air*, pp. 295–7, and Charleton, *Physiologia*, p. 59.

[124] Boyle, *Spring of the Air*, p. 289, and Charleton, *Physiologia*, p. 59; both report this figure as that determined by Galileo.

[125] Boyle discusses Mersenne's figure of 1,356 in *Spring of the Air*, p. 291, and disputes that figure as too high. Boyle does, however, give the figure 1 to 2,000 as the degree to which air can be expanded in the Magdeburg experiment. Perhaps Newton has confused the figures.

Gravity and levity

another's streams with much difficulty and pressure, and so be compacted and the descending stream will keep them compacted, by continually pressing them to the Earth until they arise to the place from whence they came. There they will attain their former liberty.

The gravity of a body in diverse places as at the top and bottom of a hill, in different latitudes, etc., may be measured by an instrument of this form

The weight of water is to the weight of quicksilver as 1 is to 14. Water is 400 (perhaps 2,000) times heavier than air and 19 times heavier than water.

Question: what proportion the weights of two bodies, such as gold and silver, have in diverse mediums as in a vacuum, air, water, etc.: which, if known, the weight of the air or water in a vacuum, or the quantity of gold to silver, can be determined. For example: if in air the gold a is equiponderant to the silver z, their weight being called b, and in water the gold a is equiponderant to the silver $2z$ and c be the weight of so much water in the air as is equal to the gold a; then cz/a is the weight of so much water as is equal to the silver z. Since the gold and silver are weighed in water, their weights are diminished by the weight of the water whose place they contain. Therefore, $b - c$ is the weight of the gold a in the water, and $b - cz/a$ is the weight of the silver z in it. Since a is equiponderant to $2z$ in water, therefore $b - c = 2b - (2cz)/(a)$, or $(ab + ac)/(2c) = z$. That

= z. y^t is $2c : b+c :: a : z$. Or if (c) y^e weight of water in y^e aire is sought $y^n \frac{ab}{2z-a} = c$. Thus might y^e absolute weights of bodys i:e: their weights in vacuo bee found. as of air or y^e Bodys ˰or fire˩ in a hot furnace w^{th}out flame. ̶&̶c̶

Try whither flame will descend in Torricellius vacuū.[126]

68 121ᵛ

In y^e descention of a body There is to be considered y^e force wch it receives every moment from its gravity (wch must bee ̶s̶o̶e̶ ̶m̶u̶c̶h̶ ̶y̶^e̶ ̶l̶e̶s̶s̶e̶ ̶b̶y̶ ̶h̶o̶w̶ ̶m̶u̶c̶h̶ ̶t̶h̶e̶ ̶s̶w̶i̶f̶t̶e̶r̶ ̶a̶ ̶b̶o̶d̶y̶ ̶i̶s̶ ̶m̶o̶v̶e̶d̶)̶ ˰least in a swiftest body)˩ & ye opposition it receives from ye aire (wch increaseth in ˰pportion to its swiftnesse). To make an experiment concerning this increase of motion ̶l̶e̶t̶ When ye Globe a is falne from e to f let ye Globe b begin to move at g soe y^t both y^e globes fall together at h.

According to Galilæus a iron ball of 100ˡ Florentine (y^t is 78ˡ at London of Adverdupois weight) descends an 100 braces Florentine or cubits (or 49,01 Ells, perhaps 66^{yds}) in 5″ of an hower.[127]

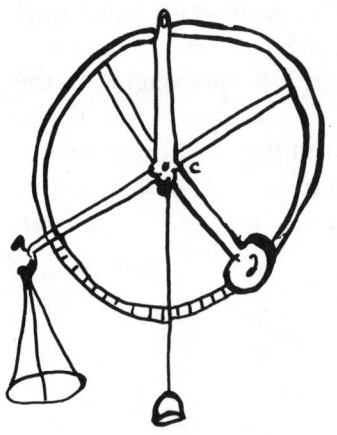

By this figure it may appeare how to weigh w^{th}out altering y^e weights. And to tell exactly y^e weight of bodys at y^e first triall.

But it will bee best to fix y^e wheele & make y^e armes cd & ac very long especially cd. This ballance may bee of excellent use

---

[126] See footnote 58.

[127] Galileo, *Systeme of the World: in Four Dialogues*, in *Mathematical Collections and Translations in Two Tomes*, edited by Thomas Salusbury (London, 1661), Vol. I, Part I, p. 200.

is, $2c : b + c :: a : z$. Or, if c, the weight of water in the air is sought, then $(ab)/(2z - a) = c$. Thus might the absolute weights of bodies, i.e., their weights in a vacuum, be found. Similarly can the weight of air, bodies in a hot furnace minus flame, and fire be determined.

Try whether flame will descend in Torricelli's vacuum.

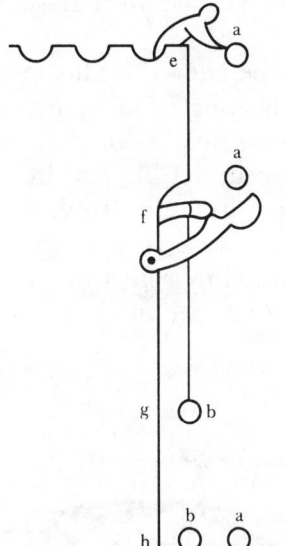

In the descension of a body there is to be considered the force which it receives every moment from its gravity – which must be least in a swiftest body – and the opposition it receives from the air – which increases in proportion to its swiftness. To make an experiment concerning this increase of motion: when the globe a has fallen from e to f let the globe b begin to move at g, so that both globes fall together at h.

68 121ᵛ

According to Galileo an iron ball of 100 lb. Florentine (that is 78 lb. at London avoirdupois weight) descends 100 Florentine braces or cubits (or 49.01 Ells, perhaps 66 yds.) in 5 seconds of an hour.

By this figure it may appear how to weigh without altering the weights, and to tell exactly the weight of bodies at the first trial. But it will be best to fix the wheel and make the arms cd and ac very long, especially cd. This balance may be of excel-

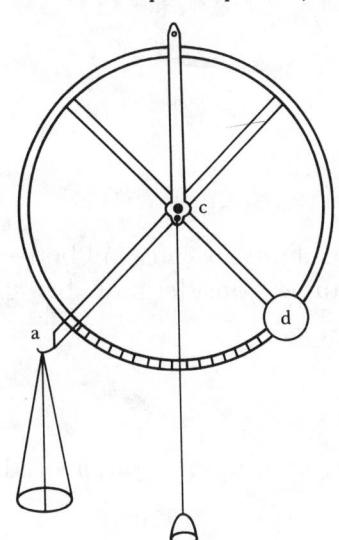

for finding y^c ~~proportions~~ ˻severall weights˼ of alloyed or mixt bodys ~~by their weight in aire & water & in~~ ˻by their weight in severall mediums as in aire & water˼ (as of gold & coppor,) or to compare y^c quantity of any two ⟨?⟩ bodys (as gold & stone) by their difference of weight in divers mediums. to compare y^c weights of ~~equall~~ bodys, viz: to find w^t proportion the weights of those bodys would have were they equall.

Try whither y^c weight of a body may be altered by heate or cold, by dilatation or condension, beating, poudering, transfering to serverall places or sev'all ⟨?⟩ heights or placing a ˻hot or˼ heavy body over it, ˻or under it,˼ or by ~~placing~~ magnetisme. whither leade or its dust spread abroade, whither a plate flat ways or edg ways is heaviest. Whither y^c rays of gravity may bee stopped by refecting or refracting y^m, if so a perpetuall motion may bee made one of these two ways

The gravity of bodys is as their solidity, because all bodys descend equall spaces in equall times consideration being had to the Resistance of y^c aire &c.

69 122^r   Of colours

Try if two Prismas y^c one casting blew upon y^c other's red doe not produce a white.

lent use for finding the several weights of alloyed or mixed bodies by their weight in several mediums as in air and water (such as gold and copper), or to compare the quantity of any two bodies (such as gold and stone) by their difference of weight in diverse mediums, or to compare the weights of bodies, that is to find what proportion the weights of those bodies would have were they equal.

Try whether the weight of a body may be altered by heat or cold, dilation or condensation, beating, powdering, transferring to several places or several heights, or placing a hot or heavy body over it or under it, or by magnetism. Whether lead or its dust spread abroad is heaviest. Whether a plate flat ways or edge ways is heaviest. Whether the rays of gravity may be stopped by reflecting or refracting them. If so a perpetual motion may be made in one of these two ways.

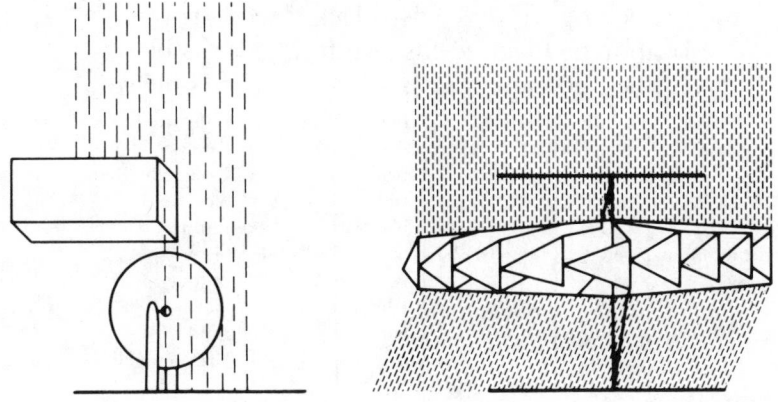

The gravity of bodies is as their solidity, because all bodies descend equal spaces in equal times, consideration being had to the resistance of the air.

Of colors

69 122^r

Try if two prisms, the one casting blue upon the other's red, do not make a white.

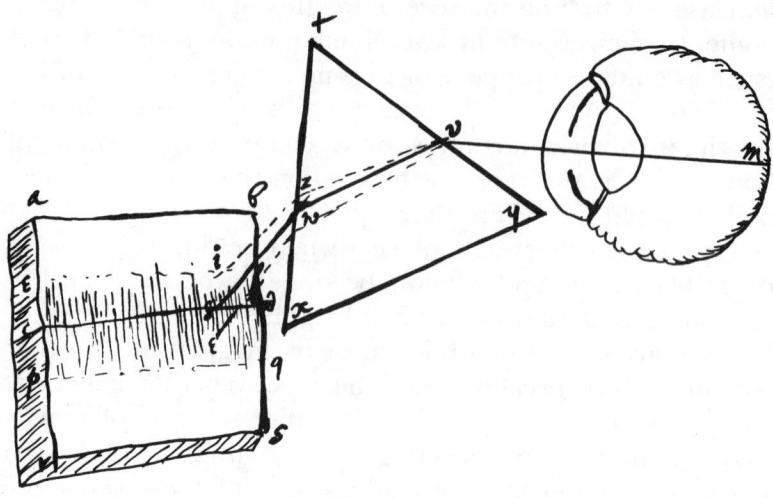

If abdc be white & cdsr black yⁿ eodc is red.
If abdc be black & cdsr white yⁿ eodc is blew.
If abdc be blew & cdsr white yⁿ eodc is blewer.

If (abdc) be		&cdsr	& (cdsr) be		yⁿ eodc is	
	white	&cdsr		blew	yⁿ c	~~yello~~ ⌐Red.⌐
	black			blew		~~Greene~~ ⌐blewer.⌐
	blew			black		Greene or Red.
	black			red		blew.
	red			black		redder.
	red		white	yⁿ eodc	blew.
	white			red	is	redder.
	white			whiter		blew.
	whiter			white		redd.
	black			blacker		Greene or dirke
						red.
	blacker			black		blew.

The more uniformely the globuli move yᶜ optick nerves yᶜ
more bodys seme to be coloured red yellow blew greene &c
but yᶜ more variously they move them the more bodys appeare white black or greys.

70 122ᵛ 1 Note yᵗ slowly moved rays are refracted more then swift
ones.
2ᵈly If adbc be shaddow & cdsr white then yᶜ slowly moved
rays ⌐comeing from cdqp⌐ will be refracted as if they had
come from eodc soe yᵗ yᶜ slowly moved ⌐ray⌐ being seperated from yᶜ swift ones ⌐by refraction,⌐ there ⟨?⟩ ariset 2

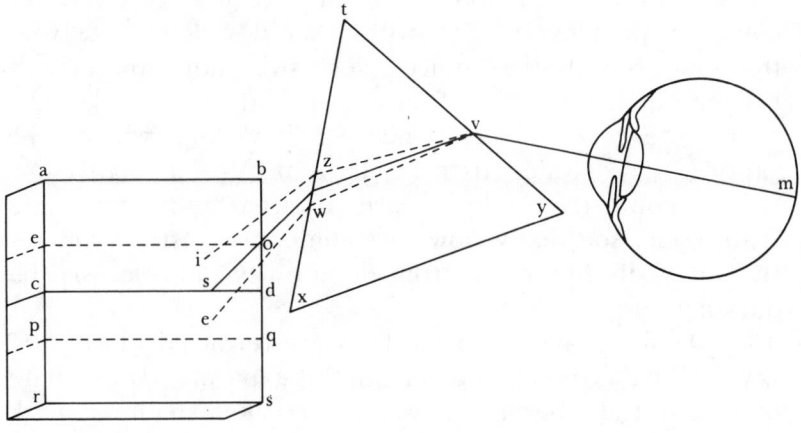

If *abdc* is white and *cdsr* black, then *eodc* is red.
If *abdc* is black and *cdsr* white, then *eodc* is blue.
If *abdc* is blue and *cdsr* white, then *eodc* is bluer.

	If *abdc* is		and *cdsr* is		then *eodc* is
	white		blue		red
	black		blue		bluer
	blue		black		green or red
	black		red		blue
	red		black		redder
	red		white		blue
	white		red		redder
	white		whiter		blue
	whiter		vhite		red
	black		blacker		green or dark red
	blacker		black		blue

The more uniformly the globuli move the optic nerves, the more bodies seem to be colored red, yellow, blue, green, etc. But the more variously they move them, the more bodies appear white, black, or gray.

1. Note that slowly moved rays are refracted more than swift ones. 70 122ᵛ

2. If *abdc* be shadow and *cdsr* white, then the slowly moved rays coming from *cdqp* will be refracted as if they come from *eodc*. So that the slowly moved ray being separated from the swift ones by refraction, two kinds of colors

~~colours~~ kinds of colours viz: from y^c slow ones blew, sky colour, & purples. from y^c swift ones red, yellow   & from them wch are neither moved very swift nor slow ariseth greene but from y^c slow & swiftly moved rays mingled ariseth white grey & black. whence it is y^t cdqp will not ~~be~~ ⌞appeare⌟ red unless qsrp be darke because as many slow rays as come from cdqp & are refracted as if they came from eodc; soe many slow rays come from qsrp & are refracted as if they came from dqpc unless unlesse qsrp be darker y^n dqpc.

3^{d}ly   That y^c rays wch make blew are refracted more y^n y^c rays wch make red appeares from this experim^{nt}. If one hafe of y^c thred abc be blew & y^c other red & a shade or black

body be put behind it y^n lookeing on y^c thred through a prism one halfe of y^c thred shall appeare higher y^n y^c other. & not ⌞both⌟ in one direct line, by reason of unequall refractions in y^c 2 ~~halfe lines~~ differing colours.[128]

4   Hence rednes yellownes &c are made in bodys by ~~st~~ stoping y^c Slowly moved rays w^{th}out ⌞much⌟ hindering of y^c motion of y^c swifter rays. & blew ⌞greene⌟ & purple by diminishing y^c motion of y^c swifter rays ⌞& not of y^c slower⌟. Or in some bodys ⌞all⌟ these colours may arise by diminishing y^c motion of all y^c rays in greater or lesse geometricall proportion, for y^n there will be lesse difference in theire motions y^n otherwise.

71 123^r 5   If y^c particles in ~~surface~~ a body have not so greate an elastick power as to returne back y^c whole motion of a ray, then y^t body may be lighter or darker colored according as y^c elastick ~~particle~~ virtues of that bodys parts is more or lesse.

6   If there be loose particles in y^c pores of a body wch by makeing y^m very narrow or by ⌞otherwise⌟ hindering y^c elastick power of y^c subtil matter whereby y^c motions of y^c rays are conserved y^t body may have some colour, &c.

---

[128] See Newton, ULC, Add. 3975 (Appendix, p. 467).

arise, viz: from the slow ones blue, sky color, and purples; from the swift ones red, yellow; and from those which are moved neither very swift nor slow arises green; but from the slow and swiftly moved mingled rays arises white, gray, and black. Whence it is that *cdqp* will not appear red unless *pqsr* is dark, because as many slow rays as come from *cdqp*, and are refracted as if they came from *eodc;* so many slow rays come from *pqsr* and are refracted as if they came from *cdqp*, unless *pqsr* is darker than *cdqp*.

3. That the rays which make blue are refracted more than the rays which make red appears from this experiment: If one half of the thread *abc* is blue and the other red, and a shade or black body be put behind it, then look-

a ———————— b ———————————— c   ing on the thread through a prism one half of the thread shall appear higher than the other, and not both in one direct line, by reason of unequal refractions in the two differing colors.

4. Hence redness, yellowness, etc., are made in bodies by stopping the slowly moved rays without much hindering of the motion of the swifter rays; and blue, green, and purple by diminishing the motion of the swifter rays and not of the slower. Or in some bodies all these colors may arise by diminishing the motion of all the rays in a greater or less geometrical proportion, for then there will be less difference in their motions than otherwise.

5. If the particles in a body have not so great an elastic power as to return back the whole motion of a ray, then that body may be lighter or darker colored according as the elastic virtues of that body's parts is more or less.   71 123ᵛ

6. If there be loose particles in the pores of a body which by making them very narrow, or by otherwise hindering the elastic power of the subtle matter whereby the motions of the rays are conserved, that body may have some color.

7 If a pore be too much straitned as at ce, so y^t y^e globulus must part y^e matter towards b from y^e matter to-  wards d so y^t there be no subtil matter on either side towards c or e, ~~y^n after~~ ⌐when⌐ some of y^t narrow poore is behind it, there will be noe matter on either side it to presse it towards its hinder ⌐pts so much as it is pressed before likewise it is forced to croude ~~all~~ before it all ye matter in ye pore ~~towar~~ so y^t its ⌐motion⌐ must needes be diminished if it had force to passe through but if it had not force enough to part ye matter & get through y^n it would be reflected back wthout looseing any considerable ⌐pte of its motion ⌐as rays passing out of glasse into ye aire whē their force is too much diminished by their obliquity they are reflected by ye aire wthout any loss of their motion.⌐ & bodys full of such straight passages in its pores must be of darke colours as blew glass whose pores may be straightned by ~~y^e~~ loose ⌐& too greate⌐ particles of ye tincture lying wthin y^m. But if this pore were something bigger so as to let ye slowly moved ~~particles~~ Os passe through wth losse of most or all theire motion, but letting ye swift ones scape freelier, y^n ye colour will be red, yellow &c: as in glasse it

72 123^v may be whose poores are full of smaller particles of ye tincture ȳ those ⌐pores⌐ of blew glasse are.

8   Though 2 rays be equally swift yet if one ray be lesse y^n y^e other that ray shall have ⌐so much a lesse⌐ effect on y^e sensorium as it has lesse motion y^n y^e other &c.

Whence supposeing y^t there are loose particles in y^e pores of a body ~~twice as big as~~ ⌐bearing proportiō to⌐ y^e greater rays, as $9 : 12$: & y^e less globulus is in proportion to y^e greater as $2 : 9$. y^e greater globulus by impinging on such a particle will loose a $\frac{6}{7}$ ~~$\frac{(?)}{}$~~ ptes of its motion y^e less glob: will loose $\frac{2}{7}$ parts of its motion & y^e remaining motion of y^e glob: Will have almost such a proportion to one another as their quantity have. viz. $\frac{5}{7} : \frac{1}{7} :: 9 : 1\frac{4}{5}$ wch is almost 2 y^e lesse glob. & such a body may produce blews & purples.

But if y^e particles on wch y^e globuli reflect are equall to y^e lesse globulus it shall loose its motion & y^e greater glob:

7. If a pore be too much straitened as at *ce,* so that the globulus must part the matter 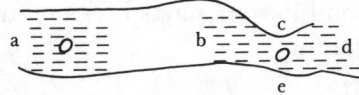 toward *b* from the matter toward *d,* so that there be no subtle matter on either side toward *c* or *e,* then when some of that narrow pore is behind it, there will be no matter on either side of it to press it toward its hinder parts, so much as it is pressed before. Likewise it is forced to crowd before it all the matter in the pore, so that its motion must needs be diminished if it had force to pass through, but if it had not force enough to part the matter and get through, then it would be reflected back without losing any considerable part of its motion; as rays passing out of glass into the air when their force is too much hindered by their obliquity are reflected by the air without any loss of their motion. Bodies full of such straitened passages in their pores must be of dark colors, as blue glass whose pores may be straitened by loose and too great particles of the tincture lying within them. But if this pore was rather bigger so as to let the slowly moved particles designated *o* in the diagram above pass through with loss of most or all their motion, but letting the swift ones escape more easily, then the color will be red, yellow, etc.: as, for example in a glass

whose pores are full of smaller particles of the tincture than are those of blue glass.

72 123ᵛ

8. Though two rays be equally swift yet if one ray be less than the other that ray shall have so much less an effect on the sensorium as it has less motion than the other. Whence, supposing that there are loose particles in the pores of a body bearing a proportion to the greater rays as $9:12$; and the lesser globulus is in proportion to the greater as $2:9$, the greater globulus, by impinging on such a particle, will lose a $\frac{6}{7}$ part of its motion and the lesser globulus will lose $\frac{2}{7}$ of its motion, and the remaining motion of the globuli will have almost such a proportion to one another as their quantities, that is, $\frac{5}{7} : \frac{1}{7} :: 9 : 1\frac{4}{5}$ which is almost twice the lesser globuli, and such a body may produce blues and purples. But if the particles on which the globuli reflect are equal to the lesser globuli they shall lose their motion and the greater globuli

shall loos $\frac{2}{11}$ parts of its motion & such a body may be red or yellow.

9   If I presse my eye on yᵉ left side (when I looke towards my right hand) as at a, yⁿ I see a wch is circle of red as at c but w^{th}in yᵉ red is blew for yᵉ capillamenta are more pressed at n & o & round about yᵉ finger yⁿ at a towards yᵉ midst of yᵉ finger. yᵗ ꝑte of yᵉ apparitiō at q is more lanquid because yᵉ capillamenta at o are duller & if yᵉ finger move towards e two much it vanisheth at q & appeareth semicircular. but if I put my finger at e or s yᵉ apparition wholly vanisheth. By puting a brasse plate betwixt my eye & yᵉ bone nigher to yᵉ midst of yᵉ tunica retina yⁿ I could put my finger I mad a very vivid impression.

73 124ʳ   But of an ellipticall figure because yᵉ edge of yᵉ plate wth wch I prest my eye was long & not round like my finger If I was in yᵉ darke ⌞& yᵉ impression be very strong⌟ towards yᵉ outside appeared a broade circle of purple next blew yⁿ grene, yⁿ yellow, red ⌞like flame⌟, yellow, greene, blew, purple wch growing from a very Darke by degrees to a lighter (?) blew ends in a greene in yᵉ midst. The colours I suppose next yᵉ flameing red looke something yᵉ darker by reason of its splendor. And if yᵉ experiment bee done in yᵉ light so yᵗ though my eyes were shut yᵉᵗ some ray got through my eye lids yᵉ ⌞outmost⌟ purple would appeare of a colour inclining to blacknes by reason I suppose of yᵉ bordering light. also & if yᵉ pressure was not very strong yᵉ greene ⌞& purple⌟ at a would not be perceived in yᵉ darke but it would be something lighter in yᵉ light yⁿ yᵉ in other places of yᵉ eye as at v or w, & all yᵉ other outmost colours w^{th}out strong pression bee but like blew. but if yᵉ presion be strong in yᵉ light yᵉ apparition will but little vary from wᵗ it is in yᵉ darke. beyond all yᵉ colours as betweene o & e yᵉ

shall lose $\frac{2}{11}$ part of their motion, and such a body may be red or yellow.

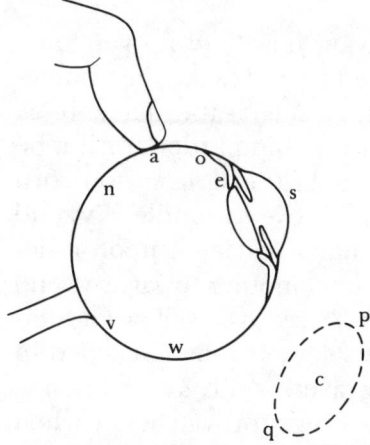

9. If I press my eye on the left side (when I look toward my right side) as at a, then I see a red circle as at c, but within the red is blue, for the capillamenta are more pressed at n and o and round about the finger than at a toward the midst of the finger. That part of the apparition at q is more faint because the capillamenta at o are duller, and if the finger move toward e too much the apparition vanishes at q and appears semicircular. But, if I put my finger at e or s the apparition wholly vanishes. By putting a brass plate between my eye and the bone nearer to the midst of the tunica retina than I could put my finger, I made a very vivid impression,

but of an elliptical figure because the edge of the plate with which I pressed my eye was long and not round like my finger. If I was in the dark and the pressure on the eye very strong, toward the outside appeared a broad circle of purple, next blue, then green, then yellow, red like flame, yellow, green, blue, and purple, which growing from a very dark by degrees to a lighter blue ends in a green in the midst. I suppose the colors next to the flaming red look darker by reason of its splendor. And, if the experiment be done in the light so that, though my eyes were shut some rays got through my eyelids, the outmost purple would appear of a color inclining to blackness by reason, I suppose, of the bordering light. If the pressure was not very strong, the green and purple at a would not be perceived in the dark. But in the light the apparition would be somewhat lighter than in other places of the eye, as at v or w. All the other outmost colors, without strong pression, be but blue. But if the pression be strong in the light, the apparition will but vary little from what it is in the dark. Beyond all the

73 124^r

light is stronger y^n in other places as at v or w, because ye pressure helpes ye motion from wthout but is not strong enough to turne it to colour.[129]

10 Heate a peice of steele glowing hot, hold it over some water untill it turne from a white to a red heate then immediately quench it y^t ~~it bee neither too brittle nor soft by beeing quenched too~~ ‚so far as you would harden for if it be quenched sooner it‚ will bee too brittle ˙if later too soft. Then makeing ye end bright hold it over a candle y^t ye end may be halfe an inch out of ye flame, or lay it upon a hot iron, & these colours will follow one another towards ye end of it; viz: bright yellow, deeper & reddish yellow or sanguine, a fainter blew, & a deeper blew, If it be quenched in tallow when tis yellow tis fit for gravers, drills &c: If when ye light blew is on it, it is fit ‚for‚ springs for watches If when ye deepe blew is on it, it will bee very soft.[130]

124^v 11 The colours succede in order according to theire more or lesse reflection of light viz: white, redd, yellow, blew, purple, greene, black. Red, & purple paint theire colours far more manifestly y^n blew, or greene.[131]
Light reflected from a yellow to a blew body makes a greene.
12 The Sunne shineing through coloured paper or glasses ‚as also ye mixture of divers colours‚ exhibits these colours viz yellow & blew make red. Yellow & red make Orange ‚~~or scarlet~~‚ colour. Purple & red make scarlet. Red & greene a darke orange Tauny.[132] Red & blew make purple. Red & white by mixture make a Carnation.[133]
12 ye yellow colour (made by a prisme) falling upon a blew makes a greene. blew falling upon red makes a greene. It would be tried w^t colours ye mixture of colours falling from 2 prismes would make.[134]
13 Leade melted very hot & haveing ye scum taken of represents these colours viz: Blew, yellow, purple, blew;

---

[129] See Newton, ULC, Add. 3975 (Appendix, p. 482).
[130] Boyle, *Colours*, pp. 6–8.   [131] *Ibid.*, pp. 186–8.
[132] *Ibid.*, pp. 190–1.   [133] *Ibid.*, p. 220.
[134] *Ibid.*, pp. 225–7. Boyle indicates that he did the experiment with two prisms.

colors, as between *o* and *e,* the light is stronger than in other places, as at *v* or *w;* because the pressure helps the motion from without but is not strong enough to turn it to color.

10. Heat a piece of steel glowing hot and hold it over some water until it turns from a white to a red heat, then quench it so far as you would harden it, for if it be quenched sooner it will be too brittle if later too soft. Then, either hold it over a candle that the end may be half an inch out of the flame or lay it upon a hot iron to make the end bright and these colors will follow one another toward the end of it, viz., bright yellow, deeper and reddish yellow or sanguine, a fainter blue, and a deeper blue. If it be quenched in tallow when it is yellow it is fit for gravers, drills, etc.; if when it is light blue, it is fit for springs for watches; if when it is deep blue, it will be very soft.

11. The colors succeed in order according to their more or less reflection of light, viz.: white, red, yellow, blue, purple, green, and black. Red and purple paint their colors far more manifestly than blue or green. Light reflected from a yellow to a blue body makes a green.

12. The Sun shining through a colored paper or glasses, as also the mixture of diverse colors, exhibits these colors: yellow and blue make red, yellow and red make orange, purple and red make scarlet, red and green make a dark orange tawny, red and blue make purple, red and white by mixture make a carnation. The yellow color made by a prism falling upon blue makes a green, and blue falling upon red makes a green. It should be tried what colors the mixture of colors falling from two prisms would make.

13. Lead melted very hot and having the scum taken off manifests these colors: blue, yellow, purple, blue; green,

greene, purple, blew, yellow, red; purple, blew, yellow & blew, yellow, blew, purple, Greene mixt, yellow, red, blew, greene, yellow, red, purple, greene.[135]

14 Motes in ye Sunne in some positions appeare of divers colors.[136]

15 Put as much common Sublimate into hot faire water as it can dissolve filter y^e solution through cap paper y^t it may be lympid. & into 2 spoonefulls of it, put about 5 drops of good limpid spirits of urine & it will be white like milke, to wch if you put in some rectified Aqua=fortis it will be transparent. more fresh spirit of urine will make it looke white but not so white as before.[137]

16 Make a strong Infusion of broken galls in faire water. filter it into a cleane violl ad more of y^e same liquor to it till it be transparent into it shake a convenient quantity of cleare but very strong solution of vitrioll & it will be black, then drop a little ⌐cleare & strong⌐ oyle of vitrioll into it shakeing ye vessell well & ye liquor will become transparent wch againe will become black by ye affusion of a small quantity of a strong solution of salt of tartar.

Note, y^t Corrosive liquors (as oyle of Vitrioll) may clarify a liquor by seperating & dividing its ⌐pts. But precipitating liquors (as salt of tartar) by uniteing theire parts make them conspicuous & ye liquor coloured.[138]

75 125^r vide pag 43.

Of Imagination

I gather y^t my Phantasie & y^e ☉ had ye same operation uppon y^e ~~optick~~ spirits in my optick nerve & y^t ye same motions are caused in my braines by both.

4 Opening my eye & lookeing in ye darke upon ye like imaginations there appeared ye like pantasme as when I shut it.[139]

[135] *Ibid.*, pp. 23–6.   [136] *Ibid.*, pp. 69–70.
[137] *Ibid.*, pp. 133–4.   [138] *Ibid.*, pp. 135–7.
[139] These afterimage experiments are also described in a letter Newton wrote to John Locke, June 30, 1691, item 365 in *The Correspondence of Isaac Newton*, edited by H. W. Turnbull, Vol. III (Cambridge University Press, 1961), pp. 152–4. They are also briefly mentioned in Newton, ULC, Add. 3975, item 63, folio 16 (see Appendix, p. 483).

purple, blue, yellow, red; purple, blue, yellow and blue, yellow, blue, purple, green mixt, yellow, red, blue, green, yellow, red, purple, green.

14. In some positions motes in the Sun appear of diverse colors.

15. Put as much common sublimate into hot fair water as it can dissolve. Filter the solution through cap paper, that it may be limpid, and into two spoonfuls of it, put about five drops of good limpid spirits of urine and it will be white like milk, to which if you put in some rectified aqua fortis it will be transparent; more fresh spirit of urine will make it look white, but not so white as before.

16. Make a strong infusion of broken galls in fair water, and filter it into a clean vial. Add more of the same liquid to it till it is transparent, into it shake a convenient quantity of clear but very strong solution of vitriol, and it will be black. Then drop a little clear and strong oil of virtriol into it, shaking the vessel well, and the liquid will become transparent; which will again become black by the affusion of a small quantity of a strong solution of salt of tartar.

Note that corrosive liquids (as oil of vitriol) may clarify a liquid by separating and dividing its parts. But precipitating liquids, as salt of tartar, by uniting their parts make them conspicuous and the liquid colored.

Of imagination   *vide* page 43[20]   75 125ʳ

I gather that my fantasy and the Sun had the same operation upon the spirits in my optic nerve, and that the same motions are caused in my brain by both. (4) Opening my eye and looking in the dark upon the like imaginations, there appeared the like phantasma as when I shut it.

[20] Page 395

5 Lookeing uppon white paper there appeared (by meanes of a strong phantasie) first a spot something darker y^n ye paper wch grew blacker & blacker until there seemed to be a dusky red speading almost over all ye pape sometime this spot would be red (?) & sometime blew.

6 lookeing on a bright cloude there appeared ye same phantasm as when I looked on ye white paper ˌonely for ye most ˌpt blackerˌ untill at last I was able to make this spot glitter ~~in ye cloude~~ ˌamidst ye dusky redˌ whither I looke on ye paper or cloude like y^e ☉ in a cloude so bight y^t my eys watered.

7 Imploying my selfe in other exercises for two or 3 howers ˌan hower before ☉ sed hee being wholy clouded.ˌ when I thought my eye was pretty well restored I repeated all ye former experiment adding this to y^m y^t though I shut ye distemperd eye & opened y^t wth wch I looked not on y^e ☉ y^t I could see y^e ☉ pictured on ye cloudes or other white objects almost as plaine as if I had looked wth my distempered ey ye other being shut ˌ& every where about ☉ appeared a dusky red & blacknesse.ˌ & wth doeing thus I made such impress on ye optick nerve ye let me looke wth wch eye I would ☉ offered itselfe to my vew unless I set my fantasie to worke on other things wch wth much difficulty I could doe.

8 If after I had thus seene ☉s image wth my left wel eye I shut it & opened my right eye all objects would appeare coloured as when I had new seene ☉ But I could not perceive any such motion in ye spirits of my left eye for all objects appeared in theire right colours to it unlesse

76 125^v when I fixed my eye (?) on y^m for y^n appeared ☉.

9 when ye impresion of ☉ was not too strong upon my eye I could easily imagine severall shapes ~~to be where I usually appre~~ as if I saw them in y^e ☉s place, whence perhaps may be gathered y^t ye tenderest sight argues ye clearest fantasie of things visible. & hence something of ye nature of madnesse & dreames may be gathered.

10 When I had beene thus affected 2 days ye same ˌwhite wallˌ If I looked not over nigh it where it was shadowed looked blew, where it was lesse shadowed looked red inclin-

(5) Looking upon white paper there appeared (by means of a strong fantasy) first a spot somewhat darker than the paper, which grew blacker and blacker until there seemed to be a dusky red spreading almost over all the paper. Sometimes this spot would be red and sometimes blue. (6) Looking on a bright cloud there appeared the same phantasm as when I looked on the white paper (only for the most part blacker), until at last I was able to make this spot glitter (amidst the dusky red) like the Sun in a cloud so bright that my eyes watered whether I looked on the paper or cloud. (7) Employing myself in other exercises for two or three hours, when I thought my eye was pretty well restored an hour before the Sun set (it being wholly clouded), I repeated all the former experiments adding this to them; that though I shut the distempered eye and opened that which had not looked on the Sun, nevertheless I could see the Sun pictured on the clouds or other white objects almost as plain as if I had looked with my distempered eye the other being shut; and everywhere about the Sun appeared a dusky red and blackness. With doing thus I made such an impress on the optic nerve that, let me look with which eye I would, the Sun offered itself to my view, unless I set my fancy to work on other things which with much difficulty I could do. (8) If after I had thus seen the Sun's image with my well left eye I shut it and opened my right eye, all objects would appear colored as when I had first seen the Sun. But I could not perceive any such motion in the spirits of my left, for all objects appeared in their right colors, except

when I fixed my eye on them for then the Sun appeared.   76 125ᵛ
(9) When the impression of the Sun was not too strong upon my eye, I could easily imagine several shapes as if I saw them in the Sun's place, whence perhaps may be gathered that the tenderest sight argues the clearest fantasy of things visible, and hence something of the nature of madness and dreams may be gathered. (10) When I had been thus affected two days, if I looked at a white wall, though not too closely, where it was shadowed it looked blue, where it was less shadowed it looked red though in-

ing to white where it was lightest by ye rays of y^e ☉ reflected fm a wall uppon it it looked white, at y^t time I had beene in a darke rome for 2 or 3 howers & my eyes were made tender thereby so ye motiō made in y^m would be easlier conserved & consequequently more unifome.

77 126^r [Blank folio]

78 126^v [Blank folio]

79 127^r [Blank folio]

80 127^v [Blank folio]

81 128^r
Of God

Were men & beasts &c made by fortuitous jumblings of attomes there would be many parts uselesse in them ~~or else uselesse standing~~ here a lumpe of flesh there a member too much Some kinds of beasts might have had but one eye some more y^n two & ye two eyes.

82 128^v see pag 32

Of Light

the candle in the glasse appears & disappeare.

83 129^r
Of ye Creation ~~is use.~~

The word ~~בראא~~[140] wch Gen i.i. is interpreted to create something out of nothing is used Gen ye 1st v. 21 where tis saide God created greate Whales ˌ&cˌ but ye matter out of w^{ch} ~~it~~ they were ~~begun~~ ˌcreatedˌ did exist before neither is it ment of creating ye soule or forme of ye whale, for y^t is not ye whale alone. & there may be but one kind of irrationall soule wch joyned wth severall kinds of bodys make severall kinds of beasts, for setting aside ye different shape of theire body beasts differ from one another but in som qualitys w^{ch} ~~w^{ch}~~ are called instincts of nature. now as in men whose soules are of one kind some love hate feare &c one

---

[140] Pronounced ~~bara~~.

clining to white, where it was lightest by the rays of the Sun
reflected from another wall upon it it looked white. At that
time I had been in a dark room for two or three hours and
my eyes were made tender thereby, so the motion made in
them would be more easily conserved and consequently
more uniform.

[Blank folio]   77 126^r

[Blank folio]   78 126^v

[Blank folio]   79 127^r

[Blank folio]   80 127^v

Of God   81 128^r

Were men and beasts made by fortuitous jumblings of
atoms there would be many useless parts in them, here a
lump of flesh, there a member too much. Some kinds of
beasts might have had but one eye, some more than two,
and others two eyes.

Of light   see page 32[21]   82 128^v

candle in the glass appear and disappear.

Of the creation   83 129^r

The word בָּרָא , which at Genesis, Chapter 1, verse 1, is
interpreted to create something out of nothing, is used at
Genesis, Chapter 1, verse 21, whence it is said God created
great whales. But the matter out of which they were created
did exist before. Neither is it meant of creating the soul or
form of the whale, for that is not the whale alone, and there
may be but one kind of irrational soul which joined with
several kinds of bodies makes several kinds of beasts, for
setting aside the different shapes of their bodies beasts differ from one another but in some qualities which are called
instincts of nature. Now, as in men whose souls are of one
kind, some love, hate, fear, etc., one thing, some another.

[21] Page 381

thing some another & few men are of ye same temper wch diversity arises from ~~(?)~~ theire bodys (for all theire soules are alike) so why may not ye severall tempers or instincts of divers kinds of beasts arise from ye different tempers & modes of theire bodys they differing one from another more y^n one mans body from anothers. To suppose then that God did create divers kind of soules for divers kinds of beasts is to suppose God did more than He needed.[141] How y^n can ye soule of ye whale be called ye whalle since before it be joyned wth ye whale tis as much ye soule of a horse & this creating y^n of whales & severall other creatures must be noe but modifying matter into ye body of a whale & infusi an irrationall soule into it. Eccles: 33 vrs 10 Adam was created of ye Earth.[142]

Whither Moses his saying Gen ye 1st y^t ye eveing & ye morning were ye first day &c do prove y^t God created time. Coll 1. 16. or heb 1 ch 2v τὸς αἰῶνας ἐποίησεν expoūded, he made ye worlds.[143] prove y^t God created time.

84 129^v   [Blank folio]

85 130^r   ## Of ye soule.

Were ye soule nothing but modified matter & did memory consist in action (for it can thus consist in nothing else) wee could never call things into o^r memory for so long as y^t action continews we must thinke of & remember y^t phantasme & when y^t action ceaseth & not before y^n wee may cease to thinke of & remember y^t p⌊h⌋antasme but how shall we call this thing into memory ye action being done & we haveing no principle wth in us to begin such a motion againe w^thin us, &c.

If sence consisted in reaction we should perceive things double or we should never se any thing before us but ther would be some apparition behind us for let this perceiving

---

[141] This sentence appears in the manuscript written in Sheltonian shorthand. See plate 3 opposite page 317. See also R. S. Westfall, "Short-writing and the State of Newton's Conscience 1662," *Notes and Records of the Royal Society* xviii. 1:10–16.

[142] There is no Eccles. ch 33 verse 10. For meaning, 3.20 and 12.7 are closest, though neither explicitly mentions Adam.

[143] From Ἡ Καινὴ Διαθήκη, *Novum Testamentum* (London, 1653).

Few men are of the same temper, which diversity arises from their bodies, for all their souls are alike; so why may not the several tempers or instincts of diverse kinds of beasts arise from the different tempers and modes of their bodies, they differing one from another more than one man's body from another's. To suppose then that God did create diverse kinds of souls for diverse kinds of beasts is to suppose God did more than he needed. How then can the soul of the whale be called the whale, since before it is joined with the whale it is as much the soul of a horse. This creating, then, of whales and several other creatures, must be nothing but modifying matter into the body of a whale and infusing an irrational soul into it. Ecclesiastes, Chapter 33, verse 10: Adam was created of the earth.

Whether Moses saying that the evening and the morning were the first day, Genesis, Chapter 1, proves that God created time. As expanded at Colossians, Chapter 1, verse 16, or Hebrews, Chapter 1, verse 2, τοὺς αἰῶνας ἐποίησεν, he made the worlds, proves that God created time.

[Blank folio]   84 129ᵛ

Of the soul   85 130ʳ

Were the soul nothing but modified matter and did memory consist in action (for it can thus consist in nothing else) we could never call things into our memory, for so long as that action continues we must think of and remember that phantasm, and when that action ceases, and not before then, we may cease to think of and remember that phantasm; but how shall we call this thing into memory the action being done and we having no principle within us to begin such a motion again within us.

If sense consisted in reaction we should perceive things double or we should never see anything before us, but there would be some apparition behind us; for let this perceiving body be what it will supose ye conarion it cannot be so much pressed on one side ⌐by y^e spirits⌐ but it will press upon y^e spirits on y^e other side & consequently they will presse upon it &c.

Hobbs. part 4th chap 1st. Motion is never y^e weake for y^e object being taken away for y^n dreames would not be so cleare as sence. but to men wakeing things past appeare ob= then things present because y^e organs being moved by other p^{r}sent objects ~~y^t arg~~ at y^e same time ~~motion~~ ⌐those phantasmes.⌐ are lesse predominant. &c.[144]

Resp: Then we should never forget any thing. 2 Phantasmes are prædominant from y^e strength of the motion causing y^m if y^n ~~there be no other cause of sence y^n motion~~ but if ye motion causing present & past phantasmes be alike strong ye effect must be ~~ye same~~ equall & so there would be no differences betwix sence & phantasie. all things wch wee ever perceived would be alike in our phantasie & wee should thinke of ~~almost~~ an immense multitude of objects at once. &c.

Of ye soule

Memory is a faculty of ye soule (in some measure) for ⌐else⌐ how can divers sounds, or words excite her to divers ~~unles by memory she app~~ ⌐thoughts⌐ or 3 4 5 or more words beget ye same thought in her. Perhaps shee remembers by ye helpe of characters in ye Braine, but y^n how doth shee remember ye signification of those characters.

Quær 1 Why objects appeare ⌐not⌐ inverst, Resp: The mind or soule cannot judge ye image in ye Braine to be inversed unlesse shee perceived externall things ~~immediatly~~ wth wch shee might compare y^t Image.

2. Why doe appeare to bee wthout our body? ~~but paines, hunger, thirst &c wth~~ Resp: Because ~~y^c~~ in ye image of things delineated in the braine by sight, ye bodys image is placed in ye midst of ye images of other things, is moved at o^r command towars & from those other images. &c:

---

[144] Thomas Hobbes, *Elements of Philosophy. The First Section, Concerning Body* (London, 1656), pp. 295–6. It is in fact Chapter XXV, but this is the first chapter of Part IV.

body be what it will, for instance suppose the conarion, it cannot be so much pressed on one side by the spirits, but it will press upon the spirits on the other side and consequently they will press upon it.

Hobbes, Part 4th, Chapter 1st. Motion is never the weaker for the object being taken away, for then dreams would not be so clear as sense. But to waking men things past appear obscurer than things present, because the organs being moved by other present objects at the same time, those phantasms are less predominant.

Response: (1) Then we should never forget anything. (2) Phantasms are predominantly from the strength of the motion causing them, but if the motion causing present and past phantasms be alike strong, the effect must be equal, and so there would be no difference between sense and fantasy. All things which we ever perceived would be alike in our fantasy and we should think of an immense multitude of objects at once.

Of the soul

86 130ᵛ

Memory is a faculty of the soul in some measure, for else how can diverse sounds or words excite her to diverse thoughts, or 3, or 4, or 5, or more words beget the same thought in her. Perhaps she remembers by the help of characters in the brain, but then how does she remember the signification of those characters.

Query 1. Why objects appear not inverted? Response: The mind or soul cannot judge the image in the brain to be inverted, unless she perceived external things immediately with which she might compare their image.

2. Why objects appear to be outside our body? but pains, hunger, thirst, etc. with Response: Because in the image of things delineated in the brain by sight, the body's image is placed in the midst of the images of other things and is moved at our command toward and from those other images, etc.

3. But why are not these objects then judged to bee in the braine. Resp: Because y^c image of y^c braine is not painted there, nor is y^c Braine perceived by y^c soule it not being in motion, & probably y^c soule perceives noe bodys but by y^c helpe of their motion. But were y^c Braine perceived together wth those images in it wee should thinke wee saw a body like the braine encompasing ⌊& comprehending our selves⌋ y^c starrs & all other visible objects. &c.

87 131ʳ   Of Quantity

If Extension is indefinite onely ⌊in greatness⌋ & not infinite y^n a point is but indefininitely little & yet we cannot comprehend any thing lesse. To say y^t extension is but indefinite (I meane all y^c extension wch exists & not soe much onely as we can fasy) because we cannot perceive its limits, is as much as to say God~~s perfection & oʳˢ differ but (?) indefinitely because wee~~ is but indefinitely perfect because wee canot apprehend his whole perfection.[145]

88 131ᵛ   [Blank folio]

89 132ʳ   Of Sleepe & Dreams &c

How is it y^t y^c Soule so often remembers her dreames by chanch otherwise not knowing shee had dreamed, & thence whither she be perpetually employed in sleepe. whither dreames are of y^c body or soule. Why are they patched up of many fragments & incoherent passages.

90 132ᵛ   [Blank folio]

91 133ʳ   Of colours   vide pag 69

17 Substances belonging to y^c vegetable or Animall kingdome when lightly burned are black, when throughly burned are white. As Ivory being skilfully burnt affords painters one of y^c deepest blacks they have &c. But minerals are to bee excepted from this rule, For Allablaster if

---

[145] Comment on Descartes, *Principia*, Part II, Article XXI, pp. 31–2.

3. But why are these objects not then judged to be in the brain? Response: Because the image of the brain is not painted there, nor is the brain perceived by the soul, it not being in motion, and probably the soul perceives no bodies but by the help of their motion. But were the brain perceived together with those images in it we should think we saw a body like the brain encompassing and comprehending, ourselves, the stars, and all other visible objects.

Of quantity 87 131ʳ

If extension is only indefinite in greatness and not infinite, then a point is but indefinitely little; and yet we cannot comprehend anything less. To say that extension is but indefinite (I mean all the extension which exists and not so much only as we can fancy) because we cannot perceive its limits, is as much as to say, God is but indefinitely perfect because we cannot apprehend his whole perfection.

[Blank folio] 88 131ᵛ

Of sleep and dreams 89 132ʳ

How is it that the soul so often remembers her dreams by chance not otherwise knowing she had dreamed, and thence whether in sleep she be perpetually employed. Whether dreams are of the body or soul. Why are dreams patched up of many fragments and incoherent passages.

[Blank folio] 90 132ᵛ

Of colors *vide* page 69[22] 91 133ʳ

17. Substances belonging to the vegetable or animal kingdom when slightly burned are black, when thoroughly burned are white. As ivory being skillfully burned affords painters one of the deepest blacks they have. But minerals are to be excepted from this rule, for alabaster if never so

[22] Page 431

never so much burnt will turne no darker than yellow. Leade being calcined ⌐wth a strong fire⌐ turnes into minium which is red, & this miniū by burning turnes darker but never to a white colour. Blew, but unsophisticated Vitriol when tis burnt a little by a slow heate to friability, is white being further burnt turnes Grey, Yellow, red, & when perustum it turnes to a purple.[146]

18   Take Rectified oyle of Vitriol mixt by degrees wth a convenient quantity of y^c Essentiall oyle of wormewodd (wch was drawke over wth store of water in a limbec) & warily distill y^c mixture in a retort, there will bee left behind a greate quantity of dry & very black matte.

Or becaus y^c Essentiall oyle of Winter-Savory is cleare y^n y^t of wormewodd mix it by degrees wth about an equall quantity of oyle of Vitrioll these two cleare liquors distilled as before leave a good quantity of black matter.[147]

19   Gold & silver melted into a lumpe & dissolved by Aqua fortis y^c pouder of gold falling to y^c bottome appeares not yellow but black though neither y^c gold silver nor Aquafortis be so, & silver rubbed on other bodys colours y^m black.[148]

20   Most bodys precipitated from y^c liquor into wch they were dissolved are white, but not all.[149]

21   The scrapeings of black horne lookes white.[150]

22   Sulphur adust is not y^c cause of blacknesse as Chimists hold, for common sulphur be either melted or sublimed turnes onely red or yellow. And ⌐y^c plant⌐ Camphire though very inflamable & consequently sulphureous by burning turnes to noe colour but white &c. But y^n w^t causeth blackness in sulphur adust.[151]

23.   A Candle looked on through blew glasse appeares greene.[152]

24.   Pouder of blew bise mixed wth a greater quantity of yellow orpiment makes a greene but y^c particles by a microscope are discovered to retaine their blewnes & yellownesse.[153]

[146] Boyle, *Colours*, pp. 138–41.   [147] *Ibid.*, pp. 144–5.
[148] *Ibid.*, pp. 148–51.   [149] *Ibid.*, p. 171.   [150] *Ibid.*, p. 175.
[151] *Ibid.*, pp. 177–9.   [152] *Ibid.*, p. 197.
[153] *Ibid.*, p. 233 and pp. 238–9.

much burned will turn no darker than yellow. Lead being calcinated with a strong fire turns into minium which is red, and this minium by burning turns darker, but never to a white color. Blue, but unsophisticated, vitriol when it is burned a little by a slow heat to friability is white. Being further burned it turns grey, yellow, red, and when it turns *perustum* it is purple.

18. Take rectified oil of vitriol mixed by degrees with a convenient quantity of the essential oil of wormwood (which was drawk over with a store of water in an alembic) and carefully distill the mixture in a retort. There will be left behind a great quantity of dry and very black matter.

Or, because the essential oil of winter-savory is clearer than that of wormwood, mix it by degrees with about an equal quantity of oil of vitriol. These two clear liquids distilled as before leave a good quantity of black matter.

19. Gold and silver melted into a lump and dissolved by aqua-fortis, the powder of gold falling to the bottom appears not yellow but black, though neither the gold, silver, nor aqua-fortis be so. Silver rubbed on other bodies colors them black.

20. Most bodies precipitated from the liquid into which they were dissolved are white, but not all.

21. The scrapings of black horn look white.

22. Sulfur adust is not the cause of blackness as chemists hold, for common sulfur being either melted or sublimated turns only red or yellow. The camphor plant though very inflammable, and consequently sulfurous, by burning turns to no color but white. But then what causes blackness in sulfur adust?

23. A candle looked at through blue glass appears green.

24. Powder of bice mixed with a greater quantity of yellow orpiment makes a green, but the particles by a microscope are discovered to retain their blueness and yellowness.

25   The steame of clear Aq: Fortis, or spirit of niter is red, &c.[154]

26.   A feather or black ribband put twixt my eye & yᶜ setting sunne makes glorious colours.[155]

27   An Acid Spirit ⌐& juices & salts⌐ (as Spirit of salt, or of Vinegar, or ⌐of⌐ Vitriol, or Lemmon juice, or Oyle of Vitrioll, or Aqua fortis   ) being droped into diversly coloured liquors & especially blew ones ~~turne yᵐ to red ones~~ (as syrrup of violets impregnated wᵗʰ yᶜ tincture of yᶜ flowers, ⌐juice of⌐ blew bottles, or coneweede, juice of ripe privet berrys,   ) it turnes them commonly to a red colour. But ⌐Sulphureous salts which are either Urinous & volatile salts of Animall⌐ ~~Volatile Salts~~ ⌐substances⌐ (as Spirit of Hartshorne, of Urine, of blood, of Sal=Armoniack   ) or ⌐Lixiviate⌐ Unctuous or Alcalizate ⌐& fixed⌐ Salts ⌐made by incineration⌐ (as yᶜ solution of Salt of Tartar, of pot ashes, of common wood ashes, of limewater, Oyle of Tartar,   ) doe change yᵐ to a Greene. as yᶜ red juice of bucthorne berrys.[156]

28   Yet ⌐either⌐ a Lixivious liquor, or urinous salt being poured on a solution of blew vitrioll in faire water makes it yellow & yᶜ precipitated corpuscles retained yᶜ ⌐yellow⌐ ~~29~~ colour when they were falne to yᶜ bottome.[157]

29   A just quantity of Oyle of Tartar poured into a strong solution of french verdigrease turnes it from greene to blew; a Lixivium of pot ashes turnes it to a lighter blew, & spirit of Urin, or Harts-horne, make other blews.[158]

30   One graine of Cochineel dissolved in spirits of urin & yⁿ by degrees in faire water, imparted a discernable colour to 125000 graines of faire water.[159]

31   Most of yᶜ Tinctures wᶜʰ chimists draw ⌐wᶜʰ abound wᵗʰ minerall or Vegetable Sulphur⌐ turne red; & ⌐both⌐ Acid & Alcalizate salts in most sulphureous or oyly bodys produce a red. & blew is more commonly turned to red yⁿ red to blew.[160]

---

[154] *Ibid.*, p. 230.   [155] *Ibid.*, p. 245.   [156] *Ibid.*, pp. 245–8.
[157] *Ibid.*, p. 251.   [158] *Ibid.*, pp. 252–3.   [159] *Ibid.*, pp. 256–7.
[160] *Ibid.*, pp. 275–7.

25. The steam of clear aqua-fortis, or spirit of niter, is red.

92 133ᵛ

26. A feather or black ribbon put between my eye and the setting Sun makes glorious colors.

27. An acid spirit, juices, and salts (as spirit of salt, of vinegar, of vitriol, lemon juice, oil of vitriol, or aqua-fortis) being dropped into diversely colored liquids, and especially blue ones (as syrup of violets impregnated with the tincture of the flowers, juice of blue bottles, cone weed, juice of ripe privet berries) turns them commonly to a red color. But sulfurous salts which are either urinous and volatile salts of animal substances (as spirit of hartshorn, urine, blood, sal ammoniac) or lixiviate, unctuous, or alcalizate and fixed salts made by incineration (as the solution of salt of tartar, potashes, common wood ashes, limewater, oil of tartar) do change them to a green, as the red juice of buckthorn berries.

28. Yet, either a lixivious liquid, or urinous salt being poured on a solution of blue vitriol in fair water makes it yellow, and the precipitated corpuscles retained the yellow color when they had fallen to the bottom.

29. A just quantity of oil of tartar poured into a strong solution of French verdigris turns it from green to blue; a lixivium of potashes turns it to a lighter blue, and spirit of urine, and hartshorn, make other blues.

30. One grain of cochineal dissolved in spirits of urine and then by degree in fair water, imparted a discernible color to 125,000 grains of fair water.

31. Most of the tinctures which abound with mineral or vegetable sulfur which chemists draw turn red. Both acid and alcalizate salts in most sulfurous or oily bodies produce a red. Blue is more commonly turned to red than red to blue.

32 Fully satiate good common sublimate wth water, filter it through paper y^t it be cleare, put a Spoonefull of it into a cleare glasse drop in 3 or 4 drops of oyle of tartar well filtred & it will be of an orange colour. but 4 or 5 drops of oyle of Vitriol dispersed about ye glasse by shaking it, makes ye liquor pellucid againe.[161]

93 134^r 32 Some tinctures (as y^t of Amber made wth Sirits of wine) appeare red or yellow as ye vessells they fill are slender or broade but cochineel dissolved as before, & others liquors never looke otherwise y^n Red ⌞or of a carnation⌟ &c.[162]

33 White bodys are comonly sulphureous.

34 Oyle of Tartar generally precipitates Metalline bodys corroded wth acid salts.[163]

35 Tinge water wth ~~faire~~ red rose leaves into wch drop a little Minium disolved in spirit of vinegar & it will be of a muddy greene, but drop ⌞in⌟ a little Oyle of Vitrioll wch though an acid Menstruum yet it will præcipitate ye leade in ye forme of a white pouder to ye bottome leaving ye rest of ye liquor above of a good red almost like a Rubie.[164]

36 Oyle or spirit of Turpentine will not mix wth water & it & water shaken together apeare white.[165]

37 Some very corrosive ⌞& acid⌟ liquors will præcipite some others, as oyle of Vitrioll præcipitates ⌞divers⌟ bodys dissolved in Aqua fortis or ⌞spirit of⌟ wine vinegar, wch precipitated bodys are usually ⌞very⌟ white.[166]

38 Bodys will scarcely be precipitated by Alcalizate Salts y^t are not first dissolved in acid Menstruum.[167]

39 A leafe of Gold held betwixt y^e eye & y^e light appeares blew.[168]

40 Acid salts rather dilute yellow & white juice y^n turne $\overline{y}$ to red,[169]

41 Gentle heates in chimicall operations rather produce rednesse y^n other colours in digested menstruums not onely sulphureous (as spirit of wine) vid: sec: 31. but saline as Spirit of Vinegar.[170]

[161] *Ibid.*, pp. 303–4. [162] *Ibid.*, p. 278. [163] *Ibid.*, p. 306.
[164] *Ibid.*, pp. 381–2. [165] *Ibid.*, p. 110. [166] *Ibid.*, pp. 168–9.
[167] *Ibid.*, pp. 136–7. [168] *Ibid.*, pp. 198–9. [169] *Ibid.*, pp. 263–4.
[170] *Ibid.*, p. 287.

32. Fully satiate good common sublimate with water. Filter it through paper so that it be clear. Put a spoonful of it into a clear glass and put in three or four drops of oil of tartar well filtered and it will be of an orange color. But four or five drops of oil of vitriol dispersed about the glass by shaking it makes the liquid pellucid again.

32. Some tinctures, as that of amber made with spirits of wine, appear red or yellow as the vessels they fill are slender or broad, but cochineal dissolved as before, and other liquids never look otherwise than red or of a carnation. 93 134ʳ

33. White bodies are commonly sulfurous.

34. Oil of tartar generally precipitates metalline bodies corroded with acid salts.

35. Tinge water with red rose leaves, into which drop a little minium dissolved in spirit of vinegar, and it will be of a muddy green, but drop in a little oil of vitriol which, though an acid menstruum, will precipitate the lead in the form of a white powder to the bottom, leaving the rest of the liquid above of a good red, almost like a ruby.

36. Oil of turpentine will not mix with water, and it and water shaken together appear white.

37. Some very corrosive and acid liquids will precipitate some others, as oil of vitriol precipitates diverse bodies dissolved in aqua-fortis or spirit of wine vinegar, which bodies precipitated are usually very white.

38. Bodies will scarcely be precipitated by alcalizate salts that are not first dissolved in acid menstruum.

39. A leaf of gold held between the eye and the light appears blue.

40. Acid salts rather dilute yellow and white juices rather than turn them to red.

41. Gentle heating in chemical operations produces redness in digested menstruums rather than other colors, not only with sulfurous substances, as spirit of wine (*vide* section 31),[23] but also with saline substances, as spirit of vinegar.

[23] Page 457

42   Alcalizate salts are wont to precipitate w^t acid salts dissolve.[171]

43   An Acid salt doth seldome restore a Vegitable substance to ye colour of wch an Alcalizate deprived it.[172]

44   The acidity of spirit of vinegar is destroyed by working on Minium, (or perhaps on crabs claws).[173]

45   Put some solution of Minium into a spoonefull of ye fresh tincture of Logwood to turne it ⌐deeply⌐ purple.

94 134^v precipitate ½ y^e leade wth spirit of sal-armonick wch precipitaco̅ ⌐as well as y^e supernatant liquor⌐ looks purple (by reason by reason of y^e predominance of y^e tinged particles over y^e white), y^n power in some spirit of salt warily & y^e precip: by Sal=Ar: is of a violet colour at y^e botome, y^e precip: by spirit of salt is white ⌐& carnacon⌐ in y^e midst. & y^e top yellow or red. vide sec 35.[174]

46   That y^e colour of a body bee altered by a ⌐cleare⌐ liquor tis commonly required y^t it have salt yet faire water powred on y^e grey & friable calx of powdred Vitriol melted by a gentle heate till it colour change & ye liquor being set by in a close violl for some days it will tune to a vitriolate colour.[175]

47   take Lignum Nephriticum (ye infusion of wch in faire water is good against y^e ~~kidne~~ stone of ye kidneys.) put a handfull of thin slices of it into 3 or 4 pounds pure spring water after it hath infused there a night put ye water into a cleare violl, & if you see ye light through it it appeares of a golden colour (excepting ⌐sometimes⌐ a sky coloured circle at ye top), but if ye infusion was too strong ye liquor will then appeare darke & reddish. But if ~~ye liq~~ your eye is twixt ye liquor & light it appeares ceruleous: &c Acid salts destroy ye blew colour & sulphureous saltes restore it againe, wthout making any change in ye golden colour. Which may bee usefull to ye finding whither bodys abound more wth acid or sulphureous Salts.[176]

48   The same may done by ~~Spirit~~ Sirrup of Violets (?) impregnated wth ye tincture of ye flowers. For an acid salt

[171] *Ibid.*, p. 373.   [172] *Ibid.*, p. 377.   [173] *Ibid.*, p. 380.

[174] *Ibid.*, pp. 384–5.   [175] *Ibid.*, p. 327.   [176] *Ibid.*, pp. 199–205.

42. Alcalizate salts are wont to precipitate what acid salts dissolve.

43. An acid salt does seldom restore a vegetable substance to the color of which an alcalizate deprived it.

44. The acidity of spirit of vinegar is destroyed by working on minium, or perhaps on crab's claws.

45. Put some solution of minium into a spoonful of the fresh tincture of logwood to turn it deeply purple.

Precipitate one half the lead with spirit of sal ammoniac, which precipitate, as well as the supernatant liquid looks purple (by reason of the predominance of the tinged particles over the white), then carefully pour in some spirit of salt, and the precipitate by sal ammoniac is of a violet color at the bottom, and the precipitate by spirit of salt is white and carnation in the midst, and at the top yellow or red. *vide* section 35.[24]   94 134ᵛ

46. That the color of a body be altered by a clear liquid it is commonly required that it have salt. Yet fair water poured on the gray and friable calx of powdered vitriol (which was melted by a gentle heat till its color changed), and the liquid being set by in a closed vial for some days it will turn to a vitriolate color.

47. Take lignum nephriticum (the infusion of which in fair water is good against kidney stones), and put a handful of thin slices of it into three or four pounds of pure spring water. After it has infused there for a night put the water into a clear vial, and if you see the light through it, it appears of a golden color (excepting sometimes a sky colored circle at the top); but if the infusion was too strong the liquid will then appear dark and reddish. But if your eye is between the liquid and light it appears ceruleous. Acid salts destroy the blue color, and sulfurous salts restore it again without making any change in the golden color. Which may be useful for finding whether bodies abound more with acid or sulfurous salts.

48. The same may be done by syrup of violets impregnated with the tincture of the flowers. For an acid salt turns

[24] Page 459

turnes it from blew to red but sulphureous one from blew to greene. vide sec 27.[177]

49 Haveing found y^t a sulphureous salt is predominant in a body it may be knowne whither y^t salt be Urinous (i.e. volatile salts of animal

95 135^r or other substances wch are contrary to acid ones) or Alcalizate by this experiment. Into ye liquor described in ye 32^d sec. that is into sublimate disolved in faire water put an Alcaly & it turnes it to an Orange Tauny, but urinous (or salsuginous) salts turne it to a pure white.

50 ye same liquor (viz oyle of Vitriol) powred into a lixivium in wch crude antimony has beene newly boyled turnes it from a cleare to ₐaₐ yellow colour, wch sec: 32 turned a yellow to a cleare one.[178]

51 There be flat peices of ₐaₐ certaine kind of glase wch exhibits ye Phenominon of lignum Nephriticū.[179]

83^r An idea as it respects ye object wthout ye mind is but a bare denomination or a meere nothing, but as it respects ye mind, or as objectively in it, (i.e. in it after y^t manner wch objects use to bee in it) tis a reall entity, viz a mode of ye intellect As ye impression is a denominācō to ye seale, but a mode to ye wax. And as y^t impressiō by how much ye more artificiall, so much ye more artificiall must bee ye seale or y^t wch is equivalent to it, so an Idea by how much ye perfecter so much ye perfecter must its cause be, whither it be y^e object, or something y^t eminently conteines it, or another Idea.

pag 53, 71, 72, 73.[180]

A necessary being is y^e cause of it selfe or its existence after y^e same manner y^t a mountaine is y^e cause of a valley or a △ y^e cause y^t its 3 angles are = to 2 right ones (wch is not from power or excellency, but y^e peculiarity of theire

[177] *Ibid.*, comments on sec. 27. pp. 245ff. on the use of color change as a test of acidity.

[178] Entries 49 and 50 are apparently comments on and conclusions drawn from previous entries.   [179] Boyle, *Colours,* pp. 216–17.

[180] Descartes, *Opera philosophica, Responsio ad Primos Objectiones* and *Responsio ad Secundas Objectiones;* page 53 of the former and pages 71–3 of the latter are the sources of this passage.

it from blue to red, but sulfurous ones from blue to green. *Vide* section 27.[25]

49. Having found that a sulfurous salt is predominant in a body, it may be known whether that salt be urinous (i.e., volatile salts of animal

or other substances which are contrary to acid ones) or al- 95 135ʳ calizate, by this experiment: into the liquid described in the 32nd section,[26] that is, into a sublimate dissolved in fair water, put an alkali, and it turns it to an orange tawny; but urinous (or salsuginous) salts turn it to a pure white.

50. The same liquid (viz., oil of vitriol, which in section 32^{27} turned a yellow liquid to a clear one) poured into a lixivium in which crude antimony has been newly boiled turns it from a clear to a yellow color.

51. There are flat pieces of a certain kind of glass which exhibit the phenomenon of lignum nephriticum.

An idea with respect to the object outside the mind is but a 83ʳ bare denomination or a mere nothing, but with respect to the mind, or as the idea is objectively in it (i.e., in it after that manner which objects use to be in it), it is a real entity, namely a mode of the intellect; as the impression is a denomination to the seal, but a mode to the wax. And as that impression by how much the more artificial, so much the more artificial must be the seal or that which is equivalent to it, so an idea by how much the more perfect so much the more perfect must its cause be, whether it be the object, or something that eminently contains it, or another idea. Pages 53, 71, 72, and 73.

A necessary being is the cause of itself or its existence after the same manner that a mountain is the cause of a valley, or a triangle the cause that its three angles are equal to two right ones (which is not from power or excellency,

---

[25] Page 457   [26] Page 459   [27] *Ibid.*

natures) pag 55, 56, 57, 58. 127, 129, 130, 131, 132, 133, 134.[181]

Of ye Idea of ⌐an⌐ infinite thing how clearly & distinctly conceivable &c pag 59.[182]

The argument drawne from necessary existence included in Gods essence explained. pag 60, 61, 62. 80.[183]

Of ye distinction twixt ye mind & ye body. pag: 63, 69, 70. 121, 122, 123, 124, 125.[184]

83^v Ax: ~~That thing~~ Tis a contradiction to say, that thing doth not exist, ~~wch may bee conceived~~ whose existence implys noe contradiction, & being supposed to exist must necessarily exist. The reason is y^t an immediate cause & effect must bee in ye same time & there fore ye præexistence of a thing ~~must~~ ⌐can⌐ bee no cause of its post existence (as also because y^e ~~former~~ ⌐after⌐ time depends not on ye former time). Tis onely from the essence of it that a thing can ~~by it owne~~ perpetuate its existence wthout extrinsecall helpe. W^{ch} essence being sufficient to continue it must bee sufficient to cause it there being ye like reason of boath.[185]

---

[181] These are references to Descartes's replies to the objections made regarding necessary being. Pages 55–8 refer to *Responsio ad Primos Objectiones*, and pages 127 and 129–34 refer to *Responsiones Quartae*, the section entitled "De Deo."

[182] This entry was made somewhat later than the preceding one. The reference to page 59 is to the *Responsio ad Primos Objectiones* of Descartes.

[183] Pages 60–2 of Descartes, *Responsio ad Primos Objectiones;* page 80 is in Descartes, *Responsio ad Secundas Objectiones.*

[184] Page 63 refers to the *Responsio ad Primos Objectiones;* pages 69 and 70 refer to *Responsio ad Secundas Objectiones;* pages 121–5 are all to the *Responsiones Quartae,* the section entitled "Responsio ad primam partem, de natura mentis humanae."

[185] Newton's gloss on the principles of the ontological argument refer to Descartes's formulation of the argument in the fifth *Meditation.*

but the peculiarity of their natures). Pages 55, 56, 57, 58, 127, 129, 130, 131, 132, 133, and 134.

Of the idea of an infinite thing how clearly and distinctly conceivable, etc. Page 59.

The argument drawn from necessary existence included in God's essence explained. Pages 60, 61, 62, and 80.

Of the distinction between the mind and the body. Pages 63, 69, 70, 121, 122, 123, 124, and 125.

Axiom: It is a contradiction to say, that thing does not exist   83ᵛ
whose existence implies no contradiction, and being supposed to exist must necessarily exist. The reason is that an immediate cause and effect must be in the same time and therefore the preexistence of a thing can be no cause of its past existence (also because the after time does not depend on the former time). It is only from the essence of it that a thing can perpetuate its existence without extrinsic help, which essence, being sufficient to continue it, must be sufficient to cause it, being the same reason of both.

APPENDIX

ULC. ADD. 3975[1]
FOLIOS 1–22

Of Colours

1   The rays reflected from Leafe Gold are Yellow but those transmitted are blew, as appeares by holding a leafe of Gold twixt yor eye & a Candle.[2]

2   Lignum Nephriticum sliced & about a handfull infused in 3 or 4 pints of faire water for a night ye liquor (looked on in a cleare violl) reflects blew rays & transmits yellow ones. And if ye liquor being too much impregnated appeares (w^n looked through) of a darker red it may bee diluted wth faire water till it appeare of a Gold in Colour.[3]

3   The flat peices of some kinds of Glase will exhibit ye same Phænomena wth Lignum Nephriticum.[4] And these Phænomena of Gold & Lignum Nephriticum are represented by ye Prisme in ye 37th experiment as also in ye 22^d & 24th Experiment.

4   But Generally bodys wch appeare of any colour to ye eye, appeare of ye same colour in all positions; Nay Gold if it bee not soe very thin as to bee transparent appeares onely yellow & perhaps ye yellow colour of Lignum Ne-

---

[1] Add. 3975, folios 1–22 consists almost entirely of entries made in 1665–6. It exhibits the emergence of Newton's views about the nature of colors as they developed from the *Questiones*. Here we see Newton taking various notes from the *Questiones* and marshaling them as evidence for a new position and as indicators of new directions for research. We have not attempted a full analysis of Add. 3975, folios 1–22, but have treated it in its relation to the *Questiones*. The reader is referred to those sections of the Commentary on physiology of perception and optics for a futher discussion of Add. 3975.

[2] See folio 93 134^r, item 39, of the *Questiones*, p. 458.

[3] See folio 94 134^v, item 47, of the *Questiones*, p. 460.

[4] See folio 95 135^r, item 51, of the *Questiones*, p. 462.

phriticum would vanish if ye tincture bee strong & y^e
liquor of a greate thicknesse. And perhaps there are
many coloured bodys wch if made so thin as to bee transparent would appeare of one colour when looked upon
& of another when looked through. Perhaps Motes in y^e
Sun doe so, for they appeare coloured.[5] And

5   The tincture of Lignum Nephriticum may bee deprived of its blew colour wthout any alteration made in y^e
Yellow by putting a little of any acid salt into it (as spirit
of Salt, of vinegar, of Vitrioll, Lemon juice, oyle of Vitrioll, Aqua fortis &c). And Sulphureous Salts (whither
Urinous (i.e. Volatile salts of Animal substances) as spirit
of hartshorne of Urin, of blood, of Sal Armoniack; Or
Lixiviate Unctuous Alcalizate & fixed salts made by incineration as ye Solution of Salt of Tartar of potashes, of
common wood ashes, of lime water, Oyle of Tarter &c)
doe restore ye blew colour wthout making any change in
the yellow.[6]

Of Colours.   2

Experiments wth y^e Prisme

6   On a black peice of paper
I drew a line opq, whereof
one half op was a good blew
y^e other pq a good deepe
red (chosen by Prob of Colours). And looking on it
through y^e Prisme adf, it ap-

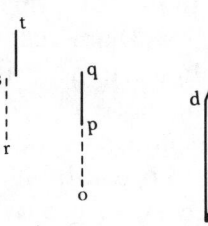
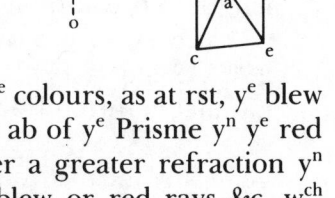

peared broken in two betwixt y^e colours, as at rst, y^e blew
parte rs being nearer y^e vertex ab of y^e Prisme y^n y^e red
parts st. Soe y^t blew rays suffer a greater refraction y^n
red ones. Note [I[7] call those blew or red rays &c, w^{ch}
make ye Phantome of such colours.

The same Experiment may bee tryed wth a thred of
two colours held against ye darke.[8]

[5] See folio 74 124^v, item 14, of the *Questiones*, p. 442.
[6] See folio 94 134^v, item 47, of the *Questiones*, p. 460.
[7] All square brackets in Add. 3975, folios 1–22, are Newton's own.
[8] See folio 70 122^v, item 3, of the *Questiones*, p. 434.

7   Taking a Pris-
me, (whose angle
fbd was about
60^{dr} into a darke
roome into wch y^e
sun shone only at
one little round
hole k. And lay-
ing it close to y^e
hole k in such

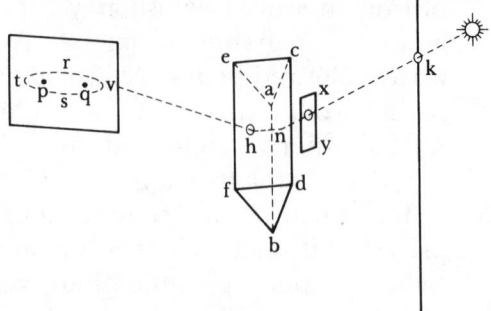

manner y^t y^e rays, being equally refracted at (n & h)
their going in & out of it, cast colours rstv on y^e opposite
wall. The colours should have beene in a round circle
were all y^e rays alike refracted, but their forme was
oblong terminated at theire sides r & s wth streight lines;
theire bredth rs being $2\frac{1}{3}$ inches, theire length to about 7
or eight inches, & y^e centers of y^e red & blew, (q & p)
being distant about $2\frac{3}{4}$ or 3 inches. The distance of y^e
wall trsv from y^e Prisme being 260^{inches}.

8   Setting y^e Prisme in y^e midst twixt y^e hole k & y^e opposite wall, in y^e same posture, & laying a boarde xy betwixt y^e hole k & y^e Prisme close to y^e Prisme, in w^{ch}
board there was a small hole as big as the hole k (viz: $\frac{1}{8}$ of
an inch in Diameter) soe y^t y^e rays passing through both
those holes to y^e Prisme might all bee almost parallell
(wanting lesse y^n 7^{minutes}, whereas in y^e former experiment some rays were inclined 31^{min}). Then was the
length & breadth of y^e colours on y^e wall every way lesse
y^n halfe y^e former by about 2 inches viz rs = $\frac{3}{8}^{inch}$, tv =
$2\frac{3}{4}^{inch}$. & pq = $1\frac{1}{4}^{inch}$. Soe y^t y^e Red & blew rays wch were
parallel before refraction may bee esteemed to be

generally inclined one to another after refraction (some   3
more some lesse y^n) 34^{min}. And y^t some of them are
inclined more y^n a degree, in this case. And therefore if
theire sines of incidence (out of glass into air) be ye same,
theire sines of refraction will generally bee in ye proportion of 285 to 286 & for ye most extreamely red & blew
rays, they will bee as 130 to 131+, For by ye experiment
if their angle of incidence out of ye glasse into ye aire bee

30^d, The angle refraction of ye red rays being 48^{gr} 35′: ye angle refraction of ye blew rays will bee 48^{gr}, 52′. generally: but if ye rays bee extreamly red & blew ye angle of refraction of ye blew rays may bee more y^n 49^{gr}, 5′.

9   In ye 7th Experiment ye colours appeared in this order. but in ye 8th exp^{er}: where ye rays were more distinct & unmixed

10   Painting a good blew & red colour on a peice of paper neither of wch was much more luminous y^n ye other (for carrying y^m gadually into ye darke, both grew faint alike almost & disappeared together) if ye Prismaticall blew fell upon ye colours they both appeared perfectly blew but y^e red paint afforded much y^e fainter & darker blew, but if y^e Prismaticall red fell on y^e colours they both appeared perfectly red but y^e painted blew afforded much y^e fainter Red. The Prisme was ordered as in y^e 8th experiment. Note y^t y^e purer y^e $\left\{ \begin{array}{c} \text{Red} \\ \text{Blew} \end{array} \right\}$ is y^e lesse tis visible wth $\left\{ \begin{array}{c} \text{blew} \\ \text{Red} \end{array} \right\}$ rays.

Of Colours   4

11   If y^e plate abc-dsr bee painted wth any two colours & abcd bee y^e lighter colour, y^e partition edge

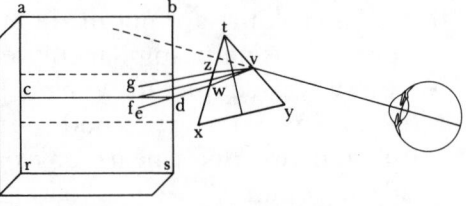

of y^e Colours, cd will appeare through y^e prisme txy of a redd colour, but if cres bee y^e lighter colour, their common edge cd will through a prisme looke blew.

12   And this will happen though y^e colours differ not in species but only in degrees, as if acdb bee black,& cdsr darkness or blacker y^n abdc ye edge dc will bee red & much more conspicuous y^n ye black, wch is strange.[9]

13   But if in a darke roome (as in Experim^{nt} 10) ye prismaticall blew or redd fall on a paper abdc ye edges of ye paper will not appeare otherwise coloured through

---

[9] See folios 69 122^r and 70 122^v of the *Questiones*, pp. 430–34.

another Prisme y^n to ye naked eye, viz: of ye same colour wth ye rest of ye paper. [For ye first Prisme perfectly seperats ye blew & red rays whereas I believe all ye colours proper to bodys are a little mixed.]

14   Prismaticall colours appeare in ye eye in a contrary order to y^t in wch they fall on ye paper.

15   If a foure square vessell abcd bee made wth
two parallell sides of well
pollished glasse AC BD,
& bee filled wth water;
And if ye sunns rays
passing into a darke
roome through y^e hole k
doe fall very obliquely
on y^e glasse sides of y^e
vessell y^e rays at their
egresse shall paint col-
ours on y^e paper EF on
wch they fall. [The blew
& red rays being seper-
ated by y^e first refraction.]

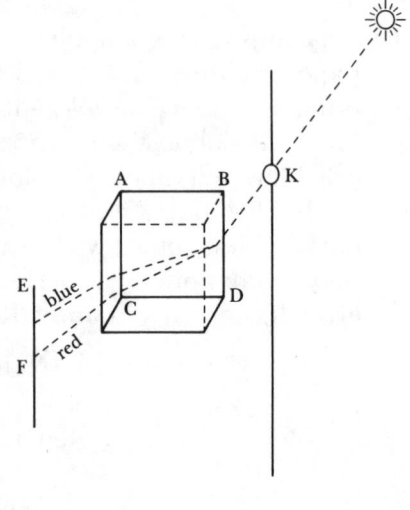

16   The colours are not made broader (as they would be were y^e prisme triangular) by removing y^e paper farther from y^e vessell. [becaus y^e blew & red rays become parallell againe after y^e second refraction] if the rays pass through two holes neare or close to ye vessell on either side ye colours.

Of Colours.   5

17   The window k being opened y^t ye Sun or other terminated light shine in freely, If I limited ye rays by an
opace body held twixt ye wall & ye vessel ye edge of y^t
bodys shaddow shaddow would not appeare coloured.
But if ye said body were on y^t side ye vessell towards y^e
sun its shaddow would be coloured on its edges.

18   But in ye Triangular Prisme whither ye said body bee
held on ye one side or on ye other the edges of its shaddow appeares coloured.

19   If you looke upon some uniformely luminous body (as on ye cleare sky or a sheet of white paper &c) through a triangular prisme & hold ye said opace body on ye further side of ye Prisme soe as to obscure parte of ye said luminous body: the farther ye said opace body is held from ye Prisme, ye more its edges will bee coloured, & ye nearer, ye lesse untill ye colours almost vanish when ye said body is held close close to ye Prisme.

20   But if instead of ye triangular Prisme you use ye said 4 square vessell ABDC, held obliquely y^t ye rays may bee much refracted in passing through it to yor eyes when ye opake body is placed as neare to ye vessell as you can distinctly see it, yor eye being close to ye vessell, ye edges of ye said body will appeare coloured, wch colours are diminished by removing ye body farther from ye vessell, & quite vanish when ye distance of ye said body is very greate. Thus ye Sun, by reason of his distance, appeares not coloured on his edges w^n looked on through ye said vessell, & yet in ye 15th Experiment hee trajects colours on a peice of paper.

21   The colours made by this vessel appeare immediatly to ye eye in ye same order in wch they fall on paper, but by y^e △Prisme y^t order is divers.

Note, That ye more ye glasse sides of ye vessell ABCD are distant, y^e better it is; y^t distance should not bee lesse y^n 6 or 8 inches to make y^e Phænomena con-spicuous. Some of y^e Phænomena may bee tryed by tying two Prismes thus together: But y^e distance of theire sides is two little to exhibit y^m.

Of Colours   6

22   If y^e sun S shine upon y^e Prisme def, some of his rays being transmitted through y^e base ef will make colours on y^e wall cb at b, others will bee re-flected to y^e wall at c mak-

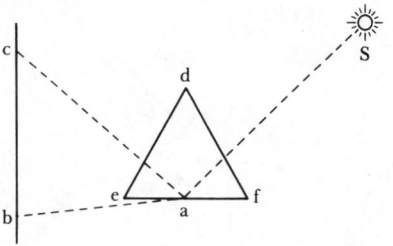

ing only a white w^{th}out colours; Now if y^e Prisme bee soe inclined as that y^e rays ab bee refracted more & more obliquly, y^e blew colour will at last vanish from b; soe y^t y^e red alone being refracted to b, y^e blew will bee reflected to c & make ye white coloure there to appeare a little blewish. But if ye Prisme bee yet more inclined, ye red colour at b will vanish too & being reflected to c will will make y^e blewish colour turne white againe.

23   If in y^e open aire you looke at y^e Image of y^e sky reflected from y^e bases of y^e Prism ef, holding yor eye O almost perpendicular to y^e basis

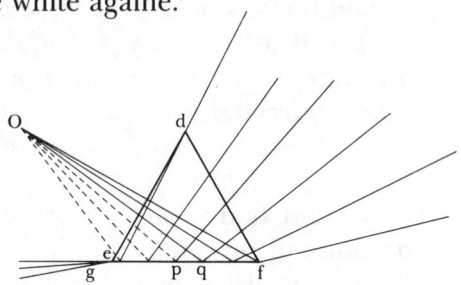

you will see one part of y^e sky ep (being as it were shaded wth a thin curtaine) to appeare darker y^n y^e other qf. [For all y^e rays wch can come to y^e eye from qf, fall soe obliquly on y^e basis as to bee all reflected to ye eye. Whereas those wch can come to ye eye from ep are so direct to ye basis as to bee most of y^m transmitted to g]: & ye partition of those two parts of ye sky, pq, appeares blew; [For ye rays wch can come to ye eye from pq, are so inclined to ye basis y^t all ye blew rays are reflected to ye eye whilst most of y^e red rays are transmitted through to g, as in experim^{nt} 22].

24   Tying two Prismes basis to basis def & bef together: I so held y^m in y^e sun beames, transmitted through a hole into a darke roome, y^t they

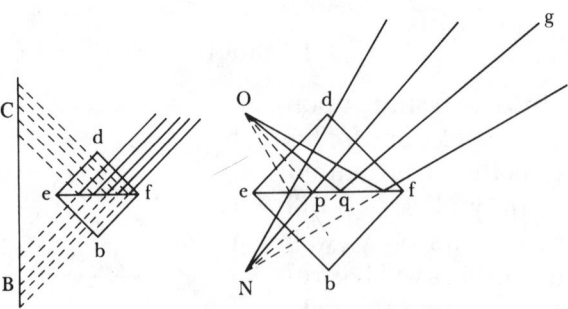

Of Colours.　　　　　　7

falling pretty directly upon y^e base ef (in fig 1) were most of y^m transmitted to B on y^e paper CB; though some of y^m were reflected to c by y^e filme of aire ef betwixt y^e Prismes. But both C & D were white. Then I inclined ye Basis (ef) of ye Prismes more & more to ye rays untill B changed from white to Red, & ye white at C became blewish; & inclining ye Prisme a little more ye Red at B vanished, & ye blewish colour at c became white againe. As in ye 22th Experiment.

25　　If I held ye said Prismes in ye open air as in ye 23^d experiment, holding my eye at O (in ye 2^d fig) to see ye reflected sky ye Phænomena were ye same as in y^t 23^d experiment; ep appearing darker y^n qf, & pq being blew. But if I held my eye at N to see ye sky through ye base of ye Prismes ef (or rather through ye plate of aire betwixt those bases) there appeared ye contrary Phænomena but much more faine; ep being very light, qf very darke, & pq very red. [The reason was given in ye 23^d Experiment.]

Note, That ye 22th & 24th (& all such like experiments y^t require y^t ye rays coming from a luminous body be all wholly or almost parallell) would bee more conspicuous were ye suns Diameter lesse, & therefore for such like experiments his rays may bee straitned through two small holes at a good distance assunder, as was done in ye 8th Experiment.

Also ye 23^d & 25^t Experiment (most other such like in wch the rays passe immediatly from ye prisme to ye eye) would bee more conspicuous were ye Pupill lesse y^n it is, And therefore it would bee convenient to look through a small hole at ye Prisme.

26　　The colours in ye partion pq appeared to ye Eye O in this order

Of Colours.　　　　　　8

27　　The two Prismes being tyed hard together then in trying ye 24th Experiment, there appeared a white spot

in ye midst of ye red colour C. And after y^e
base ef of ye Prismes was more more inclined to y^e rays, so y^t y^e red colour vanished & y^t (by y^e laws of Refraction) noe
light could penetrate y^e filme of aire ef, Yet
y^e white spot remained at B & y^e darke one
in y^e midst of y^e light at C.

28   Holding my eye at O or N (in trying y^e 25^t Exper:)
very obliquely to y^e basis ef; To my eye at O appeared a
black spot (R) in y^e midst of y^e white basis (or filme of aire)
ef, & to my eye at N appeared a white spot (R) in ye midst
of ye black basis (or plate of aire) ef; though wch spot (as
through a hole in ye midst of a black body) I could distinctly see any object, but could discerne nothing though
any other parte of ye appearingly black basis ef.

29   By variously pressing ye Prismes together at one end
more y^n at another I could make ye said spot R run from
one place to another; & ye harder I pressed ye prismes
together, ye greater ye spot would appeare to bee. [Soe y^t
I conceive ye Prismes (their sides being a little convex &
not perfectly plaine) pressed away ye interjacent aire at R
& becoming contiguous in y^t spot, transmitted ye Rays in
y^t place as if they had beene one continuous peice of
glasse; soe y^t ye spot R may bee called a hole made in y^e
plate of aire (ef)].

32   The colours of ye circles (in ye 30th & 31st Experiment) appeared more distinct at C y^n at B, & to ye Eye O
y^n to ye eye N. There being I conceive some colourlesse
light reflected wth ye coloured light to O, & C, but much
more colourlesse light transmitted to N & B; wch must
needs whiten & blend the colours.

Of Colours

9

30   In ye 27th Experiment when ye colour white or red
was trajected on B, there would apeare severall circles
of colours about ye white spot at B & also about y^e
darke one at C. But these colours vanished together w^{th}

ye red ·colour at B: Growing greater & distincter untill they vanished.

31   Likewise in ye 28th Expr: when ye spot was on y^t side ye partition pq next ye eye, it appeared to my eye both at O & N, encompassed with divers circles of colours. W^ch circles would grow greater & distincter by how much ye coloured partition pq came nearer & nearer to y^m (y^t is by how much y^e base ef was more & more oblique to y^e rays) & soe vanished by degrees as y^e said limb pq came to y^m. Before they began to vanish they appeared round or Ellipticall thus

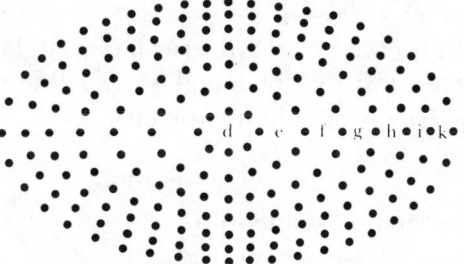

But in their vanishing (especially if looked through a hole much smaller then my pupill) they appeared incurved thus.

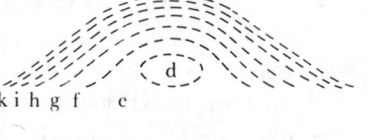

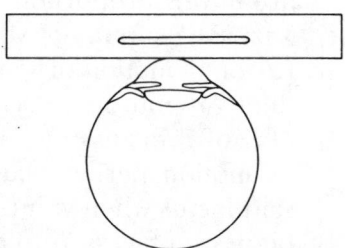

But I could see y^e most circles when I looked on y^m through a long slender slit, held parallel to y^e coloured limb pq, when y^e circles halfe disappeared: for y^n I have numbered 25 circles esteeming each consecution of red & blew to bee one circle & could perceive there were many more so close together y^t I could not number y^m; whereas wth my naked eye I could not discern above nine or ten as.

33   The circles are y^c broadest nearest to y^c center & so beeing narrower & narrower doe (I conceive by y^e exactest measure I could make) increase in number as y^e interjacent aire doth in thicknesse. (Sit cd = radio curvitatis vitri; efghik circuli color-

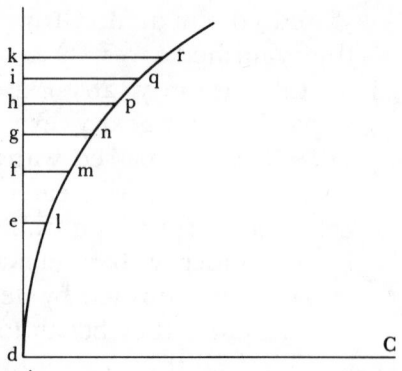

um; & el = $\frac{fm}{2} = \frac{gn}{3} = \frac{hp}{4} = \frac{iq}{5} = \frac{kr}{6}$ = crassitiei æris). And this I observed by a Sphæricall object glasse of a Prospective tyed fast to a plaine glasse, so as to make y^c said spot wth y^e circles of colours appeare.

Of colours.   10

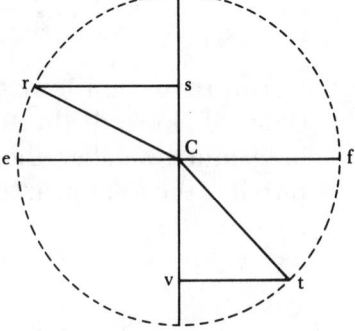

34   By the fore named Prospective glasse I observed (though not very exactly) y^t y^e more obliquely y^e ray tc was incident to y^e filme of aire ef twixt y^e glasses, y^e greater y^e coloured circles are in this proportion: Viz: as ye summe of ye factus of ye motion of ye incident ray into its velocity perpendicularly towards ye aire ef & of ye factus of ye motion of y^c said ray in ye aire ef into its motion perpendicularly through ye said aire is to ye said factus when ye incident ray is perpendicular, soe is ye bignesse of ye coloured circles w^n ye incident ray is perpendicular, to ye bignesse of ye same circles w^n ye incident ray is oblique, soe is dd × cv + ee × cs to dd + ee × ct. But ye spot in ye midst is not made greater or lesse by y^e obliquity of y^e rays rather y^e contrary.

35   When y^e rays were perpendicular to y^e aire ef, y^e diameter of 5 of y^e circles was one parte, whereas 400 was y^e radius dC of y^e glasses curvity. the said raius being 25^{inches} Soe y^t (el) y^e thicknesse of y^e aire for one circle was $\frac{1}{64000}^{inch}$, or 0,000015625. [wch is y^e space of a pulse of

y^e vibrating medium.] by measuring it since more exactly I find $\frac{1}{83000}$ = to y^e said thicknesse.

36   According as y^e glasses are pressed more or lesse together y^e coloured circles doe become greater or lesse. & as they are pressed more & more together new circles doe arise in y^e midst untill at last y^e said pellucid spot R doth appeare.

37   The circles of colours appeare in this order from ye center to ye eye O Or on ye paper at C viz Darke (or pelluced), white, yellow, greene, blew, purple, Red, Yellow, greene blew purple, Red, Yellow, Greene, blew &c. But to ye eye N or on ye Paper at B they appeare in this order Light (or pellucid) black, blew, Greene, yellow, Red, purple, blew, greene soe y^t those circles wch appeare Red to ye eye O, appeare blew to ye eye N & thos wch appeare blew to ye eye O appeare of ye contrary colour red to ye Eye N.

Of Colours.   11

38   Those circles wch appeare Red to ye eye O & blew to ye eye N are almost as broade againe as those wch appeare blew to ye eye O & Red to ye eye N.

39   Holding ye said circles in a darke roome in ye blew rays made by a Prisme (as ye 10th Experiment) all ye said circles appeared blew but those wch in ye discoloured light appeared red appeared of a blew much more diluted y^n ye others. And if ye Red Prismaticall rays fell upon those circles all ye circles appeared red but those circles wch in ye clear light appeared blew, in y^e Prismaticall red rays appeared of a much darker & obscurer red y^n y^e others.

40   Whither these circles were held in y^e Prismaticall blew or red rays they still appeared of y^e same bignesse.

41   Putting water betwixt y^e two Prismes instead of y^e filme of aire; There appeared all y^e Phæno= of y^e said circles, & also of y^e 22, 23, 24, & 25^t Experiments &c. Onely somwhat more obscurely because there is lesse refraction made out of glase into water y^n into aire; & yet

42   The coloured circles appeared as big when there was a filme of water as when there was a filme of aire betwixt ye Prismes.

43   If you make y^c pellucid spot R nimbly to run to &
fro, There will appeare another spot S to
follow it, wch spot S exhibits such Phænomena as it ought to doe were it a Spot of aire, viz: To
y^c eye O it appeares white next y^c spot R & y^n Red &c,
But to y^c eye N it appeares black next y^c spot R & y^n blew
&c: wch colours it ought to have were it a filme of aire (by
exper. 37). But it is not a filme of aire because if y^c Spot
R rests a little, the water creeps into y^c said spot S &
makes it vanish. It seems therefore y^t y^c water cannot
nimbly enough follow y^c Spot R, but leaves

Of Colours   12

y^c space S empty to bee possessed by Æther alone, untill
y^c water have time to creepe into it.

44   Refracting y^c Rays through a Prisme into a darke
rome (as in y^c 7th Experiment) And holding another
Prisme about 5 or 6 yards from y^c former to refract y^c
rays againe I found First y^t y^c blew rays did suffer a
greater Refraction by y^c second Prisme then y^c Red
ones.

45   And secondly y^t y^c purely Red rays refracted by y^c
second Prisme made no other colours but Red & y^c
purely blew ones noe other colours but blew ones.

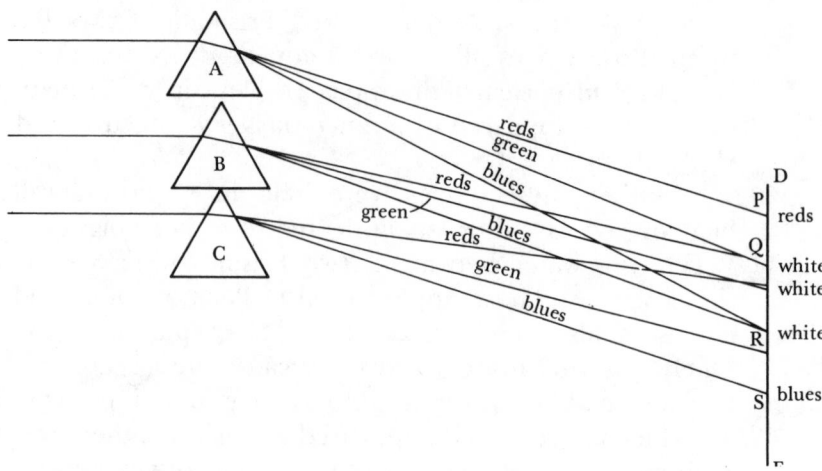

46   If three or more Prismes A, B, C, bee held in yᵉ sun soe yᵗ yᵉ Red colour of yᵉ Prisme B falls upon yᵉ Greene or yellow colour of Prisme A & yᵉ Red colour of yᵉ Prisme C falls on yᵉ Greene or yellow colour of yᵉ Prisme B; yᵉ Said colours falling upon yᵉ Paper DE at P, Q, R, S. There will appeare a Red colour at P & a blew one at S but betwixt Q & R where yᵉ Reds, yellows, Greenes, blews, & Purples of yᵉ severall Prismes are blended to-gether there appeares a white.[10]

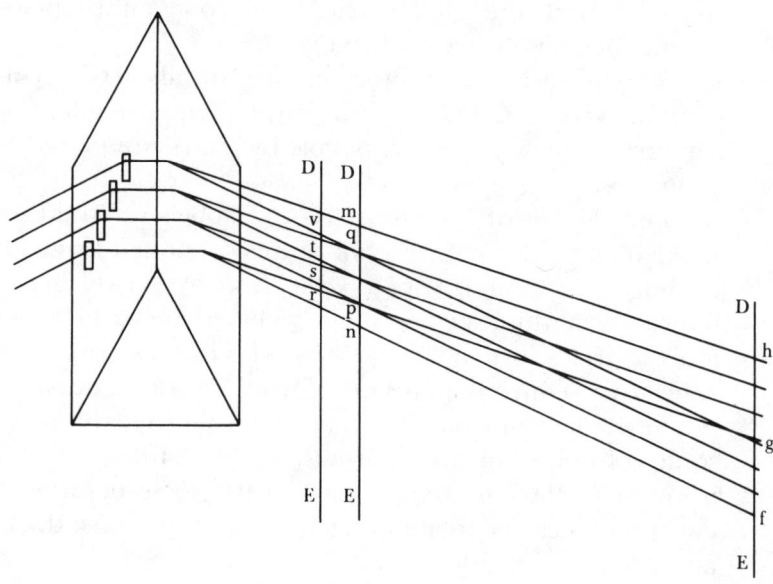

47   Of if you cleane a peice of

Of Colours   13

Paper on one side of yᵉ Prisme wᵗʰ severall slits a, b, c, d, in it parallel to yᵉ edges of yᵉ Prisme soe yᵗ yᵉ light passing through those slits make colours on yᵉ Paper DE; If yᵉ said paper be held neare to yᵉ Prisme there will appeare for each slit a, b, c, d, a coloured line r, s, t, v. The paper being held farther of untill yᵉ said coloured

[10] See folio 74 124ᵛ, item 12, and the first sentence of folio 69 122ʳ of the *Questiones*, p. 440 and p. 430.

lines bee blended together, there will appeare white twixt p & q where those colours are blended; at m there appeares Reds & at n blews. But if y^c paper bee still held farther of the white colour (pq) will appeare narrower & narrower untill it vanish. & then gh on one side appeares Red & gf on y^c other side is blew.

49   A single superficies of Glasse reflects many rays whither they passe out of glasse into aire or out of aire into Glasse & yet two surfaces of Glasse when contiguous (by y^c 27th 28th & 29th Experiment) reflect y^c Rays noe more then if the glasses had beene one entire peice wthout such superficies betwixt y^m.

48   As white was made by a mixture of all sorts of colours (in y^c 46th & 47th Experiment) Greene is made by a mixture of blew & yellow, purple by a mixture of red & yellow, &c.

50   Thin Flakes of Muscovy Glasse, Bubbles wch children make of sope & water, y^c thin skums of molten leade, of cooling iron, water wiped very thin on glasse, glasse blowne very thin, &c represent y^c Phænomena of y^c coloured circles in y^c 30th & 31st Exper &c. To wch may bee referred coloured motes in y^c Sun or in liquors, or pouders, or sollid bodys; y^c slender coloured threds of some cobwebbs, of silke wormes, & of flax finely dressed (though y^c flax in spining looseth its glosse because y^c flat thredds cleave together again into two greate a thicknesse see Exper 49).

Of Colours   14

51   If y^c Sun S shine upon a large glasse Globe abd filled wth water And if you hold your eye very neare to y^c globe, y^c rays bp will appeare coloured redd & y^c farther you hold yor eye from y^c glasse y^c

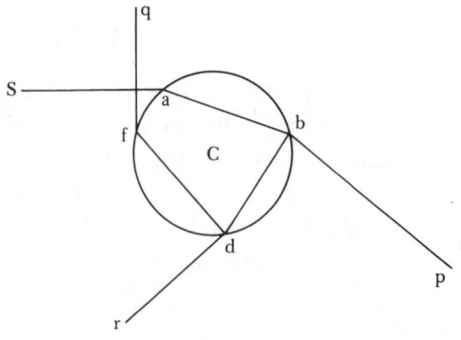

lesse they appeares coloured, untill y^e colour vanish. But y^e Rays rd & fq appeare coloured at w^t distance so ever yor eye bee placed from y^e Globe. The like you may observe by letting y^e colours fall on a peice of paper.

52   Though one termination of light trajected through ye Prisme will not make both blews & reds; yet in this globe it doth (see Cartesij Meteora cap 8 sec 9) For ye rays rd & fq make all sorts of blews & reds; indeed by ye rays bp ye red is very distinct but ye blew is scarce discernable.

53   The colours of ye Rainbow must bee explicated by ye rays rd & fq (vide Cartesij Meteor Cap 8 sec. 1, 2, 3, 9, 10, 11, 12, 15) For ye bow may bee mad by drops of water forcibly cast up into ye aire.

54   The spot R (mentioned in Experim^{nt} ye 52^d) grows lesse & lesse by how much ye rays fall more & more obliquely on ye intermediate filme of aire ef. [wch seems to intimate y^t ye thinness of ye intermediate filme of aire (or rather Æther) augments its refraction, untill (when ye glasses become contiguous) it bee equall to y^t of glasse].

55   The surfaces of Glasse doe not reflect soe much light when ye glasse is in water as when it is in aire & ye lesse any two mediums differ in refraction ye lesse their intermediate surface reflects light [wch intimates y^t tis not ye superficies of Glasse or any smoth pellucid body y^t reflects light but rather ye cause is ye diversity of Æther in Glasse & aire or in any contiguous bodys. though ye parts of ye Glasse must necessarily reflect some rays.

Of Colours   15

56   The pouders of Pelluced bodys is white soe is a cluster of small bubles of aire, ye scrapings of black or cleare horne, &c:[11] [because of ye multitude of reflecting surface] soe are bodys wch are full of flaws, or those whose parts lye not very close together (as Metalls, Marble, ye Oculus Mundi Stone &c) [whose pores betwixt their parts admit a grosser Æther into y^m y^n ye pores in their parts], hence

[11] See folio 91 133^r, items 20 and 21, of the *Questiones*, p. 454.

57   Most bodys (viz: those into which water will soake as paper, wood, marble, yᵉ Oculus Mundi Stone, &c) become more darke & transparent being soaked in water. [for yᵉ water fills up yᵉ reflecting pores].

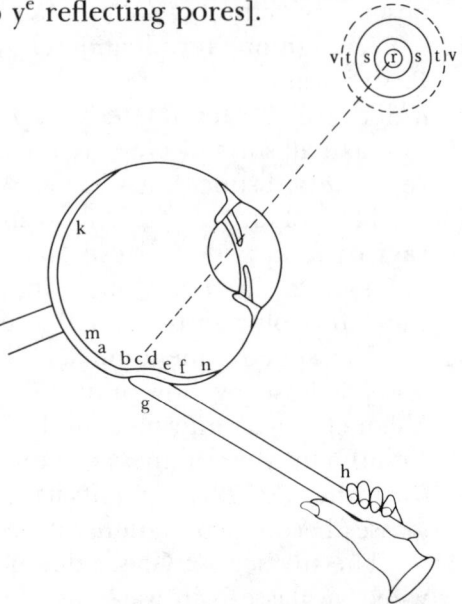

58   I tooke a bodkin gh & put it betwixt my eye & yᵉ bone as neare to yᵉ backside of my eye as I could: & pressing my eye wᵗʰ yᵉ end of it (soe as to make yᵉ curvature a,bcdef in my eye) there appeared severall white darke & coloured circles r, s, t, &c. Which circles were plainest when I continued to rub my eye wᵗʰ yᵉ point of yᵉ bodkine, but if I held my eye & yᵉ bodkin still, though I continued to presse my eye wᵗʰ it yet yᵉ circles would grow faint & often disappeare untill I renewed yᵐ by moving my eye or yᵉ bodkin.¹²

59   If yᵉ experiment were done in a light roome so yᵗ though my eyes were shut some light would get through their lidds There appeared a greate broade blewish darke circle outmost (as ts), & wᵗʰin that another light spot srs whose colour was much like yᵗ in yᵉ rest of yᵉ eye as at k. Within wᶜʰ spot appeared still another blew spot r,

Of Colours   16

Espetially if I pressed my eye hard & wᵗʰ a small pointed bodkin. & outmost at vt appeared a verge of light.

---

¹² See folios 72 123ᵛ and 73 124ʳ, item 9, of the *Questiones*, pp. 438–40.

60   But on ye contrary if I tryed ye Experiment in very darke roome ye circle ts apeared of a Reddish light, sr of a darkish blew & ye middle spot r appeared lighter againe; & there seemed to be a circle of darke blew tv w^{th}out ye circle ts ye outmost of all. [I conceive (in ye 60th experiment) where ye curvature of ye Retina at ma & fn began & was but little ye blew colour tv was caused; at ab & ef where ye Retina was most concave, ye bright circle ts was caused: at bc, & de where ye Retina was not much incurved nor strained ye dark blew circle sr was caused & at cd where ye Retina was stretched & made convex ye light spot r was caused. In ye 59th Experiment ye spirits were perhaps strained out of ye Retina at ab, ef, & cd or otherways made incapable of being acted upon by light & soe made a lesse appearance of light y^n ye rest of ye Retina].

61   That ye same circle ts wch appeared light in ye darke, appeared darke in ye light I found by suddenly letting in light into a darke darke roome for y^n ye bright circles would imediatly turne into darke ones & ye darke ones into bright ones.

62   I could sometimes perceive vivid colours of blew & red, made by ye said pressure & perhaps a criticall eye might have discerned this order of colours in ye 60th experiment viz from ye center greene, blew, purple, darke, purple, blew, greene, yellow, red like flame, yellow, greene, blew, broade purple, darke.

63   Looking on a very light object as ye Sun or his image reflected; for a while after there would remaine an impression of colours in my eye: viz: white objects looked red & soe did all objects in ye light but if I went into a dark roome ye Phantasma was blew.[13]

64   That vision is made in the retina appeares because colours are made by pressing the bakside of the eye: but when ye eye turns towards ye pressure soe y^t it is pressed before ye colours cease.

---

[13] See folios 43 109^r, 75 125^r, and 76 125^v of the *Questiones*, pp. 394–96, and pp. 442–46.

Of Colours.[14]

The Tunica Retina grows not from y^c sides of y^c opticks nerve (as y^c other two wch rise one from y^c dura, y^c other from y^c Pia mater) but it grows from y^c middle of y^c nerve sticking to it all over the extremitys of its marrow. Which marrow if the nerve bee any where cut cross wise twixt y^c eye & y^c union of the nerves, appeares full of small spots or pimples, wch are a little prominent, especially if the nerve be pressed or warmed at a candle. And these shoot into y^c very eye & may bee seene wth in side where y^c retina grows to y^c nerve: and they also continue to y^c very juncture EFGH. but at the juncture they end on a suddein into a more tender white pap like the interior part of the braine & soe y^c nerve continuing after y^c juncture into y^c braine filld wth a white tender pap in wch can bee seene noe distinction of parts as betwixt y^c said juncture & y^c eye.

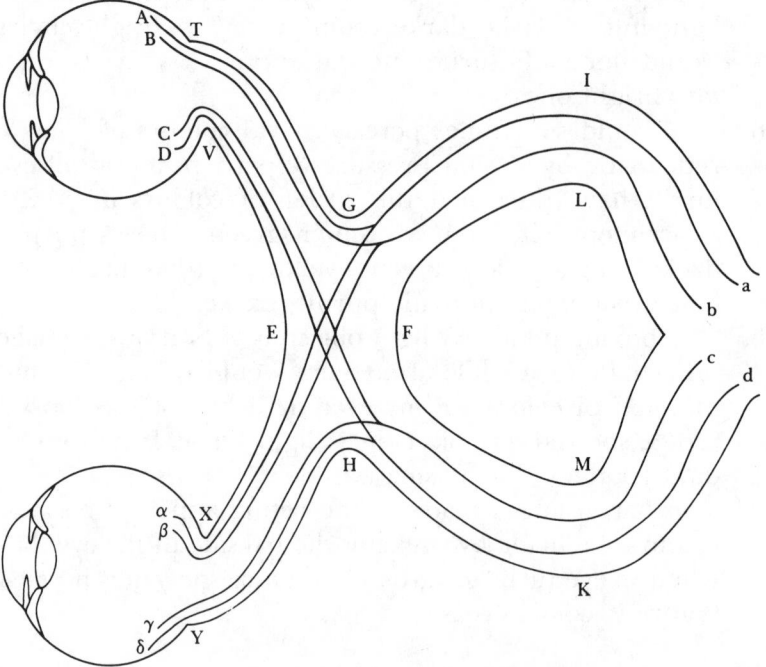

[11] This section of Add. 3975 was published by Sir David Brewster as No. VII of his Appendix to Volume 1 of *Memoirs of the Life, Writings, and Discoveries of Sir Isaac Newton* (Edinburgh, 1855). Brewster apparently never saw the manuscript of Add. 3975 and reproduces a version published by Joseph Harris in his *Treatise of Optics* (London, 1775).

Now I conceive that every point in the retina of one eye hath its correspondent point in ye other, from wch two very slender pipes, filld wth a most lympid liquor doe wthout either interruption or any other uneavenesse or irregularity in their processe, goe along the optick nerves to ye juncture EFGH where they meete either twixt GF or FH, & there unite into one pipe as big as both of them, & so continue in one passing either twixt IL or MK into ye braine where they are terminated. perhaps at ye next meeting of ye nerves twixt ye Cerebrum & cerebellum, in ye same order that their extremitys were scituate in the Retinals. And so there are a vast multitud of these slender pipes wch flow from the braine the one halfe through the right side nerve IL till they come at the junctur GF where they are each divided into two branches the one passing by G & T to ye right side of ye right eye AB the other halfe shooting through y^e juncture[15] EF & soe passing by X to y^e right side of the left eye αβ. And in like manner other halfe shooting through the left side nerve MK divide them selves at FH & their branches passing by EV to the right ey & by HY to the left, compose that $\frac{1}{2}$ of the Retina

in both eys wch is towards y^e left side, CD, & γδ.[16]   18

Hence it appears 1 why y^e two images of both eyes make but one image abcd in the braine. 2 Why when one eye is distorted objects appear double, For if y^e image of any object bee made upon A in the one ey & β in the other, y^t object shall have two images in the brain at a & b. Therefore the pictures of any object ought to bee made upon the corresponding points of ye two Retinas, if upon A in ye right ey, then upon α in ye left. If upon B then also upon β. And soe shall ye motions concurr after they have past ye juncture GH & make one image at a or b more vivid then one ey alone could doe. 3 Why though one thing may appeare in two places by distorting the eys yet two things cannot appear in one place. If the picture of one thing fall upon A & another upon α, they may both procee to p but noe farther, they cannot both be carried on ye same pipes pa into ye braine, that wch is [s]trongest or most helped by fantacy will

---

[15] Brewster's version has "space" for "juncture."
[16] See folio 35 105^r of the *Questiones*, p. 386.

there prevaile & blot out ye other. 4 Why a blew seene by one eye & a yellow by the other at ye same time produces a greene unlesse ye fantasy make one colour prædominant.[17] 5^{thy} Why if one of the branches of ye nerve beyond y^e juncture as at as at GF or FH should bee cut: That halfe of both eys toward y^e wounded nerve would bee blind, the other halfe remaing perfect.[18] 6^{thy} Why the juncture is almost as broad again twixt G & H then twixt E & F, becaus all the tubuli of both eys pass twixt G & H & but $\frac{1}{2}$ of them twixt E & F. It is not quite so broad again because y^e tubuli crossing in G are joining &c: also y^e thicknes of the quicks &c[19] 7^{thly} why the nerve GILF buts not directly upon the nerve XEHY, but deviates deviates a little towards TV because its Tubuli are to passe only into that side of the nerve EHYX towards EX. The like of FMKH 8^{thly} why the marrow of the nerve TVEG grows soft on a suddein when it comes at the juncture EF & more suddenly on that side towards G then towards E. And the like of the nerve EXYH. For it being necessary that the nerve TVEG should bee stretcht & bended severall ways by the motion of the eye: Therefore the tubuli are involved or wrought up w^{th}in the substance of severall tough skins wch being foulded up together compose ye marrow of ye nerve, pretty sollid & flexible least ye tubuli should be prejuced by the severall motions of the nerve. And those small pimples or prominences wch appeare in the nerve cut cross wise I conceive to bee made by the foldings of these crasser skins. But the nerve at ye juncture EGFH being well guarded from all violence &

motion by the bones into wch it is closely adapted: tis not necessary the said membranes substance should be continued any further then EG therefore the tubuli there on a suddein unsheath themselves those on ye inner side of the nerves towards VE & XE may severally crosse twixt EF & bee united wth their correspondents on the other sides YH & TG. Now because ye inner tubuli must first crosse before they can convene wth the outmost tubuli of the opposite

19

[17] Item 4 is omitted from the Brewster version.
[18] This item is numbered 4 in the Brewster version.
[19] This sentence is omitted from the Brewster version.

nerve hence it is that y^e nerves grow soft sooner on y^e inner side at E then on y^e outer side at G & H.

9^{thly} why y^e two nerves meet a second time in the brains, because y^e two half images caried along IL & MK may bee united into one complete image in the sensory. Note y^t y^e nerves at their $\overset{\text{meeting}^{20}}{\text{contact}}$ are round about disjoyned from y^e rest of the braine, nor are they soe thick there as a little before their meeting. But by their externall figure they seeme as if the capillamenta concentered like y^e radij of a hemisphere to a point in y^e lower part of the juncture. And tis probable y^t the visive faculty is there for else why doe the nerves swell there to so great a bulke as it were preparing for their last office, why doe they run directly crosse from eitherside the braine to meet there if the designe was to have ye motions coveyed by the shortest cut from ye eye to ye sensorium before they grew too weak. If they were to proceed further they might have gone a shorter cut & in a lesse channell. There is indeed a marrow shoots from under them toward ye cerebrellum to wch they are united but ye greatest part of their substance if not all of it lys above this marrow & also shoots cross beyond it to ye center of the brain where they meet. Lastly the substance here is most pure, ye scituation in ye mist of the brain, constituting ye upper part of that small passage twixt all ye ventricles. where all superfluous humors have the greatest advantages to slide away that they may not incumber y^t p^rcious organ.

Light seldom striks upon ye parts of grosse bodys (as may bee seen in its passing through them), its reflection & refraction is made by ye diversity of æthers, & therefore it effect on the Retina can only bee to make this vibrate W^{ch} motion then must be either carried in ye optick nervs to ye sensorium or produce other motions that are carried thither. Not ye latter for water is too grosse for such subtile impressions & as for animal spiritts

though I lyed a peice of ye optick nerve at one end & 20 warmed it in ye middle so see if any aery substance by that

[20] In the Brewster version, "meeting" appears, but not "contact."

meanes would disclose it selfe in bubbles at the other end, I could not spy the least bubble; a little moisture only & y^c marrow it selfe squeezed out. And indeed they that know how difficultly aire enters small pores of bodys, have reason to suspect y^t an aery body though much finer then aire can pevade easily[21] & wthout violence (as it ought to doe) y^c small pores of the braine & nerves, I should say of water, because those pores are filled wth water, & if it could it would bee too subtil to bee imprisoned by y^c dura mater & Skull & might passe for æther. However, w^t need of such spirits much Motion is ever lost by communication especially twixt bodys of different constitutions: and therefore it can noe way bee conveyed to y^c sensorium so entirely as by the æther it selfe. Nay granting mee but that there are pipes fill'd wth a pure transparent liquor passing from y^c ey to y^c sensorium & y^c vibrating motion of y^c æther will of necessity run along thither. For nothing interrupts that motion but reflecting surfaces, & therefore also y^t motion cannot stray through y^c reflecting surfaces of y^c pipe but must run along (like a sound in a trunk) intire to y^c sensorium. And that vision bee thus made is very conformable to the sense of hearing wch is made by like vibrations.[22]

From y^c whitenes of the brain & nerves the thicknesse of its vessells may be determined & their cavitys guessed at. And its pretty to consider how these agree wth the utmost distinctnesse in vision. As also wth y^c extent of nature in conveying distinctly y^c motions of the Aether.

[Blank Folio]   21

Of colours   22

If rays be incident out of glasse upon a film of air terminated twixt two glasses, the thicknesse of a vibration is $\frac{1}{81000}$, or $\frac{1}{80000}$ part of an inch.

If water was put twixt the glasses the thicknes of a vibration was $\frac{1}{100000}$inch, of $\frac{3}{4}$ of its former dimensions. viz as y^c densitys of the interjected mediums.

[21] Brewster omits "easily."
[22] Here the Brewster version ends.

If the rays were incident obliquely, the circles increase so that their diameters are as y^e secants of the rays obliquily w^{th}in the film of air, or reciprocally as their celerity w^{th}in the said film.

And the thicknesse belonging to each vibration is as the squares of these secants of celeritys, And ye lengths of ye rays belonging to each vibration as their cubes.

The first pulse ends at the first dark circle.

The thicknesse of a pulse of extream rubiform rays to that of purpuriform ones perpendicularly incident is greater then 3 to 2 & lesse then 5 to 3. viz as 9 to 14 or 13 to 20. And the thickness belonging to each coulour is 13, 14, 14½, 15½.16½.17½.18½.19. for extreame purple, intense purple, Indico, blew, green, ye terminus of green & yellow, yellow, orange, red,extream red.

Mr Boyle mentions one that by sickness became so tender sighted as in ye dark night to see & distinguish plainly ye colours of ribbans (& other objects) on purpose pinned on ye inside of his curtains against he awaked. Of ye determinate nature of Effluviums p 26, And of another y^t by a feaver became of so tender hearing as to hear plainly soft whispers at a distance wch others could not at all perceive, but when he grew well his hearing became but like y^t of other men. Ibid.

Stipic vegetables, as gall, oaken bark, red roses, Logwood, Sumach &c turn vitriol to a black precipitate.[23]

[23] See folio 91 133^r, item 18, of the *Questiones*, p. 454.

GLOSSARY

In this Glossary we make no claim to completeness. Our concern is with the meanings of the following terms only as they are used by Newton in the *Questiones*. In some cases Newton's uses are extremely obscure and may have been coined by Newton himself only to be dropped by him soon after composition of the *Questiones*. In most cases, however, Newton's uses were those of his time. We hope that this fact may serve to make this Glossary useful beyond the context of this volume.

The word or words given initially are spelled as they are found in our Expansion. Where this spelling differs from Newton's own, as recorded in the Transcription, his spelling has been given in parentheses. A folio reference to an example of the term is given at the end of each entry.

Acid spirit: Acids obtained by distillation. (92 133ᵛ)

Acute sounds: Sounds sharp or shrill in tone, high sounds. (37 106ʳ)

Aldebaran (Aldeboran): The α star in the constellation of Taurus (the Bull) of 0.8 apparent magnitude. (56 115ᵛ)

Alembic (Limbec): An apparatus used in distilling, consisting of a gourd-shaped vessel containing the substance to be distilled, surmounted by the cap that conveys the vaporous product to a receiver, where it is condensed. In Newton's usage, probably the gourd-shaped vessel alone. (91 133ʳ)

Almacantars (Almicanthers): Small circles of the sphere parallel to the horizon, cutting the meridian at equal distances, thus parallels of altitude. The horizon itself was considered the first almacantar. (57 116ʳ)

Andromeda: Constellation of the chained maiden. A constellation of the autumn and winter northern sky. (58 116^v)

Angle of contact: The angle between a tangent and circle at the point of contact. In this case the external angle of a side of the n-sided inscribed polygon as n has become indefinitely large. (5 90^r)

Animal spirits (Animall spirits): The supposed 'spirit' or principle of sensation and voluntary motion; answering to nerve fluid, nerve force, nervous action. (33 104^r)

Antimony, crude: Calcined and powdered antimony trisulfide. Also called black antimony. (95 135^r)

Antimony, Regulus of: The metallic form of antimony. (48 111^v)

Aqua-fortis: Nitric acid. (91 133^r)

Aqua-fortis, rectified: Purified nitric acid. (74 124^v)

Arcturus: The α star in the constellation of Bootes (the Herdsman), of apparent magnitude −0.1. (12 93^v)

Arsenic, white (white arsnick): Trioxide of arsenic (As_2O_3). (48 111^v)

Artificial (artificiall): Product of art or artifice. (83^r)

Asperity: Roughness or unevenness of surface. (17 96^r)

Asteria, the stone: Astroite or asterite. (40 107^v)

Beard: The tail of a comet when it appears to precede the nucleus. (54 114^v)

Bell metal (bell mettall): The substance of which bells are made; an alloy of copper and tin. (48 111^v)

Bird raying: Apparently a comet nucleus producing a halo obscuring the nucleus itself. Perhaps Newton meant to write beard raying. (56 115^v)

Blood, spirit of: Apparently distilled blood or serum. (92 133^v)

Blue bice (blew bise): Smalt, glass usually colored a deep blue by oxide of cobalt; when finely ground to make a pigment, it is called blue bice. (91 133^r)

Brine, strong (strong bryan): Water saturated, or strongly impregnated, with salt. (20 97^v)

Bull: The constellation of Taurus, a winter constellation that contains the star Aldebaran. (56 115^v)

Burning waters: Ardent spirits, inflammable or combustible spirits. (26 100^v)

Calcinated (calcined): Reduced by roasting or burning. (91 133^r)

Camphor (camphire): A whitish translucent crystalline volatile substance ($C_{10}H_{16}O$) belonging chemically to the vegetable oils. (91 133^r)

Capillamentum: One of the slender hairlike ultimate ramifications of a nerve. (35 105^r)

Cassiopeia: The constellation of a woman in a chair, containing the magnitude-2 star Shedir. A constellation of the northern sky. (58 116^v)

Ceruleous: Dark blue. (94 134^v)

Clackers: Instrument that by striking the hopper causes the corn to be shaken into the millstones of a mill. (49 112^r)

Cochineal (cochineel): Carminic acid, the anthraquinone of chineal insects (*Dactylopius*). (92 133^v)

Comet's bird: Apparently a comet's halolike tail. (12 93^v)

Common sensorium: The seat of sensation, the percipient center to which sense impressions are transmitted by the nerves. (33 104^r)

Comprehending: Including or containing. (86 130^v)

Conarion: The pineal gland. (33 104^r)

Corrosive liquids (corrosive liquors): Acids. (74 124^v)

Crookedness (crookednes): Degree of change of direction in the perimeter of a regular-sided polygon as measured by the sum of the external angles of each side. (5 90^r)

Decocted: Concentrated by boiling. (23 99^r)

Decocted herbs (decocted hearbes): Boiled-down herbs. (40 107^v)

Descension (descention): Descending, falling. (68 121^v)

Distempered (distemperd): Disordered or deranged. (75 125^r)

Drawk (drawke): To saturate with moisture. (91 133^r)

Ductility: Malleability. (16 95^v)

Equilibrio, in: In equilibrium, in balance; in this case, neither moving (much) closer to nor (much) farther away from the Sun. (32 103^v)

Fairystone (fairy stone): A fossil sea urchin or echinite. (40 107^v)

Fantasy (phantasie): The imagination. (43 109^r)

Filtration: Percolation, the slow passage of a liquid through a porous body. (31 103^r)

Fire-stone: Iron pyrite. (48 111^v)

Fishes: The constellation of Pisces, an autumn constellation of rather dim stars partly in the northern sky, partly in the southern sky. (57 116^r)

Friability: Being capable of being easily crumbled or reduced to powder. (91 133^r)

Galls, infusion of broken: Gallotanic or tannic acid, $C_{27}H_{22}O_{17}$. (74 124^v)

Glass egg: A glass receptacle shaped like an egg with a glass tube emerging for filling, etc. (20 97^v)

Globe: The celestial globe. (58 116^v)

Globule: A small spherical body. (30 102^v)

Goat (goate): The constellation of Aries, the Ram; an autumn constellation. (56 115^v)

Gratefully: So as to give pleasure. (44 109^v)

Grave sounds: Sounds low in pitch, deep in tone. (37 106^r)

Hartshorn, spirit of (spirit of hartshorne): Aqueous solution of ammonia. (92 133^v)

Hebetude: Condition of being blunt or dull; dullness or bluntness. (17 96^r)

Humor crystal (humor chrystall): The lens of the eye. (20 97^v)

Impregnated: Imbued or saturated with something; having some active ingredient diffused through it. (92 133^v)

Inane: Void or empty. (52 113^v)

Inanities (inantys): Voids, empty spaces. (19 97^r)

Indiscerpible: Incapable of being divided into parts. In Henry More's sense, an indiscerpible body or substance is absolutely resistive of the entry of another body into its place. (3 89^r)

Lamp-black (lamb black): A pigment consisting of almost pure carbon in a state of fine division. (36 105^v)

Lapis calaminaris (lapis calaminariæ): Calamine, usually zinc carbonate $ZnCO_3$. (48 111^v)

Lath: The bending part of a crossbow. (31 103^r)

Lignum Nephriticum: A wood, the infusion of which was thought to be efficacious in diseases of the kidney; from a tree of New Spain or horseradish tree. (94 134^v)

Limewater: An alkaline solution of water and hydrate of calcium (or lime) CaHO. (92 133^v)

Link (linke): A torch of short fibers of hemp or flax and pitch. (48 111^v)

Lixivious liquid (lixivious liquor): A liquid alkalide, a base. (92 133^v)

Logwood, tincture of: The heartwood of an American tree used in dyeing, also called blockwood. (93 134^r)

Lower wheels in the Great Bear: The constellation Ursa Major has also been called the Big Dipper, the Plow, and the Wagon. The 'wheels' are the stars Phecda and Merak. (55 115^r)

Lute: To coat with lute, usually a pipeclay. (18 96^v)

Menstruum, acid: Any acid that will dissolve a solid substance. (93 134^r)

Menstruums, digested: Heated menstruums. (93 134^r)

Meridian: Either the point at which a celestial body attains its highest altitude, or the great circle of the celestial sphere that passes through the celestial poles and the zenith of any place on the Earth's surface. (56 115^v)

Meteors: Phenomena of, or pertaining to, the atmosphere. (46 110^v)

Minium: Red lead. (91 133^r)

Niter, spirit of: Niter is either sodium carbonate or potassium nitrate (saltpeter). (91 133^r)

North knot of the Fishes: The constellation Pisces is partially in the northern sky and partially in the southern sky. The 'north knot' is the star Eta Piscium of that constellation. (57 116^r)

Occult qualities (occult qualityes): Hidden, nonmanifest qualities. (28 101^v)

Orifice of the stomach: The mouth of the stomach, the cardiac sphincter. (33 104^r)

Orpiment, yellow: The trisulfide of arsenic. (91 133^r)

Pellucid: Translucent, transparent. (26 100^v)

Perspicuity: Transparency, translucency. (14 94^v)

Perustum: A dark red. (91 133^r)

Phantasm (phantasme): Mental image, appearance, or representation. (85 130^r)

Phantasma (pantasme): Mere appearance, phantom, image. (75 125^r)

Plenitude: The condition of being absolutely full in quantity, measure, or degree. (2 88^v)

Plummet: A plumb bob. (9 92^r)

Potashes (pot ashes): A crude form of potassium carbonate. (92 133^v)

Potashes, lixivium of (lixivium of pot ashes): Lye. (92 133^v)

Pression: In Cartesian physics, pressure or impulse communicated to and propagated through a fluid medium. (49 112^r)

Procyon (Procion): A star in the constellation of Canis Minor of apparent magnitude 0.4. (12 93^v)

Quicksilver: Mercury. (47 111^r)

Ram, the: The constellation of Aries, an autumn constellation. (57 116^r)

Raw water: Unboiled water, unheated water. (26 100^v)

Rigel: Beta Orionis. A star of apparent magnitude 0.1. (56 115^v)

Sal ammoniac, spirit of (spirit of sal-Armoniack): Aqueous ammonia. (92 133^v)

Salt, spirits of: Hydrochloric acid, HCl. (47 111^r)

Salt, white: Common salt, sodium chloride. (47 111^r)

Salts, alcalizate: A salt that will neutralize an acid, i.e., behave like a soda (a base). (92 133^v)

Salts, fixed: Salts not easily volatized, i.e., not losing weight under the influence of fire. (92 133^v)

Salts, lixiviate: An alkaline salt, same as alcalizate salts. (92 133^v)

Salts, salsuginous: The fugitive salts of animal substances. (95 135^r)

Salts, sulfurous (sulphureous salts): Potassium sulfate 'impregnated' with sulfuric oxide. (92 133^v)

Salts, urinous: Ammonio-sodic phosphates. (92 133^v)

Salts, volatile: Ammonium carbonate. (92 133^v)

Saltpeter (saltpeeter): Potassium nitrate. (48 111^v)

Sapors: Tastes, savors. (38 106^v)

Satiate: To saturate. (92 133^v)

Scapula Pegasi: The α star of the constellation Pegasus. The star in the shoulder (or the first star of the wing in Tycho's catalog) of Pegasus. The star Markab. (54 114^v)

Semidiameter: Radius, half the diameter. (4 89^v)

Sensorium: See Common sensorium.

Septum lucidum: Now more usually septum pellucidum. The thin double partition extending vertically from the lower surface of the corpus callosum to the fornix and neighboring parts, separating the lateral ventricles of the brain. (33 104^r)

Shedir (Schedir): A star of magnitude 2 in the constellation of Cassiopeia. (58 116^v)

Siccity: Dryness, absence of moisture. (16 95^v)

Sirius: α Canis Majoris. Star of apparent magnitude −1.4. The brightest star in the sky. (12 93^v)

Spanish whiting (Spanish whiteing): Finely powdered chalk used as a pigment or for its cleansing properties. (36 105^v)

Speculum: A mirror or reflector. (23 99^r)

Spica virginis: α Virginis. A star of apparent magnitude 1.0 in the constellation Virgo. (12 93^v)

Square-ruler (suare ruler): A surveying instrument consisting of two arms at right angles to each other with tangents marked for various possible hypotenuses across points on the two arms. (9 92^r)

Style (stile): The pin, rod, or triangular plate that forms the gnomon of a sundial. (9 92^r)

Sublimate, common: A solid product of sublimation, especially in the form of a compact crystalline cake. (74 124^v)

Sublimated (sublimed): Subjected to the action of heat in a vessel so as to be converted into vapor, which is carried off, and on cooling is deposited in a solid form. (91 133^r)

Substyler (substilar): The line on which the style or gnomon stands. (9 92^r)

Subtility: Fineness, tenuity. (17 96^r)

Suculæ: A mistranslation of the Greek γάδες, the Hyades. The Hyades is a V-shaped group of relatively faint stars near the brilliant star Aldebaran in the constellation of Taurus. (57 116^r)

Sufur, mineral (minerall sulphur): Sulfur or sulfurous substances derived from minerals such as gypsum, barite, etc. (92 133^v)

Sulfur, vegetable: Sulfurous substances derived from plant life. Sulfur enters plants from the soil through bacterial action. (92 133^v)

Superficies: Surfaces. (5 90^r)

Supernatant: Swimming above, floating above, as a lighter liquid on a heavier. (94 134^v)

Sympathy: A relation between two bodily organs or parts such that disorder, or any condition, of the one induces a corresponding condition in the other. (33 104^r)

Tartar: Potassium hydrogen tartrate. (18 96^v)

Tartar, salt of: Potassium carbonate. (18 96^v)

Tincture: Coloring matter, dye, pigment. (71 123^r)

Tincture of flowers: Coloring matter derived from flowers. (92 133^v)

Tinglass (tinglasse): Bismuth. (48 111^v)

Tobacco-pipe clay: A fine white clay that forms a ductile paste with water. (18 96^v)

Tractility: The quality of being capable of being drawn out to a thread. (16 95^v)

Trepan: A surgical instrument in the form of a crownsaw for cutting out small pieces of bone, especially from the skull. (33 104^r)

Tunica retina: The retina of the eye, the lining of the back of the eye on which images are focused by the lens of the eye. (72 123^v)

Tunnel (tunnell): Funnel. (9 92^r)

Turpentine, oil of (oyle of turpentine): A volatile oil distilled from crude turpentine $C_{10}H_{16}$; also called spirit of turpentine. (20 97^v)

Urine, spirits of: The solution of ammonium carbonate, obtained by distilling putrid urine. (72 124^v)

Ventricle, 4th (4th ventrickle): The fourth ventricle is located in the hindbrain. It communicates below with the central canal of the spinal cord and above with the third ventricle in the thalamencephalon. (33 104^r)

Ventricles (ventrickles): The cavities in the brain, normally numbering four in the adult human, formed by the enlargements of the neural canal. (33 104^r)

Verdigris (verdigrease): A green or greenish blue substance obtained by the action of dilute acetic acid on thin plates of copper, used as a pigment in dyeing. (92 133^v)

Vinegar, spirit of: Vinegar, dilute acetic acid. (91 133^r)

Violets, syrup of (syrrup of violets): A syrup made from the flowers. (92 133^v)

Vitriol (vitrioll): Sulfate of metal, most often iron sulfate. (74 124^v)

Vitriol, blue (blew vitriol): Sulfate of copper. (91 133^r)

Vitriol, friable calx of powdered (friable calx of powdred vitriol): An easily crumbled or powdered substance produced by thoroughly burning or roasting (calcining) a sulfate of metal, so as to consume or drive off all its volatile parts. (94 134^v)

Vitriol, oil of (oyle of vitrioll): Concentrated sulfuric acid. (74 124^v)

Vitriol, rectified oil of (rectified oyle of vitriol): Purified sulfuric acid, nonsmokey sulfuric acid. (91 133^r)

Vitriol, unsophisticated: Unmixed, unadulterated vitriol. (91 133^r)

Vortex: A supposed rotatory movement of cosmic matter round a center or axis, regarded as accounting for the origin or phenomena of the terrestrial and other systems; a body of such matter rapidly carried round in a continuous whirl. (32 103^v)

Water, fair (faire water): Clean, pure water. (74 124^v)

Weatherglass (witherglasse): A kind of thermometer. (47 111^r)

Whale's mouth: The Whale is the constellation of Cetus. Its mouth consists of the stars Deneb Kaitos and Tau Ceti. (55 115^r)

Wine, spirit of: Aqueous alcohol. (93 134^r)

Winter-savory, essential oil of (essentiall oyle of winter-savory): Apparently oil of wintergreen, a volatile oil of specific gravity 1.1423, being a mixture of hydrocarbon $C_{10}H_{16}$ with methyl-salicylic ether. (91 133^r)

Wormwood, essential oil of (essential oyle of wormewodd): A volatile oil of specific gravity 0.9122 obtained from wormwood of a very deep blue color caused by cocrulein. (91 133^r)

SYMBOLS AND SHORTHAND
DEVICES

Newton made ample use of symbols and shorthand devices during the period of the *Questiones*. For the most part, his symbols and shortenings were consistent with the usage of his time. In some cases, however, Newton made unique use of some of these devices (e.g. his use of the right-hand margin in place of punctuation). If a sentence conveniently ended at the right-hand margin, Newton would usually omit the period. In addition, the right-hand margin sometimes plays the role of comma, colon, or semicolon as the opportunity arises.

Newton makes frequent use of tittles and nasals to shorten his words. He also uses superscripts (e.g., y^c, o^r, w^t, etc.) to shorten his words, a common practice of his day. In our Transcription we have reproduced Newton's use of these various devices, with the exception of the one line of actual shorthand in Sheltonian brevigraphy on folio 83 129^r. This original line, however, can be seen in photographic plate 3, Chapter 6, Section 2.

We have distinguished tittles and nasals in the following way. Nasals are indicated by a single bar above a letter (usually an *n* or *m*) and generally indicate that the letter beneath should be doubled. Tittles come at the ends of words, but are used in two ways. One is to complete a word whose final letter (usually an *n* or *m*) or letters (often -ing or -al endings) have been omitted. In such cases the tittle is indicated by a bar above the final letter of the word and extending one space beyond the word. The other use of the tittle is in contractions of -tion and -sion endings. In these cases we have placed a bar, usually four letters in length, above the syllable having the 'shun' sound, as in 'Filtr$\overline{acon}$' on folio 87^r.

In the following list of symbols and shorthand devices that are drawn from the Newton manuscripts transcribed in this volume we have indicated at least one example of each and indicated its location in the text. We consider it unnecessary to give every use of every device and symbol, because the Expansion essentially does that job. We have also included the symbols we have used in the Transcription to indicate deletions, insertions, and so forth.

Editors' symbols

⌞ ⌟ The material between corner brackets was inserted into the text by Newton either at the time of the writing or at some later date.

——— Deletion by Newton.

‖‖ Deletion by Newton. Used in the case of lengthy passages or where Newton made other deletions (indicated by ———) before deleting the entire passage.

[ ] Square brackets contain the editors' clarifying additions and are used sparingly.

& "and," as on folio 1 88^r

&c "etc.," as on folio 1 88^r

⊖ "Earth," as on folio 47 111^r

☽ "Moon," as on folio 46 110^v

☉ "Sun," as on folio 11 93^r

✳ Either "star," as on folio 56 115^v, or "nucleus" of a comet, as on folio 56 115^v

♒ "Aquarius," as on folio 12 93^v

♓ "Pisces," as on folio 54 114^v

♈ "Aries," as on folio 54 114^v

♉ "Taurus," as on folio 54 114^v

♎ "Libra," as on folio 54 114^v

♍ "Virgo," as on folio 54 114^v

♏ "Scorpio," as on folio 54 114^v

∠ "angle," as on folio 46 110^v

: "proportional to," as on folio 5 90^r

◎ "wheel" (in a wheel), as on folio 59 117^r

△ "triangle" or "triangular," as on folio 83^r and folio 5 of the Appendix

//	"parallel," as on folio 55 115^r
'	"minute," as on folio 54 114^v
"	"second," as on folio 54 114^v

Newton's shorthand devices

-a$\overline{\text{con}}$	"-ation" or "-asion," as in "Filtr$\overline{\text{acon}}$" on folio 87^r
3^c	"Chapter 3," as on folio 27 101^r
ca$\overline{\text{n}}$ot	"cannot," as on folio 6 90^v
3^{chap}	"Chapter 3," as on folio 27 101^r
continu$\overline{\text{u}}$	"continuum," as on folio 2 88^v
3^d	"third," as on folio 11 93^r
3$^{\underline{d}}$	"3 degrees," as on folio 12 93^v
9^d	"9 degrees," as on folio 12 93^v
Decemb^r	"December," as on folio 12 93^v
5^{degr}	"5 degrees," as on folio 46 110^v
2^{d}ly	"secondly," as on folio 70 122^v
D^r	"doctor," as on folio 2 88^v
2^{ds}	"seconds," as on folio 5 90^r
El:	"element," as on folio 11 93^r
Eq^r	"Esquire," as on folio 6 90^v
exp^{er}	"experiment," as on folio 3 of the Appendix
experim^{nt}	"experiment," as on folio 70 122^v
figu$\overline{\text{d}}$	"figured," as on folio 26 100^v
fm	"from," as on folio 76 125^v
48^{gr}	"48 degrees," as on folio 3 of the Appendix
9^h	"9 hours," as on folio 49 112^r
100^l	"100 pounds," as on folio 68 121^v
40^m	"40 minutes," as on folio 12 93^v
Math:	"mathematical," as on folio 4 89^v
30$^{min:}$	"30 minutes," as on folio 12 93^v
M^r	"Mister," as on folio 18 96^v
natu$\overline{\text{r}}$	"natural," as on folio 43 109^r
o^r	"our," as on folio 11 93^r
oth^r	"other," as on folio 31 103^r
ꝓ	"pr," as in "proposition" on folio 10 92^v
ꝓhaps	"perhaps," as on folio 37 106^r
ꝑs, ꝑts	"parts," as on folio 1 88^r
ꝑte	"part," as on folio 1 88^r
$*^s$	"stars," as on folio 57 116^r

s^d	"squared," as on folio 9 92^r
v.	"verse," as on folio 83 129^r
13^v	"verse 13," as on folio 27 101^r
vrs	"verse," as on folio 83 129^r
w^{ch}	"which," as on folio 1 88^r
w^n	"when," as on folio 18 96^v
w^t	"what," as on folio 1 88^r
w^{th}out	"without," as on folio 2 88^v
$\overline{y}$	"them," as on folio 6 90^v
66^{yds}	"66 yards," as on folio 68 121^v
y^e	"the," as on folio 1 88^r
y^m	"them," as on folio 4 89^v
y^n	"then," as on folio 1 88^r, or "than," as on folio 2 88^v
yor	"your," as on folio 32 103^v
y^t	"that," as on folio 1 88^r
y^u	"you," as on folio 48 89^v

SELECTED BIBLIOGRAPHY

Books

Adam, Charles, and Milhaud, Gerard (editors), *Descartes' Correspondence publiee avec une introduction et des notes.* (Paris, 1936–63).

Archimedes, *On Floating Bodies.* In: *The Works of Archimedes with the Method of Archimedes,* edited by T. L. Heath (New York: Dover Press, 1953).

Archimedes, *Opera quae extant* (Paris, 1615). Harrison 75.

Ball, Rouse, *Notes on the History of Trinity College, Cambridge* (London, 1899).

Barrow, Isaac, *Euclidis elementorum libri XV. breviter demonstrati* (Cambridge, 1655). Harrison 581.

Barrow, Isaac, *Lectiones habitae in scholis publicis academiae Cantabrigiensis 1664, 1665, 1666* (London, 1684).

Bate, John, *The Mysteryes of Nature and Art* (London, 1634).

Bouma, P. J., *Physical Aspect of Color* (New York: St. Martin's Press, 1971).

Boyle, Robert, *A Defence of the Doctrine touching The Spring of the Air* (London, 1662).

Boyle, Robert, *Essays of the Strange Subtilty Determinate Nature, and Great Efficacy of Effluviums* (London, 1673). Harrison 259.

Boyle, Robert, *Experiments and Considerations touching Colours* (London, 1664).

Boyle, Robert, *New Experiments Physico-Mechanicall, touching the Spring of the Air* (Oxford, 1660). Harrison 269.

Boyle, Robert, *New Experiments and Observations touching Cold, or An experimental history of cold* (London, 1665). Harrison 268.

Boyle, Robert, *The Works of the Honourable Robert Boyle,* 5 vols., edited by Thomas Birch (London, 1744).

Brewster, Sir David, *Memoirs of the Life, Writings, and Discoveries of Sir Isaac Newton,* 2 vols. (New York: Johnson Reprint Corp., 1965; reprint of the Edinburgh edition of 1855).

Brown, P. L., *Comets, Meteorites & Men* (New York: Taplinger Publications, 1974).

Buridan, John, *Quaestiones super octo, libros physicorum Aristotelis,* edited by Johannes Dullaert (Paris, 1509).

Castellus, D. Benedictus, *Discourse of the Mensuration of Running Waters.* In:

Mathematical Collections and Translations, edited by Thomas Salusbury (London, 1661).

Charleton, Walter, *Physiologia Epicuro-Gassendo-Charltoniana* (London, 1654).

Cohen, I. Bernard (editor), *Isaac Newton's Papers & Letters on Natural Philosophy* (Cambridge, Mass.: Harvard University Press, 1958).

Costello, W. T., *The Scholastic Curriculum at Early Seventeenth Century Cambridge* (Cambridge, Mass.: Harvard University Press, 1958).

Crombie, A. C., *Medieval and Early Modern Science*, 2 vols. (New York: Doubleday Anchor, 1959).

Crombie, A. C., *Robert Grosseteste and the Origins of Experimental Science.* (Oxford University Press, 1953).

Descartes, René, *Dioptrics.* In: *Opera philosophica*, 3rd edition (Amsterdam, 1656). Harrison 509.

Descartes, René, *Meditations.* In: *Opera philosophica*, 3rd edition (Amsterdam, 1656). Harrison 509.

Descartes, René, *Meteorology.* In: *Opera philosophica*, 3rd edition (Amsterdam, 1656). Harrison 509.

Descartes, René, *More geometrica dispositae.* In: *Opera philosophica*, 3rd edition (Amsterdam, 1656). Harrison 509.

Descartes, René, *Objections and Replies.* In: *Opera philosophica*, 3rd edition (Amsterdam, 1656). Harrison 509.

Descartes, René, *Oeuvres de Descartes*, 12 vols., edited by Charles Adam and Paul Tannery (Paris, 1973).

Descartes, René, *Passiones sive affectus animae, prima pars.* In: *Opera philosophica*, 3rd edition (Amsterdam, 1656). Harrison 509.

Descartes, René, *Principia philosophiae.* In: *Opera philosophica*, 3rd edition (Amsterdam, 1656). Harrison 509.

Diels, H., and Kranz, W., *Die Fragmente der Vorsokratiker.* (Dublin: Weidmann, 1968).

Digby, Kenelm, *Two Treatises: in the one of which, The nature of bodies: in the other, The nature of man's soul, is looked into* (London, 1658). Harrison 516.

Diogenes Laertius, *De vitis dogmatis et apophthegmatis* (London, 1664). Harrison 519.

Diogenes Laertius, *Lives of Emminent Philosophers*, 2 vols., translated by R. D. Hicks (Cambridge, Mass.: Harvard University Press, 1925).

Dircks, Henry, *Perpetuum Mobile; or, Search for Self-Motive Power During The 17th, 18th, and 19th Centuries* (London, 1861).

Du Val, Guillaume, *Aristotelis opera omnia, Graece et Latine* (Paris, 1654).

Epicurus, *Letter to Herodotus.* In: *De vitis dogmatis et apophthegmatis*, by Diogenes Laertius (London, 1664).

Eustachius of St. Paul, *Ethica, sive summa moralis disciplinae in tres partes divisa* (Cambridge, 1654).

Evans, J. D. G., *Aristotle's Concept of Dialectic* (Cambridge University Press, 1977).

Furley, David J., *Two Studies in the Greek Atomists* (Princeton University Press, 1967).

Galen, *On the Natural Faculties*, translated by A. J. Brock (Cambridge, Mass: Harvard University Press, 1952).

Galileo, *The Systeme of the World, in Four Diologues*. In: *Mathematical Collections and Translations*, edited by Thomas Salusbury (London, 1661).

Gaskell, Philip, and Robson, Robert, *The Library of Trinity College Cambridge: a short history* (Cambridge: University Printing House, 1971).

Gassendi, Pierre, *Animadversiones in decimum librum Diogenis Laertii* (Lyon, 1649).

Gassendi, Pierce, *De motu impresso a motore translato* (Paris, 1642).

Gassendi, Pierre, *Syntagma philosophicum* (London, 1658).

Gassendi, Pierre, *Syntagma philosophicum*. In: *Opera omnia*, 6 vols. (Stuttgart-Bad: Cannstatt, 1964).

Giussani, C., *Studi lucreziani* (Turin, 1896).

Glanvill, Joseph, *The Vanity of Dogmatizing* (London, 1661).

Grant, Edward, *Much Ado About Nothing: Theories of Space and Vacuum from the Middle Ages to the Scientific Revolution* (Cambridge University Press, 1981).

Grassi, Horatio, *On the Three Comets of the Year MDCXVIII*. In: *The Controversy on the Comets of 1618*, translated and edited by Stillman Drake and C. D. O'Malley (Philadelphia: University of Pennsylvania Press, 1960).

Grosseteste, Robert, *Commentarius in VIII libros physicorum Aristotelis*, edited by R. C. Dales (Boulder: University of Colorado Press, 1963).

Grunbaum, Adolf, "Absolute Relational Theories of Space-Time." In: *Foundations of Space-Time Theories, Minnesota Studies in the Philosophy of Science, Vol. 8* (Minneapolis: University of Minnesota Press, 1977).

Hall, A. Rupert, *Philosophers at War* (Cambridge University Press, 1980).

Hall, A. Rupert, and Hall, Marie Boas (editors), *Unpublished Scientific Papers of Isaac Newton* (Cambridge University Press, 1962).

Hall, Francis, *Tractatus de corporum inseparabilitate* (London, 1661).

Harris, Joseph, *Treatise of Optics* (London, 1775).

Harrison, John, *The Library of Isaac Newton* (Cambridge University Press, 1978).

Heath, Thomas L. (editor), *The Thirteen Books of Euclid's Elements*, 3 vols. (New York: Dover Press, 1956).

Heath, Thomas L. (editor), *The Works of Archimedes with the Method of Archimedes* (New York: Dover Press, 1953; reprint of 1912 Cambridge edition).

Herivel, John, *The Background to Newton's Principia* (Oxford University Press, 1965).

Hobbes, Thomas, *Elementorum philosophiae, sectio prima de Corpore* (London, 1655).

Hobbes, Thomas, *Elements of Philosophy. The First Section concerning Body* (London, 1656).

Hooke, Robert, *Micrographia* (London, 1665).

Keckermann, Bartholomew, *Systema physicum, septum libris adornatum*, 3rd edition (Hanover, 1612).

Kepler, Johannis, *De cometis* (Augsburg, 1619).

Kepler, Johannis, *Gesammelte Werke*, edited by Walther von Dyck and Max Caspar (Munich: C. H. Beck, 1937–).

Klein, Jacob, *Greek Mathematical Thought and the Origin of Algebra* (Cambridge, Mass.: M.I.T. Press, 1968).

Koslow, Arnold, "Ontological and Ideological Issues of the Classical Theory of Space and Time." In: *Motion and Time, Space and Matter*, edited by Peter K. Machamer and Robert G. Turnbull (Columbus: Ohio State University Press, 1976).

Koyré, Alexandre, *Newtonian Studies* (London: Chapman and Hall, 1965).

Lucretius, *De rerum natura, libri sex*, edited by H. A. J. Munro (Cambridge, 1866).

McGuire, J. E., "Space, Infinity, and Indivisibility: Newton on the Creation of Matter." In: *Contemporary Newtonian Research*, pp. 145–90 (Dordrecht: Reidel, 1982).

Magirus, J., *Physiologiae peripateticae contractio* (Cambridge, 1642).

Magnenus, Johannes, *Democritus reviviscens: sive vita & philosophia Democriti* (The Hague, 1658). Harrison 1014.

Maier, Anneliese, *Die Vorläufer Galileis Im 14. Jahrhundert* (Rome, 1949).

Maimonides, Moses, *The Guide for the Perplexed* (New York: Dover Press, 1956).

Manuel, Frank, *The Religion of Isaac Newton* (Oxford: Clarendon Press, 1974).

Mau, Jürgen, *Zum Problem des Infinitesmalen bie den antiken Atomisten* (Berlin: Akademie-Verlag, 1954).

Mersenne, Marin, *Harmonicorum libri [12] in quibus agitur de sonorum natura causis et effectibus: de consonantiis, dissonantiis . . . compositione, orbisque totius harmonicis instrumentis* (Paris, 1635).

Molland, A. G., "Mathematics in the Thought of Albertus Magnus." In: *Albertus Magnus and the Sciences: Commemorative Essays*, edited by James A. Weisheipl, pp. 464–78 (Toronto: Pontifical Institute of Medieval Studies, 1980).

More, Henry, *The Immortality of the Soul* (London, 1659). Harrison 1113.

More, Henry, *The Immortality of the Soul*. In: *A Collection of Several Philosophical Writings* (London, 1662).

Murdoch, John E., "Henry of Harclay and the Infinite." In: *Studi sul XIV secolo in memoria di Anneliese Maier*, edited by A. Maieru and A. Paravicini-Bagliani (in press).

Newton, Isaac, *The Correspondence of Isaac Newton*, edited by H. W. Turnbull, J. F. Scott, A. Rupert Hall, and Laura Tilling (Cambridge University Press, 1959–77).

Newton, Isaac, *Isaac Newton's Cambridge Lecturers on Optics, 1670–72*, edited by D. T. Whiteside. Facsimile of CUL ms. Add. 4002 (Cambridge: The University Library, 1973).

Newton, Isaac, *Opticks* (New York: Dover Press, 1952; based on the 4th edition, London, 1730).

Newton, Isaac, *Philosophiae Naturalis Principia Mathematica*, 2 vols., edited by Alexandre Koyré and I. Bernard Cohen (Cambridge, Mass.: Harvard University Press, 1972).

Oldenburg, Henry, *The Correspondence of Henry Oldenburg*, edited and translated by A. R. Holland and M. B. Hall (Madison: University of Wisconsin Press, 1966).

Owen, G. E. I., "Aristotle on Time." In: *Motion and Time, Space and Matter*, edited by Peter Machamer and Robert G. Turnbull, pp. 3–19 (Columbus: Ohio State University Press, 1976).

Plato, *The Collected Dialogues of Plato*, edited by Edith Hamilton and Huntington Cairns (New York: Pantheon Books, 1963).

Pliny, *Natural History*, 10 vols., translated by H. Rackham (Cambridge, Mass.: Harvard University Press, 1938).

Proclus, *A Commentary on the First Book of Euclid's Elements*, translated by Glenn R. Morrow (Princeton University Press, 1970).

Sabra, A. I., *Theories of Light from Descartes to Newton* (London: Oldbourne, 1967).

Sanderson, Robert, *Logicae artis compendium* (Oxford, 1631). Harrison 1442.

Simplicius, *Aristotelis physicorum, libros quattor priores commentaris.*

Sextus Empiricus, *Against the Physicists.* In: *Sextus Empiricus*, 4 vols., translated by R. G. Bury (Cambridge, Mass.: Harvard University Press, 1936).

Sextus Empiricus, *Opera quae extant* (Paris, 1621). Harrison 1503.

Snell, Willebrord, *Descriptio cometae* (Lugduni Batavorum, 1619).

Sorabji, Richard, "Atoms and Time Atoms." In: *Infinity and Continuity in Ancient and Medieval Thought*, edited by Norman Kretzmann (Ithaca: Cornell University Press, 1982).

Stahl, Daniel, *Axiomata philosophica* (Cambridge, 1645).

Usener, Herman, *Epicurea* (Rome, 1963; reprint of the 1887 edition).

Wallace, William A., *Causality and Scientific Explanation*, 2 vols. (Ann Arbor: University of Michigan Press, 1972–4).

Wallis, John, *Arithmetica infinitorum.* In: *Operum mathematicorum, Pars altera* (Oxford, 1656).

Wallis, John, *De angulo contactus.* In: *Operum mathematicorum, Pars altera* (Oxford, 1656).

Wallis, John, *De sectionibus conicis.* In: *Operum mathematicorum, Pars altera* (Oxford, 1656).

Wallis, John, *Defensio tractatus de angulo contactus et semicirculi.* In: *Opera mathematica* (Oxford, 1693–9). Harrison 1710.

Wallis, John, *Mathesis universalis.* In: *Operum mathematicorum, Pars prima* (Oxford, 1657).

Weisheipl, James A., *Nature and Gravitation* (River Forest, Ill.: Albertus Magnus Lyceum, 1955).

Westfall, Richard S. *Never At Rest: A Biography of Isaac Newton* (Cambridge University Press, 1980).

Whiston, William, *Memoirs of the Life of Mr. William Whiston by himself*, 2 vols. (London, 1749).

Whiteside, D. T., *The Mathematical Papers of Isaac Newton*, 8 vols. (Cambridge University Press, 1967–82).

Wilkins, John, *Mathematicall Magick or, The Wonders that may be Performed by Mechanicall Geometry* (London, 1648).

Wilson, Curtis, "Newton and the Eötvös Experiment." In: *Essays in Honor of Jacob Klein* (Annapolis: St. John's College Press, 1976).

Wing, Vincent, *Harmonicon celeste: or, The celestial harmony of the visible world* (London, 1651). Harrison 1744.

Wing, Vincent, *Astronomia Britannica* (London, 1669). Harrison 1743.

Articles

Bechler, Zev, " 'A Less Agreeable Matter': The Disagreeable Case of Newton and Achromatic Refraction." *British Journal for the History of Science* 8(1975):101–26.

Bechler, Zev, "Newton's Search for a Mechanistic Model of Colour Dispersion: A Suggested Interpretation." *Archive for the History of Exact Sciences* 11(1973):1–37.

Bizzi, Emilio, "The Coordination of Eye-Hand Movements." *Scientific American* 231(1974):100–6.

Cohen, I. Bernard, " 'Quantum in se est': Newton's Concept of Inertia in Relation to Descartes and Lucretius." *Notes and Records of the Royal Society of London* 19(1964):131–55.

Guerlac, Henry, "Newton's Optical Aether: His Draft of a Proposed Addition to his Opticks." *Notes and Records of the Royal Society of London* 22(1967):45–57.

Hall, A. Rupert, "Sir Isaac Newton's Notebook, 1661–65." *Cambridge Historical Journal* 9(1948):239–50.

Hall, A. R., and Hall, M. B., "Newton's Electric Spirit: Four Oddities." *Isis* 50(1959):473–6.

Hawes, Joan I., "Newton's Revival of the Aether Hypothesis and the Explanation of Gravitational Attraction." *Notes and Records of the Royal Society of London* 23(1968):200–12.

Hendry, John, "Newton's Theory of Colour." *Centaurus* 23(1980):230–51.

Herivel, J. W., "Newton's First Solution to the Problem of Kepler Motion." *British Journal for the History of Science* 2(1965):350–4.

Kitcher, Philip, "Fluxions, Limits, and Infinite Littleness. A Study of Newton's Presentation of the Calculus." *Isis* 64(1973):33–49.

Konstan, David, "Problems in Epicurean Physics." *Isis* 70(1979):394–418.

Koyré, Alexandre, and Cohen, I. B., "Newton's 'Electric & Elastic Spirit.' " *Isis* 51(1960):337.

McGuire, J. E., "Existence, Actuality and Necessity: Newton on Space and Time." *Annals of Science* 35(1978):463–508.

McGuire, J. E., "Force and Active Principles and Newton's Invisible Realm." *Ambix* 15(1968):154–208.

Millington, E. C. M. "Theories of Cohesion in the Seventeenth Century." *Annals of Science* 5(1945):253–69.

Russell, J. L., "Kepler's Laws of Planetary Motion: 1609–1666." *British Journal for the History of Science* 2(1964):1–24.

Shapiro, Alan E., "The Evolving Structure of Newton's Theory of White Light and Color." *Isis* 71(1980):211–35.

Shapiro, Alan E., "Kinematic Optics: A Study of the Wave Theory of Light in the Seventeenth Century." *Archive for the History of Exact Sciences* 11(1973):134–266.

Shapiro, Alan E., "Light, Pressure and Rectilinear Propagation: Descartes' Celestial Optics and Newton's Hydrostatics." *Studies in the History and Philosophy of Science* 5(1974):239–96.

Shapiro, Alan E., "Newton's 'Achromatic' Dispersion Law: Theoretical Background and Experimental Evidence." *Archive for the History of Exact Sciences* 21(1979–80):91–128.

Strang, Colin, and Mills, K. W., "Plato and The Instant." *Proceedings of the Aristotelian Society* 48(1974):63–96.

Tamny, Martin, "Newton and Galileo's Dialogue on the Great World Systems." *Isis* 68(1977):288–9.

Tamny, Martin, "Newton, Creation, and Perception." *Isis* 70(1979):48–58.

Westfall, Richard S., "The Foundations of Newton's Philosophy of Nature." *British Journal for the History of Science* 1(1962):171–82.

Westfall, Richard S., "Short-writing and the State of Newton's Conscience 1662." *Notes and Records of the Royal Society of London* 18(1963):10–16.

Whiteside, D. T., "Isaac Newton: Birth of a Mathematician." *Notes and Records of the Royal Society of London* 19(1964):53–62.

Whiteside, D. T., "Newton's Early Thoughts on Planetary Motion: A Fresh Look." *British Journal for the History of Science* 2(1964):117–37.

Whiteside, D. T., "Patterns of Mathematical Thought in the Later Seventeenth Century." *Archive for the History of Exact Sciences* 1(1961):179–388.

Zajonc, Arthur G., "Goethe's Theory of Color and Scientific Intuition." *American Journal of Physics* 44(1976):327–33.

INDEX

References to the Transcription, Expansion, and the Appendix (Add. 3975) are indicated by bold face numbers. In references to the *Questiones* page numbers are given for the Expansion alone, the Transcription being on the facing pages.

511